The Lighter Side of Mathematics

Proceedings of the Eugène Strens Memorial Conference on Recreational Mathematics & its History

Edited by
Richard K. Guy & Robert E. Woodrow

SPECTRUM SERIES

Published by
THE MATHEMATICAL ASSOCIATION OF AMERICA

About the cover: Background is Escher's design for the cover of *Regelmatige vlakverdeling.*

Library of Congress Catalog Card Number 90-70790

ISBN 0-88385-516-X

Printed in the United States of America

Current printing (last digit):
10 9 8 7 6 5 4 3 2 1

The Lighter Side of Mathematics

Proceedings of the Eugène Strens Memorial Conference on Recreational Mathematics & its History

Edited by
Richard K. Guy & Robert E. Woodrow

Published and Distributed by
THE MATHEMATICAL ASSOCIATION OF AMERICA

SPECTRUM SERIES

The Spectrum Series of the Mathematical Association of America was so named to reflect its purpose: to publish a broad range of books including biographies, accessible expositions of old or new mathematical ideas, reprints and revisions of excellent out-of-print books, popular works, and other monographs of high interest that will appeal to a broad range of readers, including students and teachers of mathematics, mathematical amateurs, and researchers.

Complex Numbers and Geometry, by Liang-shin Hahn
Cryptology, by Albrecht Beutelspacher
From Zero to Infinity, by Constance Reid
I Want to be a Mathematician, by Paul R. Halmos
Journey into Geometries, by Marta Sved
The Last Problem, by E. T. Bell (revised and updated by Underwood Dudley)
Lure of the Integers, by Joe Roberts
Mathematical Carnival, by Martin Gardner
Mathematical Circus, by Martin Gardner
Mathematical Cranks, by Underwood Dudley
Mathematical Magic Show, by Martin Gardner
Mathematics: Queen and Servant of Science, by E. T. Bell
Memorabilia Mathematica, by Robert Edouard Moritz
Numerical Methods that Work, by Forman Acton
Out of the Mouths of Mathematicians, by Rosemary Schmalz
Polyominoes, by George Martin
The Search for E. T. Bell, also known as John Taine, by Constance Reid
Shaping Space, edited by Marjorie Senechal and George Fleck
Student Research Projects in Calculus, by Marcus Cohen, Edward D. Gaughan, Arthur Knoebel, Douglas S. Kurtz, and David Pengelley
The Trisectors, by Underwood Dudley
The Words of Mathematics, by Steven Schwartzman

Mathematical Association of America
1529 Eighteenth Street, NW
Washington, DC 20036
800-331-1MAA FAX 202-265-2384

Preface

We didn't really mean to have a book. It would be far too much hard work to make one (as indeed it was). And it didn't jell with the idea of the conference, which was to get fun out of mathematics. And who on earth would publish it?

But there were lots of good talks, and many participants wanted to have written versions of them. And when the news got around to those who didn't come, about what a good conference we had, they wanted a written version, too. But there still would be no book had it not been for the energetic chairmanship of Warren Page of the MAA Notes Editorial Board, who not only solicited the work but also has proceeded steadily during his tenancy to improve the appearance of the MAA Notes. We apologize to him that the volume grew so large that it became more at home in the MAA Spectrum series.

And the book could not have appeared without the conference, and the conference would not have happened but for the energetic organization of Bill Sands and his several helpers, and the financial help provided by the Natural Sciences & Engineering Research Council of Canada, the Social Sciences & Humanities Research Council of Canada, The University of Calgary, its Department of Mathematics & Statistics, and by Richard Guy.

Of course, the eight-year gap between the conference and the publication of the book has meant that some things have got lost—Hendrik Lenstra's masterly melding of mathematics and amusement in his talk on exotic number systems, and Alan MacDonald's witty welcome during the opening ceremonies. And we cannot convey in a book the beauty and the interest provided by the presence of the exhibits of puzzles by Jerry Slocum and Tom Ransom and others, of Kathy Jones's games and puzzles, and the actual battles on the board games provided by Aviezri Fraenkel.

But much remains, and some of it has matured with age. Several papers have meanwhile appeared in journals and are correspondingly that much more polished. Indeed, most of the papers have been rewritten. Two of them, "The Strong Law of Small Numbers" and "Fourteen Proofs of a Result About Tiling a Rectangle," have gained for their respective authors, Richard Guy and Stan Wagon, the MAA's prestigious Lester R. Ford award for exposition.

Elwyn Berlekamp's article on Blockbusting and Domineering was the natural precursor of his more recent remarkable breakthrough in giving values to endgames in Go, a game long thought to be even more intractable than Chess.

At least two of the participants in the Conference, Ken Falconer and Angela Newing, are keen bell-ringers. Here Ken Falconer applies group theory to bell-ringing, and, by a happy coincidence, Calgary possesses one of the eight rings of bells in Canada, and he and Angela Newing were able to pursue their hobby locally, albeit five thousand miles from its traditional home.

The picturesque aspects of mathematics are beautifully displayed by Coxeter & Rigby, by

Branko Grünbaum, by Doris Schattschneider and several others. It will be no surprise that Maurits Escher (who, appropriately enough, was a friend of Eugène Strens) was mentioned, and used by way of illustration, by each of these people as well as by Athelstan Spilhaus.

We thank Don Albers and his staff for producing a fine volume while bearing with the editors' idiosyncrasies. Richard Nowakowski and John Selfridge were of great help with proof-reading, and are not to be blamed for any remaining errors.

But it would take too long to detail all the numerous and varied contributions: browse and enjoy! We hope that you have as much fun as we did.

Contents

Preface v

The Strens Collection 1

Eugène Louis Charles Marie Strens 5

Part 1: Tiling & Coloring

Frieze Patterns, Triangulated Polygons and Dichromatic Symmetry, *H. S. M. Coxeter & J. F. Rigby* 15

Is Engel's Enigma a Cubelike Puzzle? *J. A. Eidswick* 28

Metamorphoses of Polygons, *Branko Grünbaum* 35

SquaRecurves, E-Tours, Eddies, and Frenzies: Basic Families of Peano Curves on the Square Grid, *Douglas M. McKenna* 49

Fun with Tessellations, *John F. Rigby* 74

Escher: A Mathematician in Spite of Himself, *D. Schattschneider* 91

Escheresch, *Athelstan Spilhaus* 101

The Road Coloring Problem, *Daniel Ullman* 105

Fourteen Proofs of a Result About Tiling a Rectangle, *Stan Wagon* 113

Tiling R^3 with Circles and Disks, *J. B. Wilker* 129

Part 2: Games & Puzzles

Introduction to Blockbusting and Domineering, *Elwyn Berlekamp* 137

A Generating Function for the Distribution of the Scores of all Possible Bowling Games *Curtis N. Cooper & Robert E. Kennedy* 149

Is the Mean Bowling Score Awful? *Curtis N. Cooper & Robert E. Kennedy* 155

Recreation and Depth in Combinatorial Games, *Aviezri S. Fraenkel* 159

Recreational Games Displays
- Combinatorial Games, *Aviezri S. Fraenkel* 176
- Combinatorial Toys, *Kathy Jones* 195

Rubik's Cube—application or illumination of group theory? *Mogens Esrom Larsen* 202

Golomb's Twelve Pentomino Problems, *Andy Liu* 206
A New Take-Away Game, *Jim Propp* 212
Confessions of a Puzzlesmith, *Michael Stueben* 222
Puzzles Old & New: Some Historical Notes, *Jerry Slocum* 229

Part 3: People & Pursuits

The Marvelous Arbelos, *Leon Bankoff* 247
Cluster Pairs of an n-Dimensional Cube of Edge Length Two, *I. Z. Bouwer & W. W. Chernoff* 254
The Ancient English Art of Change Ringing, *Kenneth J. Falconer* 261
The Strong Law of Small Numbers, *Richard K. Guy* 265
Match Sticks in the Plane, *Heiko Harborth* 281
Misunderstanding My Mazy Mazes May Make Me Miserable, *Mogens Esrom Larsen* 289
Henry Ernest Dudeney: Britain's Greatest Puzzlist, *Angela Newing* 294
From Recreational to Foundational Mathematics, *Victor Pambuccian* 302
Alphamagic Squares, *Lee C. F. Sallows* 305
Alphamagic Squares: Part II, *Lee C. F. Sallows* 326
The Utility of Recreational Mathematics, *David Singmaster* 340
The Development of Recreational Mathematics in Bulgaria, *Jordan Stoyanov* 346
$V - E + F = 2$, *Herbert Taylor* 353
Tracking Titanics, *Samuel Yates* 355
List of Conference Participants 363

The Strens Collection

Richard K. Guy

As with so much else to do with the enjoyment of mathematics, first news of the Strens Collection came via $(MG)^2$, Martin Gardner's Mathematical Grapevine. When Eugène Strens died, there were obvious places in The Netherlands where his collection of chess books could be appropriately disposed, and his famous collection of Ex Libris was already properly housed. Among other things, there remained a collection of over two thousand items concerned with recreational mathematics, including Strens's own manuscripts in which he had made extensive explorations of magic squares, knight's tours and other classical areas.

His children were reluctant to see the careful collection of a lifetime broken up, but they did not have the facility to continue curating it themselves. A friend of the family, Lee Sallows, had the good thought to write to Martin Gardner, who copied the letter to several of us, including Doris Schattschneider, Donald Knuth, David Singmaster and Ron Graham. Donald Knuth would have liked Stanford to acquire the collection. David Singmaster's plans for a centre for recreational mathematics in London were not far enough advanced to enable him to house it. Martin Gardner himself seriously considered building an extra room onto his house, but realized that, within a finite time (and we all hope that this is not short), he would have to think about redisposing the collection.

I was particularly fortunate that our Director of Libraries, Alan MacDonald, was willing to house the collection as a Special Collection in The University of Calgary Library. Hendrik Lenstra, who is not only an internationally known number theorist, but also an experienced bibliophile, went with me to Breda to visit the Strens family and the collection, and to help put a value on it. The Strens children generously agreed to donate a substantial fraction of the collection, once they knew that it would be preserved as a whole, and this enabled me to purchase the balance, as I knew that there was a good chance of a matching grant from the government of the Province of Alberta.

Strens was a friend of Maurits Escher and, with the assistance of Pascal Strens, we have been able to acquire copies of four Escher prints, Earth, Air, Fire and Water, which commemorate four wedding anniversaries of Strens and his wife.

When we first acquired the Strens Collection, Alan MacDonald described it as "a pearl that will grow", and grow is indeed what it's doing. Small but interesting donations arrive almost every week, and larger contributions are not uncommon. Martin Gardner has sent more than 200 volumes. Wade Philpott's widow has presented his extensive library, and Mrs. Avetta Trigg has given us Charles W. Trigg's collection of more than a thousand items. These last two bequests contain a good deal of manuscript material, which should be of interest to investigators in the years to come. Charles Trigg in particular has been the doyen of recreational mathematics, and his mathematical activity continued from April 1932 until his death in June 1989. Recently we have negotiated with William L. Schaaf, author of the well known Bibliography, for the part donation and part purchase of his specialized library of about 250 volumes. Jordan Stoyanov, of the Bulgarian Academy of Sciences, contributes a steady stream of Eastern European literature.

How to make sure that the Strens Collection serves the interested public? It contains an enormous amount of information, not easily obtainable from any other source: and unfortunately, not always easily obtainable from the Collection itself. We encourage interested persons to visit the Collection, and to record the results of their consultations. David Singmaster and the late Victor Meally are among those who have already helped, and Donald Knuth recently spent a profitable ten days in the Collection. After a visit from Patrick Ion of *Mathematical Reviews*, Richard Nowakowski has designed two databases, NTRENSIC and XTRENSIC, which are compatible with MATHSCI, the on-line version of *Mathematical Reviews*, with which we eventually hope to amalgamate. NTRENSIC lists the items in the Collection, while XTRENSIC (presently in only an inchoate state) consists of bibliographies of a wide collection of topics, and includes references to items not yet in the Collection. Both databases enable searches to be made, and are invaluable to editors, authors and other investigators, and also enable us to answer sporadic enquiries. Very soon the information will be available on-line.

There remains an enormous amount of work to be done in attaching keywords and commentaries and reviews to the now more than 5000 entries: this can only be achieved by help from interested and dedicated people. Enquiries are welcome at any time and may be addressed to the Strens Collection, The McKimmie Library, The University of Calgary, Calgary, Alberta, Canada, T2N 1N4, or to the Honorary Curator, Richard K. Guy, Department of Mathematics & Statistics at the same address.

Eugène Louis Charles Marie Strens

Pascal Strens

Translated by Lee Sallows

Eugène Strens was born on August 5th 1899 in Roermond (Limburg), a provincial town in the south of Holland. He was the oldest son of Johanna Herten and Eugène Strens, a notary in Roermond. After him came his brother Otto, born in 1900 and his sister Jeanne in 1904.

In Roermond too he received his secondary education, attending the *Bisschoppelijk College*, a school funded by the clergy, where, after finishing the Humanities course (history, geography, languages), he went straight on to complete the Science syllabus (mathematics, chemistry, physics, biology). This is already characteristic of his intellectual grasp and catholic interests. Attention for the Humanities side had been inspired by his father who would have been happiest seeing his son study law, the Strens family having already produced a number of lawyers.

It was electrical engineering, however, that captured Eugène's imagination. During the First World War he constructed his first radio set, witnessing at the same time the arrival of the early motor cars which, at home, were already supplanting horses and horse-drawn vehicles. And so, shortly before his father's early death in 1919, he gained from the latter permission to study electrical engineering at the *Technische Hogeschool* in Delft.

Military service interrupted his training for a period of two years, which ended with his being an officer, having gained the rank of captain. Then, in 1927, he successfully completed his studies in Delft, becoming thus a qualified *ingenieur* or engineer. It was a profession he was to exercise only for a short while. After a brief interval occupied with lift-shaft installations at the coal mines in south Limburg, and three months in Paris, he worked briefly with a central-heating firm in Schiedam. After this came a further short connexion, this time with the Patent Office in The Hague.

Even by the time of his studies he had developed broad interests over many different areas and, not being dependent on employment for income—living at this time as a bachelor in a flat at The Hague—he enjoyed every opportunity to devote himself to these right up to the beginning of the war. He watched the emergence of the first air flights and followed the foundation of the KLM company in 1919, step by step. Driven by his philatelic interests, he sent letters abroad over the new air routes, including special flights to the Dutch East Indies.

Around 1928 saw the start of his enthusiasm for the "Ex Libris", the book-plates pasted inside the front covers of books, bearing the owner's name or crest. It was a graphical art form that appealed strongly to him and became the basis of his life's work: the assembly of a very large collection of Exlibris and similar graphic artwork. This patient acquisition, still complete and in the hands of the family, is one of the largest such private collections in existence, forming an excellent anthology of what Europe has put forth in this area. He maintained close contacts with fellow collectors, commissioned work from many artists, including Maurits Escher, and founded the Netherlands Exlibris Circle.

The range of his interests was by no means narrowed through this activity, however, for in the same period he undertook extensive travel, including to the U.S.A.; he remained a regular attender at theatres and concerts, and he kept closely in touch with Dutch literature.

It was in these pre-war years also that he laid the foundations for his mathematical library. And then, just a few weeks before the outbreak of the Second World War in Holland (May 10th 1940), he married Hendrika Wilhemina te Strake. Four children were to spring from this marriage: two daughters, Bianca (1942) and Marina (1947) and two sons, Pascal (1951) and Carlo (1954).

As an ex-officer being sought by the Germans, in 1943 Eugène went underground, living at a farmhouse close to Nijmegen in the east of Holland. Being separated from the larger part of his possessions, this period offered unusual opportunities for concentrated work on mathematical puzzles. Without the aid of an electronic calculator, he studied many such problems, often working out new variations on existing themes.

In 1945, after the war, he returned to the south to live in Breda, where he remained until his death. In spite of a large family, the interest in his hobbies remained undiminished.

Besides the fields of interest mentioned above, it was shortly after the war that he became particularly preoccupied with chess, collecting in the process about a thousand books on the subject. Meteorology and astronomy fascinated him too, little as his way of life was affected by the weather.

The daily routine followed a fairly strict pattern and despite the absence of a regular job, was tightly disciplined. Every morning he would rise at 7:00 a.m., listening first to the news and weather reports. Household chores would occupy him next, stoking the coal furnace and walking the dogs.

The house became filled with books—upwards of ten thousand—and was simply too small for the graphic art collection. In the "Cultural Centre" in Breda, though, he had a room of his own for the collection. Taking his *Solex*, a single-cylinder powered bicycle affectionately regarded in Holland for its quaint simplicity, every day he would ride to town to do the shopping and to work on the collection. At the beginning of the afternoon he would return for a short rest. After this it would be time to work through the large pile of daily post, including letters from his numerous correspondents abroad and many periodicals, his subscriptions numbering more than fifty. Evenings were filled with correspondence, lectures on philately, meteorology and astronomy, cultural pursuits and theatre.

Although his hobbies were those of an indoor type, the family had lots of animals: all sorts of poultry and, for the children, ponies. The latter, however, lay more in the province of his wife, Willy. Holidays tended to be active, the family going to Italy or the south of France in summer, and Austria for winter sports.

Not unnaturally, Eugène would communicate his enthusiasms to his children; a great treat for us would be those special occasions when the forbidden drawer was opened and each of us was invited to choose a game or puzzle to examine.

His eldest children attended the *Gymnasium* in Breda. Bianca went to Italy in 1961, studying art history and linguistics in Florence. Marina chose psychology, and Pascal medicine, later specializing in orthopaedic surgery.

Eugène Strens was an unassuming, gentle-hearted man of many-facetted learning. He lived a retiring life in spite of his many activities and, to an extent, was even eccentric—if only in having no job. He lived soberly, denying himself the luxury of a car. His health remained good until 1973, when he suffered a cerebral thrombosis. After that he never fully recovered, being unable to write with the hand and finding difficulty in walking. Nor would his familiar figure grace the trusty *Solex* again. His world thus contracted, and yet on a more modest scale he remained active, maintaining his various interests thanks to a mind in good order, until he passed away in February 1980.

The family he left behind greatly appreciates the transferral of his mathematical books to The University of Calgary, being especially happy that this portion is to remain intact, forming a fitting memorial to the remarkable man who was our father.

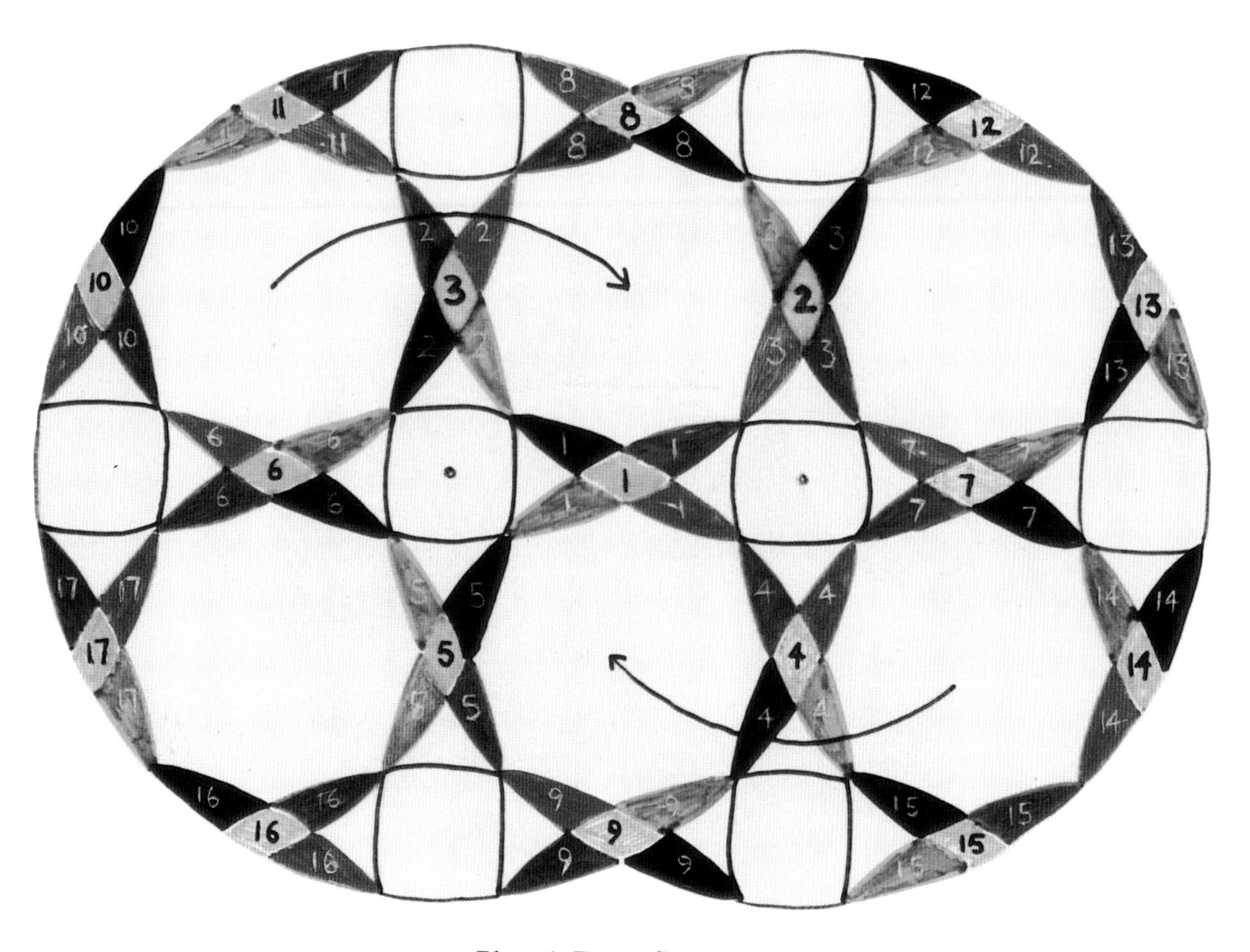

Plate 1. Butterfly puzzle.

Plate 2a

Plate 2b

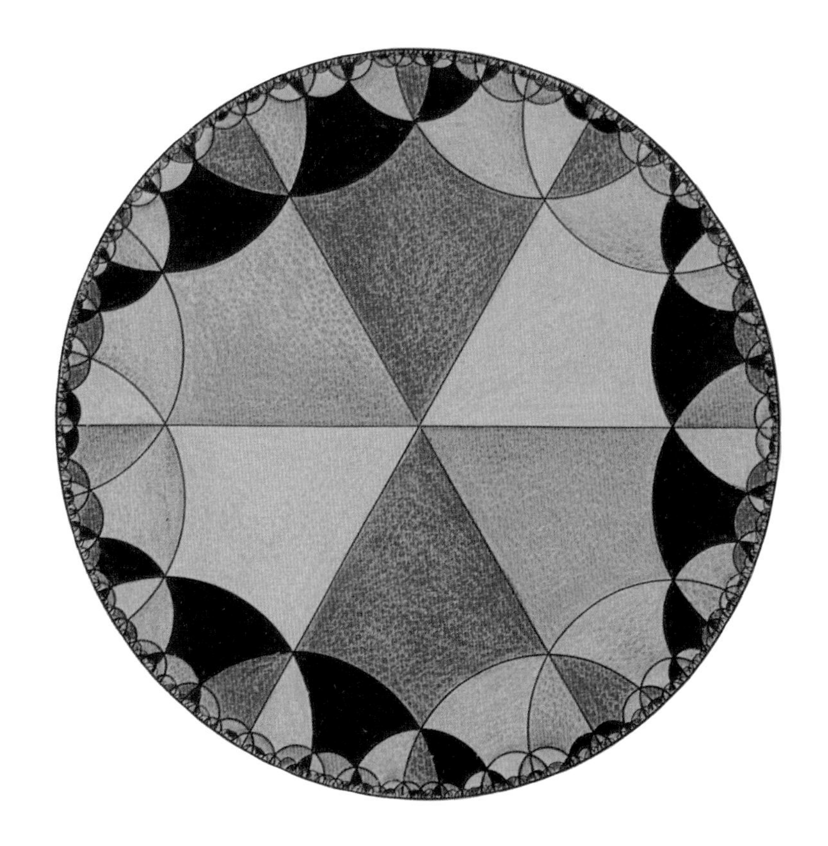

Plate 3a

Plate 3b

Plate 4a

Plate 4b

Part 1

Tiling & Coloring

Frieze Patterns, Triangulated Polygons and Dichromatic Symmetry

H.S.M. Coxeter & J.F. Rigby

The material in this article is not new, except perhaps for the final formulae in Section 3, but we have tried to present proofs of some known results in an intuitive and recreational manner.

1 Frieze Patterns

A frieze, in architecture, is "the part of the entablature between the architrave and cornice, often ornamented with figures" [**14**]; the Elgin Marbles, which formed the frieze on the Parthenon, are a well known example. But a frieze is also "a decorated band along the top of a room wall" [**14**], and in mathematics the term "frieze pattern" is reserved for any plane pattern that repeats regularly in one direction; such a pattern has **translational symmetry**, and a simple example is shown in the top row of Figure 1, where the basic translation is indicated by an arrow.

Frieze patterns can have other types of symmetry in addition, and they can be classified according to their symmetry types. The seven different symmetry-types are all shown in Figure 1, labelled according to one of the standard notations [**16**, p. 683; **13**] in which the symbols *1*, *m*, *2*, and *g* stand for translation, mirror, rotation and glide-reflexion. Type *1m* has a horizontal line of symmetry or **mirror line**. Type *m1* has vertical mirror lines; these occur at two different places in the pattern. Type *mm* has both horizontal and vertical mirror lines.

Type *12* has no mirror lines, but it has centres of 2-fold rotational symmetry, which we shall call simply **centres of symmetry**, indicated by dots; in type *12* these centres of symmetry occur at two different places in the pattern. Type *mg* has vertical mirror lines and centres of symmetry.

In type *1g* there is clearly some type of symmetry connecting the motifs in the top and bottom halves of the pattern, but the pattern has no mirror lines or centres of symmetry. We can transform a motif at the top to an adjacent motif at the bottom by means of a translation and a horizontal reflexion performed simultaneously. This type of transformation is called a **glide reflexion** (the relation between successive footprints along a snowy path), and we can indicate it by two half-arrows.

Any translation, reflexion, rotation or glide reflexion that transforms a pattern into itself is called a **symmetry** of the pattern.

Glide reflexions occur in Types *1m*, *mm*, and *mg* also, but in these types the glide reflexion is the product of two "more basic" symmetries of the pattern, either a translation and a horizontal reflexion, or a rotation and a vertical reflexion; this is not the case in Type *1g*.

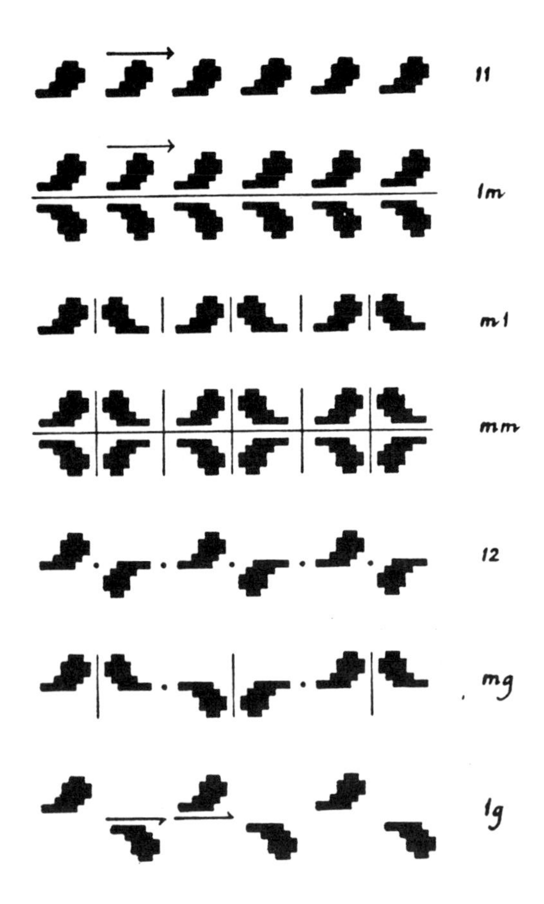

Figure 1: The seven symmetry types of frieze pattern.

Type *mm* has centres of symmetry, at the intersection of each vertical mirror line with the horizontal mirror line. Rotation through 180° about such a centre, which transforms the whole pattern to itself, can be achieved as the product of the reflexions in the two mirror lines through the centre.

We shall not attempt to show here that these are the only types of symmetry that a frieze pattern can have. It should be emphasized that symmetry types are not concerned with the style of a pattern; Figure 2 shows another pattern, from Turkey, whose symmetry-type is *m1*, having only vertical mirror lines.

Figure 2: A frieze pattern from Turkey.

2 Triangulated Polygons and Frieze Patterns

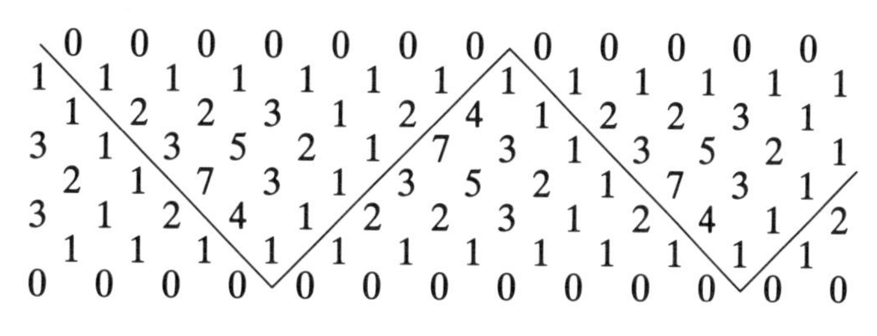

Table 1.

The first author once asked an audience of a hundred students of mathematics to look at the pattern in Table 1 and find the simple rule connecting each number with its neighbors and allowing the pattern to be extended indefinitely to the right and left. After an embarrassingly long time it was necessary to break the suspense by explaining that any four numbers forming a diamond, such as

$$\begin{matrix} & b & \\ a & & d \\ & c & \end{matrix}$$

satisfy the relation $ad - bc = 1$, which may also be written $c = (ad - 1)/b$; this is called the **unimodular rule**. Later, to test the effect of a brilliant brain, the same pattern was shown to Paul Erdős; he needed only a few seconds!

Using the same unimodular rule to construct other similar patterns with initial and final rows of zeros, it is soon observed that the diagonal sequence

$$0 \quad 1 \quad 1 \quad 1 \quad 1 \quad 2 \quad 1 \quad 0$$

starting at the top left corner of Table 1 (from which the entire pattern can be constructed) can be replaced by any sequence

$$0 \quad 1 \quad a \quad c \quad \ldots \quad 1 \quad 0,$$

where a, c, ... are positive integers each dividing the sum of its two neighbors; for instance

$$0 \quad 1 \quad 2 \quad 3 \quad 7 \quad 4 \quad 1 \quad 0.$$

All such patterns turn out to be frieze patterns with symmetry type *1g*, or in some special cases *1m*, *mg*, or *mm* as we shall see later. This remarkable periodicity is easy to prove when the number of rows is sufficiently small [**3**, pp. 90, 175]; for greater numbers of rows the proof is more tricky, and the search for it caused many restless nights [**5**; **6**].

Instead of beginning with a diagonal sequence, we could just as well begin by writing down a "suitable" periodic sequence below the first horizontal row of 1s; but what type of sequence is a "suitable" sequence to provide eventually another row of 1s? The surprisingly simple answer was supplied by J. H. Conway [**3**, pp. 87–88, 180].

A convex n-gon needs $n-3$ diagonals to **triangulate** it, that is, to divide it up into $n-2$ triangles. For instance, Figure 3 [**2**, p. 174] shows one of the four essentially different triangulations of a heptagon [**16**, p. 355; **15**]. At each vertex of the heptagon is written the number of triangles that come together at that vertex, and these numbers, repeated in their cyclic order, form the third row of numbers in Table 1. Conway found that a sequence of integers is "suitable" for forming a frieze pattern, using the unimodular rule, if and only if it arises in this way from a triangulated polygon. In this manner a triangulated n-gon yields a sequence of period n (or possibly a divisor of n) and a frieze pattern of $n + 1$ rows (including the top and bottom rows of 0s).

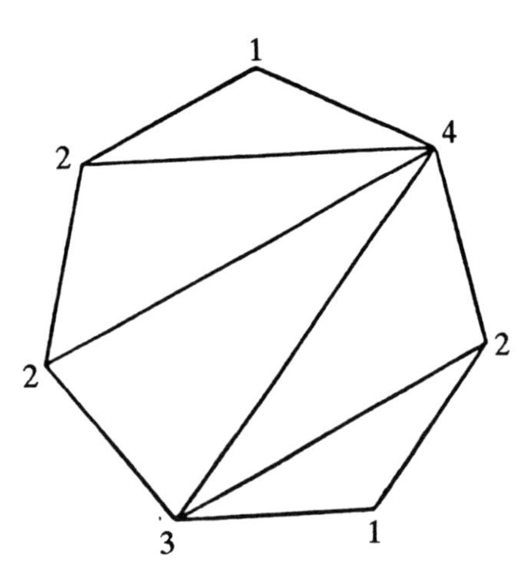

Figure 3: A triangulation of a heptagon.

Figure 4: Triangle 125 adjoined to side 14 of Figure 3.

If we remove the top triangle from the triangulated polygon in Figure 4, we obtain Figure 3 (actually, a distortion of Figure 3, but the exact shape of the polygons is not important), so Figure 4 is obtained by adjoining a triangle to Figure 3. Table 2 shows how the frieze pattern associated with Figure 4 is obtained from Table 1: the triangular portions in Table 1 have been separated to leave diagonal channels between them, and each new number in the channels is the sum of its two nearest neighbors in the separated portions. This observation leads to a proof by induction of Conway's result that every triangulated polygon has an associated frieze pattern of positive integers with glide-reflexion symmetry. We shall sketch the proof.

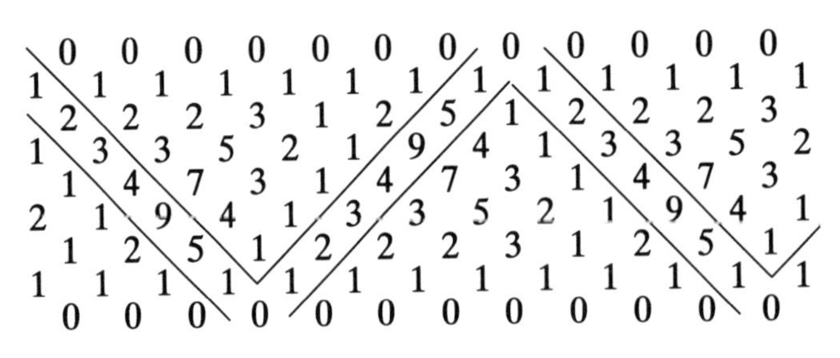

0	0	0	0	0	0	0	0	0	0	0	0	
1	1	1	1	1	1	1	1	1	1	1	1	1
2	2	2	3	1	2	5	1	2	2	2	3	
1	3	3	5	2	1	9	4	1	3	3	5	2
1	4	7	3	1	4	7	3	1	4	7	3	
2	1	9	4	1	3	3	5	2	1	9	4	1
1	2	5	1	2	2	2	3	1	2	5	1	
1	1	1	1	1	1	1	1	1	1	1	1	1
0	0	0	0	0	0	0	0	0	0	0	0	

Table 2.

Suppose, as an inductive hypothesis, we have proved that every triangulated n-gon has an associated frieze pattern of $n+1$ rows, bordered at the top and bottom by rows of 1s and 0s, with the "vertex numbers" of the polygon in the third and antepenultimate rows, satisfying the unimodular rule, and with glide-reflexion symmetry. Any given triangulated $(n + 1)$-gon may be obtained by adjoining a triangle to a suitable n-gon. This n-gon has an associated frieze pattern with a glide reflexion, which can be divided into triangular portions as in Table 1, taking into account the position where the extra triangle is to be adjoined to the polygon. Separate the triangular portions as in Table 2, and insert a new positive integer at each position in the channels by adding together the two nearest positive integers in the triangles. The new frieze pattern of n rows of positive integers has the vertex numbers of the triangulated $(n + 1)$-gon in its third and antepenultimate rows, and it has glide-reflexion symmetry. Also it satisfies the unimodular law, because if

$$\begin{matrix} & b & \\ a & & d \\ & c & \end{matrix}$$

is a unimodular diamond, then so are the two diamonds contained in

$$\begin{matrix} & b & & \\ a & & b+d & \\ & a+c & & d \\ & & c & \end{matrix}$$

since

$$a(b + d) - b(a + c) = (a + c)d - (b + d)c = ad - bc = 1.$$

Hence the new frieze pattern is the frieze pattern associated with the $(n + 1)$-gon. The induction can be started, since the unique triangulated 4-gon certainly has an associated frieze pattern; hence by induction every triangulated polygon has an associated frieze pattern with symmetry type *1g*.

A school teacher might well show a class of young children how to triangulate a convex polygon and how to construct the corresponding frieze pattern. This is a nice exercise in multiplication and division, because any mistake is liable to cause its own penalty in the form of a fraction or a negative number, whereas accuracy yields the pleasant surprise of finding that the process ends with another row of 1s followed by a row of 0s.

A triangulated polygon with 2-fold rotational symmetry, as in Figure 5a, gives a frieze pattern of type *1m*. A triangulated polygon with a line of symmetry, as in Figure 5b, gives a frieze pattern of type *mg*, and one with two perpendicular lines of symmetry (which must have rotational symmetry also), as in Figure 5c, gives a frieze pattern of type *mm*.

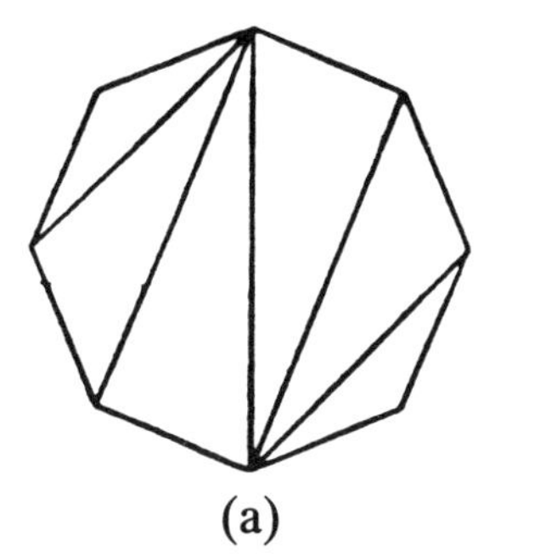
(a)

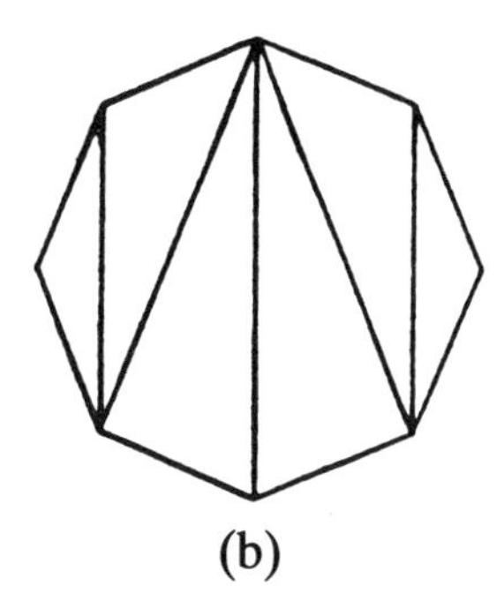
(b)

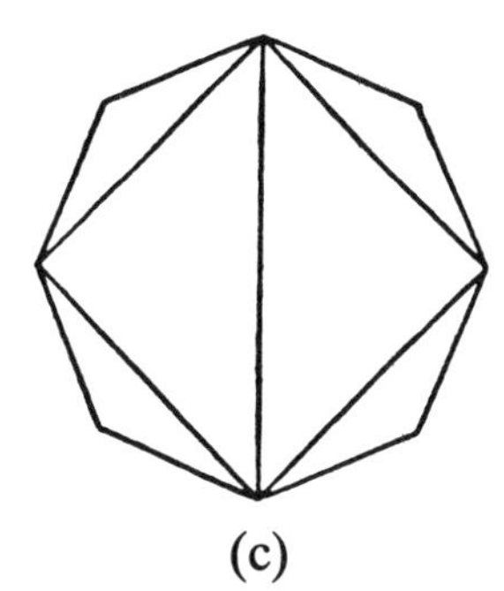
(c)

Figure 5: Symmetrical triangulations.

In Section 3 we shall show that every multiplicative pattern (i.e., every pattern satisfying the unimodular rule) with a finite number of rows and bounded by rows of 0s and 1s, and with no other 0s in it, is a frieze pattern with glide-reflexion symmetry. The fact that every such frieze pattern whose entries are positive integers is associated with a triangulated polygon (the converse of what we have just proved) is verified by simply reversing the steps in the above proof; but before we can take the general step that is equivalent to reducing Table 2 to Table 1, we have to show that every frieze pattern of positive integers contains at least one 1 in the third row. The details are given in [**3**].

In [**3**] there is a formula for the numbers in the third row of a frieze pattern in terms of the numbers in any diagonal. In Section 3 we shall give a formula for all the numbers in a frieze pattern in terms of the numbers in any diagonal.

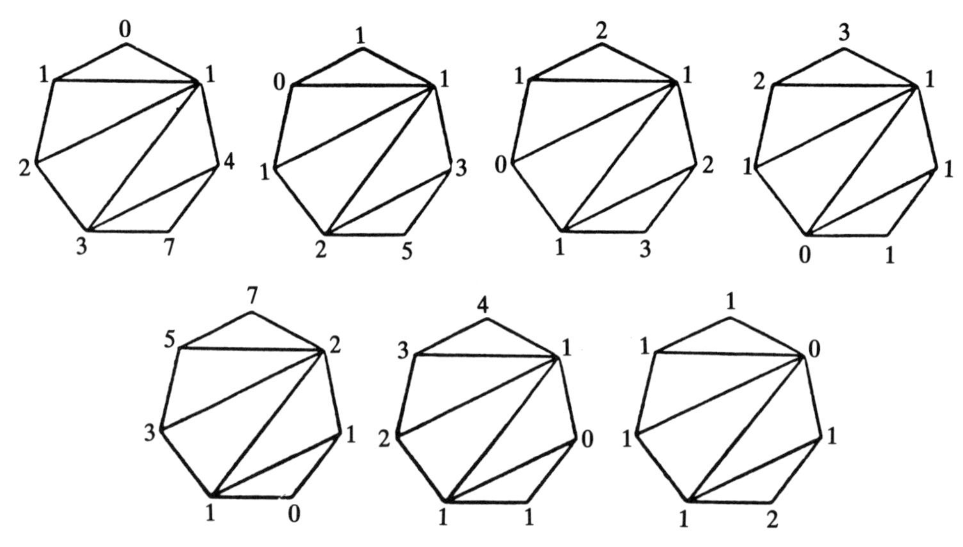

Figure 6: The seven different diagonals of Table 1.

Conway [**3**, pp. 93, 183] discovered also that the triangulated polygon provides, in a simple way, not only the third row in the frieze pattern but also the diagonal sequences in the pattern. For this purpose new numbers are assigned to the vertices, as in Figure 6, which shows the same triangulation as in Figure 3 but with different labels. Each vertex in turn is labelled 0. The number 1 is assigned to each of the two or more vertices that are joined to the vertex marked 0. Then whenever a triangle has two marked vertices, the third vertex takes their sum. The labels thus accumulated all round the polygon provide one of the diagonal sequences in the frieze pattern.

One surprising corollary of this numbering scheme is the following statement [**3**, pp. 93, 183]:

Every frieze pattern of integers either contains a 4
or consists entirely of Fibonacci numbers!

The above definition of a frieze pattern of numbers may be modified in various significant ways. Duane Broline [**1**] considered modifying the unimodular rule $ad - bc = 1$, but he had to allow the symmetry to be reduced from *1g* to *11*. Shephard's "additive" frieze patterns [**17**] will be discussed in Section 5.

In connexion with determinants and continued fractions [**6**, pp. 306–308], regular polytopes [**7**, pp. 22, 54–57, 141–147, 165–178] and "polygonometry" [**4**, pp. 204–205; **10**], it is natural to abandon the restriction to integers; but then, of course, there is no longer a connexion with triangulated polygons.

3 More About Multiplicative Frieze Patterns

If a multiplicative pattern contains 0s other than those in the first and last rows, it need not repeat regularly; an example is shown in Table 3, where a, b, c etc. can be any numbers.

	0		0		0		0	
1		1		1		1		1
	a		0		c		0	
−1		−1		−1		−1		−1
	0		b		0		d	
1		1		1		1		1
	0		0		0		0	

Table 3.

We shall henceforth assume without explicit mention that all numbers in our patterns are nonzero, except in the first and last rows.

If we consider the portion of a multiplicative pattern shown below

		a_1	
	a_2		b_1
a_3		b_2	
	b_3		

we easily deduce from the unimodular law that

$$(a_1 + a_3)/a_2 = (b_1 + b_3)/b_2. \tag{1}$$

Consider now the multiplicative pattern in Table 4.

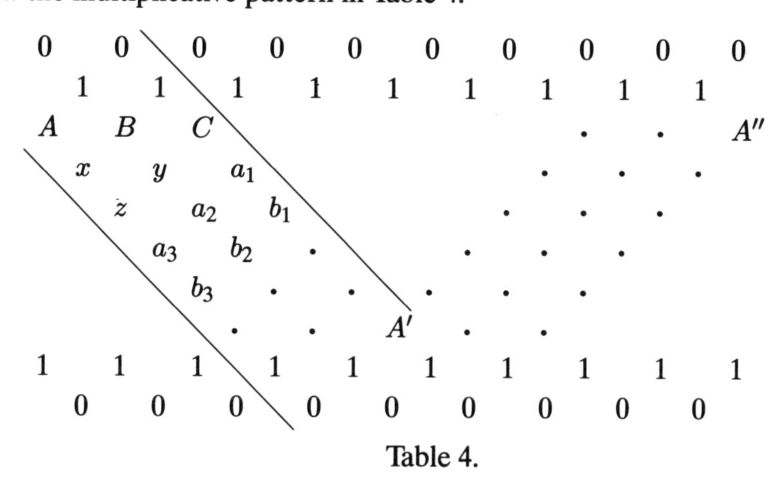

Table 4.

By applying (1) to consecutive triples in the table, we find that $(A + 0)/1 = (0 + A')/1$, i.e. $A = A'$. Similarly $A' = A''$. Hence the third row (and therefore all the other rows also) repeats regularly, so we have a frieze pattern, and because $A = A'$ we have glide-reflexion symmetry (since the pattern can equally well be built up from the bottom).

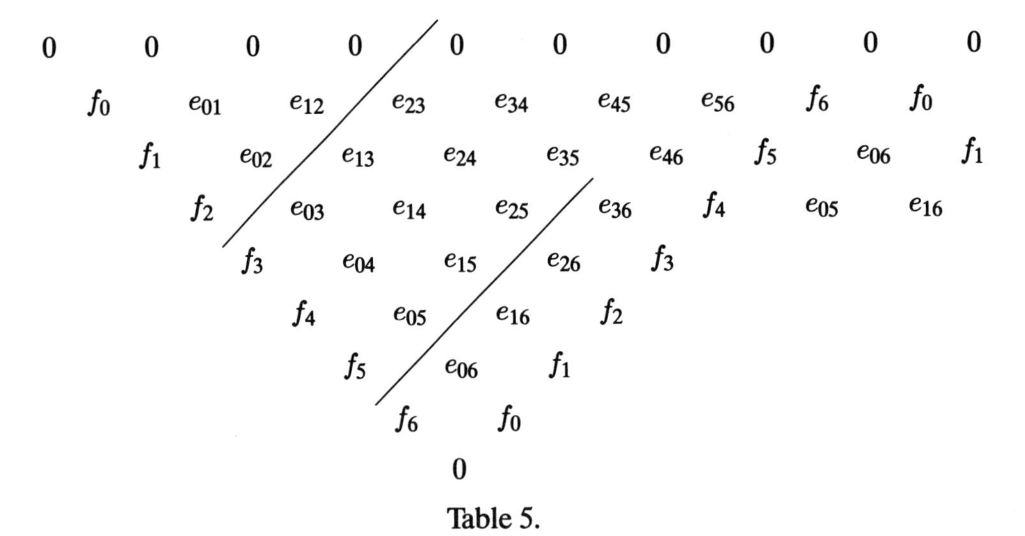

Table 5.

Table 5 now shows a typical frieze pattern; the entries in the second row are all equal to 1, and 0, f_0, f_1, f_2, ... is one of the diagonals. Applying (1) to consecutive triples along the diagonal indicated in the table, we find that

$$(f_3 + f_5)/f_4 = (0 + e_{35})/e_{34} = e_{35}$$

($= a_4$ in the notation of [3]). In the same way we obtain the general formula

$$a_s = e_{s-1,s+1} = (f_{s-1} + f_{s+1})/f_s,$$

which gives us values for all the entries in the third row except for e_{06}.

Write $u_i = 1/f_i$; then the general entry in Table 5 is given by the formula

$$e_{ij} = f_i f_j (u_i u_{i+1} + u_{i+1} u_{i+2} + \ldots + u_{j-1} u_j) \qquad (i < j).$$

To verify this, we simply have to check that with these values for the e_{ij} the unimodular law is satisfied throughout the table; this is left to the reader.

We now have a value for the "missing entry" e_{06} in the third row, which was not given by the previous formula.

4 Black-and-White Symmetry

Suppose that we color each motif in a frieze pattern either black or white (and imagine the "background" to have some third neutral color) in such a way that each symmetry of the original pattern either maps black to black and white to white or else interchanges black and white; we then have a **perfect black-and-white coloring** (or dichromatic coloring) of the frieze pattern. As an example, consider Figure 7, where our original pattern of type *mg* has been colored in three different ways. The first coloring has reflexions in vertical mirror lines that do not interchange the colors, indicated by unbroken lines, and rotations about centres of symmetry that do interchange the colors, indicated by white dots. The second has vertical mirror lines that do interchange the colors, indicated by broken lines, and centres of symmetry that do not, indicated by black dots. In the third coloring, both the mirror lines and the centres of symmetry interchange the colors.

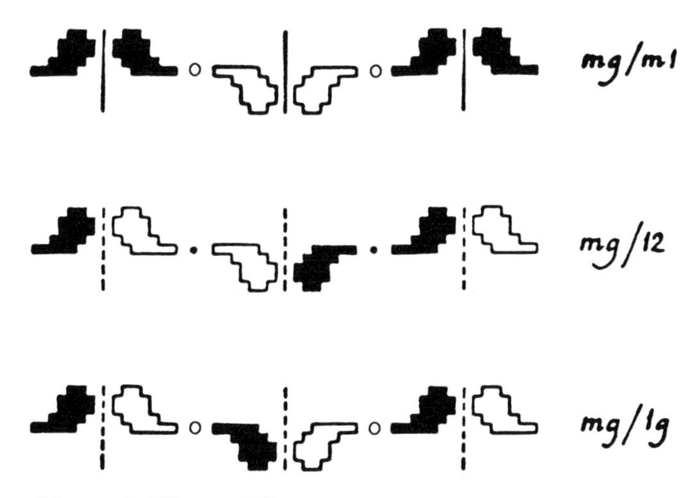

Figure 7: Three different colorings of a frieze pattern.

If in the first coloring in Figure 7 we look at the black motifs only, we see that these form a pattern of type *m1*; hence this coloring is labelled *mg/m1*. For similar reasons the other two colorings are labelled *mg/12* and *mg/1g*. This notation is an adaptation of the one devised by Schwarzenberger for colored plane symmetry patterns and described in [**9**].

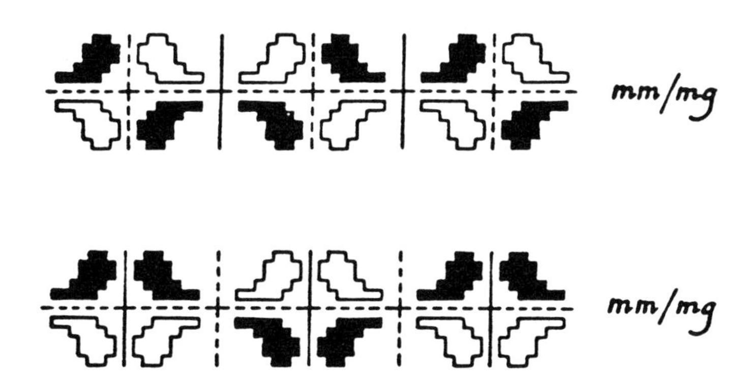

Figure 8: Two different colorings of the same type.

Two patterns have the same type of black-and-white symmetry if they have the same arrangement of broken and unbroken mirror lines or arrows, and white and black dots. For example, Figure 8 shows two different colorings of our pattern of type *mm*, but the two colorings are of the same type because they have the same arrangement of vertical mirror lines, alternately broken and unbroken, and both have a broken horizontal mirror line.

There are seventeen types of black-and-white symmetry [**8**; **17**], but Figure 9 shows only those that will interest us here, namely those possessing a glide reflexion that interchanges the colors.

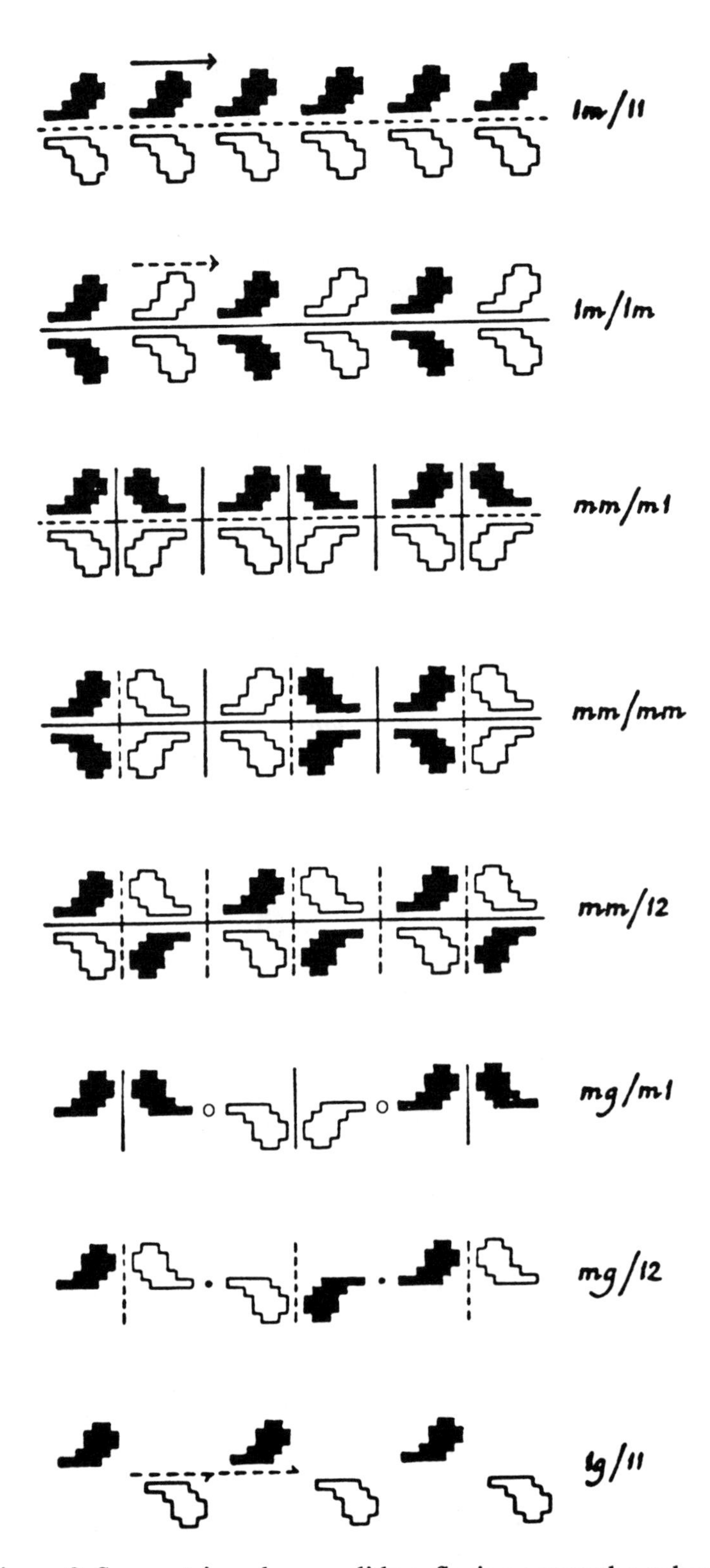

Figure 9: Symmetries where a glide reflexion swaps the colors.

5 Additive Frieze Patterns

In [**18**] Shephard considers **additive** frieze patterns, in which the rule for "completing the diamond" is

$$a + d = b + c + 1$$

and the first and last rows consist of 0s; one such pattern is shown in Table 6.

0		0		0		0		0		0		0		0		0		0		0
	1		1		4		1		5		3		1		1		4		1	
3		1		4		4		5		7		3		1		4		4		5
	2		3		3		7		6		6		2		3		3		7	
4		3		1		5		7		4		4		3		1		5		7
	4		0		2		4		4		1		4		0		2		4	
0		0		0		0		0		0		0		0		0		0		0

Table 6.

Table 7 shows the **constant** additive frieze pattern with seven rows; it is a simple exercise in induction to show that a constant additive frieze pattern with $n + 2$ rows must have $n/2$ as the constant number in the second row.

0		0		0		0		0		0
	$2\frac{1}{2}$		$2\frac{1}{2}$		$2\frac{1}{2}$		$2\frac{1}{2}$		$2\frac{1}{2}$	
4		4		4		4		4		4
	$4\frac{1}{2}$		$4\frac{1}{2}$		$4\frac{1}{2}$		$4\frac{1}{2}$		$4\frac{1}{2}$	
4		4		4		4		4		4
	$2\frac{1}{2}$		$2\frac{1}{2}$		$2\frac{1}{2}$		$2\frac{1}{2}$		$2\frac{1}{2}$	
0		0		0		0		0		0

Table 7.

If we subtract Table 7 from Table 6, and then multiply all the resulting numbers by 2 to avoid fractions, we obtain Table 8.

0		0		0		0		0		0		0		0		0		0		0
	−3		−3		3		−3		5		1		−3		−3		3		−3	
−2		−6		0		0		2		6		−2		−6		0		0		2
	−5		−3		−3		5		3		3		−5		−3		−3		5	
0		−2		−6		2		6		0		0		−2		−6		2		6
	3		−5		−1		3		3		−3		3		−5		−1		3	
0		0		0		0		0		0		0		0		0		0		0

Table 8.

Each diamond in this table satisfies the rule

$$a + d = b + c;$$

we shall call frieze patterns satisfying this rule **simple** frieze patterns, because the rule is so simple. We see that the frieze pattern in Table 8 possesses a glide reflexion that multiplies each number by

-1, thus transforming positive to negative and vice versa. This corresponds to the black-and-white symmetry type *1g/11* .

It is easy to prove that a simple frieze pattern always has this type of positive-negative symmetry. Taking a pattern with six rows as an example, suppose that one diagonal consists of the entries $0 \quad a \quad b \quad c \quad d \quad 0$, which we shall write in the form

$$0-0 \qquad a-0 \qquad b-0 \qquad c-0 \qquad d-0 \qquad 0-0.$$

Using the "simple diamond rule" we quickly verify that the complete pattern is

$0-0$		$a-a$		$b-b$		$c-c$		$d-d$		$0-0$
	$a-0$		$b-a$		$c-b$		$d-c$		$0-d$	
$a-d$		$b-0$		$c-a$		$d-b$		$0-c$		$a-d$
	$b-d$		$c-0$		$d-a$		$0-b$		$a-c$	
$b-c$		$c-d$		$d-0$		$0-a$		$a-b$		$b-c$
	$c-c$		$d-d$		$0-0$		$a-a$		$b-b$	

Table 9.

which clearly shows the sign-changing glide reflexion. This result is implicit in **[18]** but is expressed differently there.

There is a one-to-one correspondence between additive frieze patterns and simple frieze patterns (although we obscured this by multiplying by 2 to obtain Table 8); hence any additive pattern with a finite number of rows must repeat and is therefore a frieze pattern, and it possesses a hidden type of symmetry, which is displayed clearly in the corresponding simple frieze pattern as a positive-negative glide reflexion.

It is an interesting exercise to construct simple frieze patterns with extra positive-negative symmetries, corresponding to all the black-and-white symmetry types shown in Figure 9. Table 10 shows one whose symmetry type is *mg/m1*; the reader may like to construct others, some of which must perforce contain a row of 0s in the middle.

0		0		0		0		0		0		0		0		0		0		0
	5		5		−5		3		−3		−3		3		−5		5		5	
0		10		0		−2		0		−6		0		−2		0		10		0
	5		5		3		−5		−3		−3		−5		3		5		5	
8		0		8		0		−8		0		−8		0		8		0		8
	3		3		5		−3		−5		−5		−3		5		3		3	
0		6		0		2		0		−10		0		2		0		6		0
	3		3		−3		5		−5		−5		5		−3		3		3	
0		0		0		0		0		0		0		0		0		0		0

Table 10.

References

1. Duane M. Broline, Frieze patterns as matrix multiplication tables, *Proc. 9th SE Conf. Combin. Graph Theory Comput. Congress. Numer.*, **21**(1978) 151–161; MR **80i**:05022.

2. Duane M. Broline, D.W. Crowe & I.M. Isaacs, The geometry of frieze patterns, *Geometriae Dedicata*, **3**(1974) 171–176; MR **51** #210.

3. J.H. Conway & H.S.M. Coxeter, Triangulated polygons and frieze patterns, *Math. Gaz.*, **57**(1973) 87–94, 175–183; MR **57** #1254–5.

4. H.S.M. Coxeter, Twelve Geometric Essays, Southern Illinois University Press, Carbondale IL, 1968; MR **46** #9843.

5. H.S.M. Coxeter, Cyclic sequences and frieze patterns, *Vinculum (Melbourne)*, 1971.

6. H.S.M. Coxeter, Frieze patterns, *Acta Arith.*, **18**(1971) 297–310; MR **44** #3980.

7. H.S.M. Coxeter, Regular Complex Polytopes, Cambridge University Press, 1974; MR **51** #6555.

8. H.S.M. Coxeter, The seventeen black and white frieze types, *C.R. Math. Rep. Acad. Sci. Canada*, **7**(1985) 327–331; MR **87a**:20052.

9. H.S.M. Coxeter, Coloured Symmetry, in *M.C. Escher: art and science (Rome)*, North-Holland, Amsterdam, 1986, pp. 15–33.

10. H.S.M. Coxeter, A simple introduction to coloured symmetry, *Internat. J. Quantum Chem.*, **31**(1987) 455–461; MR **88g**:20101.

11. H.S.M. Coxeter, Trisecting an orthoscheme, in *Symmetry 2: Unifying Human Understanding*, ed. Istvan Hargittai, Pergamon Press, Oxford, 1989.

12. D.W. Crowe & D.K. Washburn, Groups and geometry in the ceramic art of San Ildefonso, *Algebras Groups Geom.*, **2**(1985) 263–277; MR **87k**:05055.

13. Martin Gardner, Mathematical Games, *Scientific Amer.*, **226**#4(Apr 1972) 101.

14. E.M. Kirkpatrick, ed., Chambers 20th Century Dictionary, Edinburgh, 1983.

15. J.W. Moon & Leo Moser, Triangular dissections of n-gons, *Canad. Math. Bull.*, **6**(1963) 175–178; MR **27** #4765.

16. Th. Motzkin, Relations between hypersurface cross-ratios, and a combinatorial formula for partitions of a polygon, for permanent preponderance, and for non-associative products, *Bull. Amer. Math. Soc.*, **54**(1948) 352–360; MR **9**, 489d.

17. Doris Schattschneider, In black and white: how to create perfectly colored symmetric patterns. Symmetry: unifying human understanding II, *Comput. Math. Appl. Part B* **12**(1986) 673–695; MR **87j**:05064.

18. G.C. Shephard, Additive frieze patterns and multiplication tables, *Math. Gaz.*, **60**(1976) 178–184; MR **58** #16353.

Department of Mathematics,
University of Toronto,
Toronto M5S 1A1,
Canada.

University of Wales College of Cardiff,
School of Mathematics,
Senghennydd Road,
Cardiff CF2 4AG,
Wales, United Kingdom.

Is Engel's Enigma a Cubelike Puzzle?

J.A. Eidswick

1 Introduction

At first glance, the question of the title seems to mix apples and oranges. Engel's Enigma (Figure 1(c)) is a two-dimensional puzzle; cubelike puzzles (Figure 2) are *a priori* three-dimensional. Nevertheless, any abstract study of cubelike puzzles like the one suggested in Section 2 would have to take into account "two-faces puzzles" (Figures 1(a)–(b)) and would undoubtedly recognize Engel's Enigma to be in that category. Two-faces puzzles are discussed in [1], [3], [5], and [8].

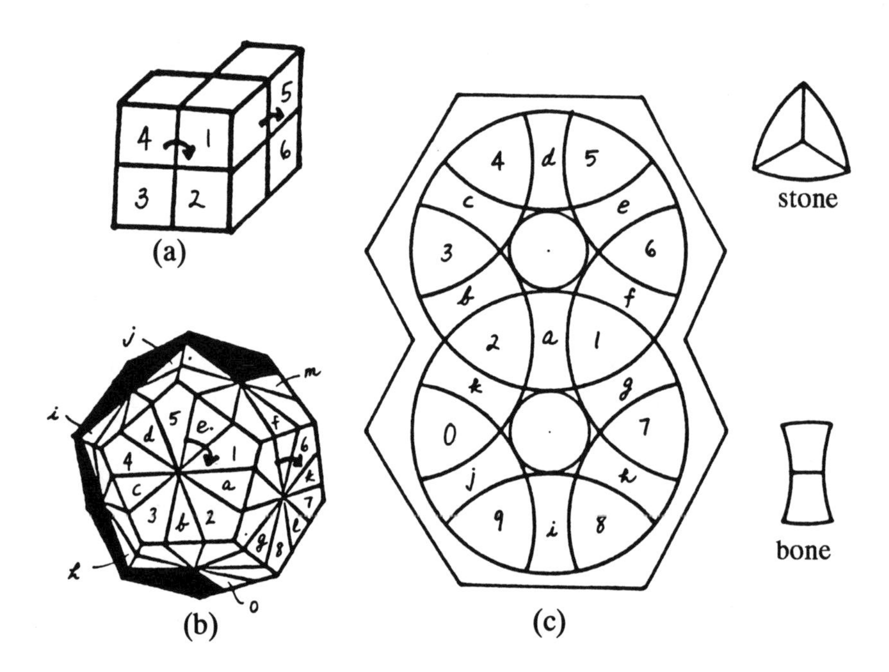

Figure 1: Engel's Enigma.

Engel's Enigma is discussed in the October 1985 issue of *Scientific American* [2]. Briefly, Engel's Enigma consists of two intersecting disks which are partitioned into pieces, called "stones" and "bones", in such a way that each disk can be rotated through multiples of 60°. After a few rotations, the puzzle will be so mixed up that even avid Rubik cube solvers will have trouble restoring it.

Engel's Enigma is just one of a family of two-dimensional puzzles called "circle puzzles". A subfamily of this family will be discussed in Section 3. Most circle puzzles exist only on paper. A booklet [4] written, published, and distributed by inventor/engineer Douglas Engel, describes these puzzles in considerable detail and challenges the mathematics community to put some order into the subject. The booklet concludes with an interesting section on puzzle patents.

2 Cubelike puzzles

Cubelike puzzles, discussed at length in [3], are, *a priori*, three-dimensional devices consisting of pieces which can be permuted by performing combinations of certain basic moves, the basic moves being rotations through a fractional multiple of 2π.

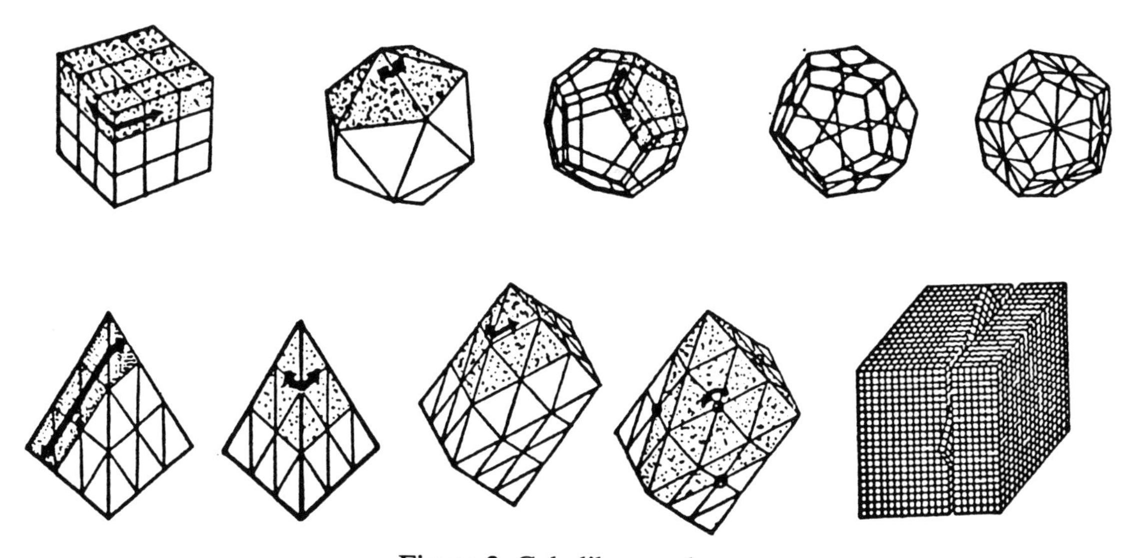

Figure 2: Cubelike puzzles.

How can such puzzles be described mathematically? One possibility is to characterize the permutation groups G which arise or could conceivably arise from such puzzles. A reasonable starting point might be the following three properties, common to the puzzles of [3] (including those in Figure 2):

(1) G is generated by a finite number of elements of the form $c_1c_2\dots c_q$, where the factors c_i are disjoint p-cycles (p, q fixed).

(2) If c_1 and c_2 are any two p-cycle factors of any two generators, then c_1 and c_2 intersect in at most two points.

(3) If g_1 and g_2 are generators, then there exist involutions ϕ and ψ such that $g_2 = \phi^{-1}g_1\phi$ and $g_2^{-1} = \psi^{-1}g_1\psi$.

Property (2) says that the cycles that make up the generators behave like circles relative to one another. Property (3) says that, in a sense, all generators look alike.

In this article the focus will be on groups satisfying (1)–(3) whose transitive constituents are primitive. Thus, in the following examples, the focus will be on permutations of pieces rather than on permutations of the points that make up the pieces. Note that from the puzzle-solver's point of view, this is a significant simplification. For a more general discussion involving wreath products, see **[3]**. For the basic theory of permutation groups, see **[9]**.

EXAMPLE 1. The cubelike puzzle group corresponding to Figure 1(a) is generated by

$$\begin{aligned} g_1 &= (1234) \quad \text{and} \\ g_2 &= (1562). \end{aligned}$$

Here $p = 4$, $q = 1$, and we may take

$$\begin{aligned} \phi &= (12)(35)(46) \quad \text{and} \\ \psi &= (36)(45). \end{aligned}$$

It can be shown that this group is isomorphic to S_5 (see, e.g., **[8]**).

EXAMPLE 2. The cubelike puzzle group corresponding to Figure 1(b) is generated by

$$\begin{aligned} g_1 &= (12345)(abcde)(fghij) \quad \text{and} \\ g_2 &= (16782)(afklg)(bemno). \end{aligned}$$

Here $p = 5$, $q = 3$, and we may take

$$\begin{aligned} \phi &= (12)(36)(47)(58)(bf)(ck)(dl)(eg)(hm)(in)(jo) \quad \text{and} \\ \psi &= (38)(47)(56)(bg)(cl)(dk)(ef)(ho)(in)(jm). \end{aligned}$$

It can be shown that this group is isomorphic to $A_8 \times A_{15}$.

EXAMPLE 3. Engel's Enigma (orientations disregarded) is generated by

$$\begin{aligned} g_1 &= (123456)(abcdef) \quad \text{and} \\ g_2 &= (178902)(aghijk). \end{aligned}$$

Here $p = 6$, $q = 2$, and we may take

$$\begin{aligned} \phi &= (12)(37)(48)(59)(60)(bg)(ch)(di)(ej)(fk) \quad \text{and} \\ \psi &= (30)(49)(58)(67)(bk)(cj)(di)(eh)(fg). \end{aligned}$$

It can be shown (see below) that this group is a subgroup of index 2 of $S_{10} \times S_{11}$.

Note that, as in **[3]**, one can simultaneously find solution algorithms and determine group structures. For example, to solve Engel's Enigma, one may use commutators to find 3-cycles of bones and stones as follows:

$$\begin{aligned} [g_1, g_2] &= g_1 g_2 g_1^{-1} g_2^{-1} \\ &= (1\,6)(2\,0)(akf) \\ [g_1, g_2]^2 &= (afk) && \text{(a 3-cycle of bones)} \\ h_1 &= [g_1, g_2]^3 = (1\,6)(2\,0) \\ h_2 &= g_1^2 h_1 g_1^{-2} = (2\,3)(4\,0) \\ [h_1, h_2] &= (2\,0\,4) && \text{(a 3-cycle of stones)} \end{aligned}$$

One may then use conjugacy to obtain all 3-cycles of bones and stones. It follows that any element of the form BS can be constructed, where B and S are, respectively, permutations of bones and stones that have the same parity. Note that the parity restriction is necessary.

A great deal of research has been devoted to groups generated by products of q p-cycles where p is prime and $q < p$ (see [6], [7]), but very little seems to be known about groups that satisfy (1)–(3). Note that properties (1)–(3) are insufficient to characterize cubelike puzzles. For example, the group generated by

$$g_1 = (123)(456) \quad \text{and} \quad g_2 = (127)(345)$$

satisfies (1)–(3), but, clearly, it could not represent a cubelike puzzle.

3 A kaleidoscope of circle puzzles

A subfamily of the family of circle puzzles will now be described.

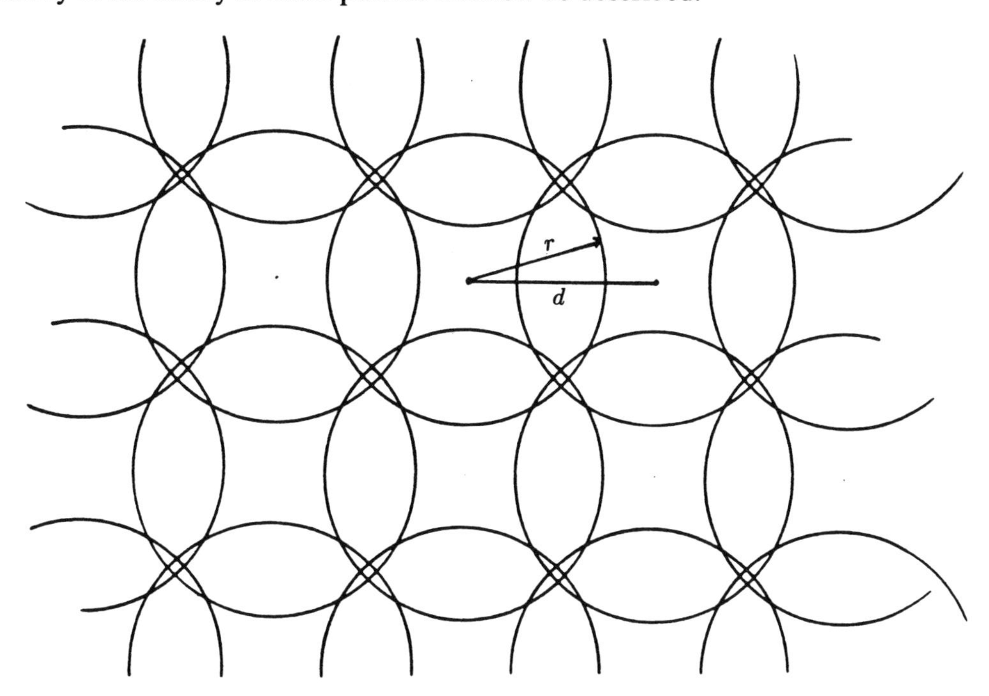

Figure 3: Circle puzzles.

On an infinite square grid with lattice points d units apart, imagine drawing circles centered at the lattice points with radii r (see Figure 3). Here we require only that $r > d/2$. Note that the most interesting cases will occur for $r > d$.

Now restrict the drawing to just two disks centered at adjacent lattice points. The result is a two-generator puzzle with $p = 4$. See Figure 4.

As r increases, one gets a kaleidoscopic effect. The puzzle changes continuously, then makes a sudden change, then repeats this behavior over and over, getting more and more complicated with increasing r. It is not hard to see that these discontinuities occur at values of r for which three or

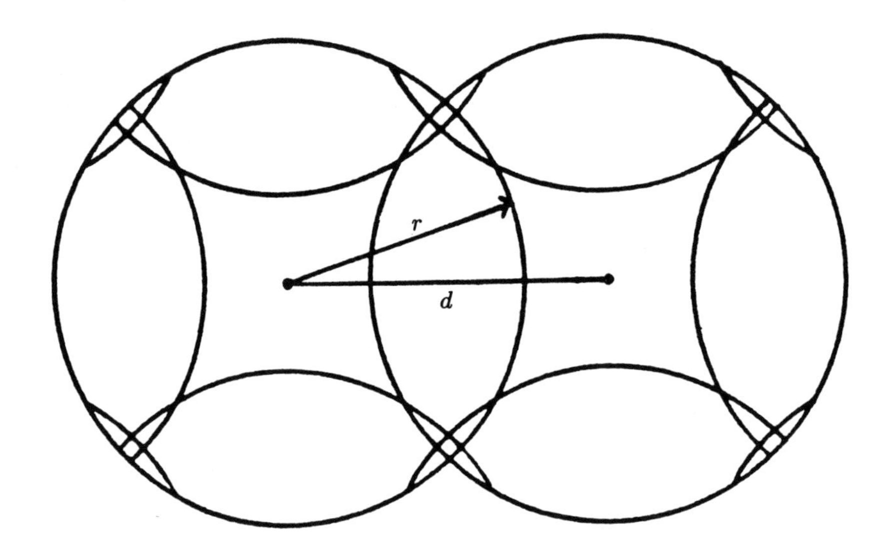

Figure 4: A two-generator puzzle.

more circles intersect at a point or two circles intersect at a point of tangency. Group-theoretically, we get a family

$$\{G(r) : \frac{d}{2} < r < \infty\}$$

of groups and an increasing sequence $\{r_k\}$ such that $G(r)$ is isomorphic to $G(s)$ whenever $r_k < r < s < r_{k+1}$.

Questions: What precisely is the sequence $\{r_k\}$? What are the **orders** n of the $G(r)$? How many **orbits** m does $G(r)$ have? Table 1 shows the first twelve distinct stages which correspond to the first nine values of r:

$$1/2, 1/\sqrt{2}, 1, \sqrt{5}/2, 5\sqrt{2}/6, 5/4, \sqrt{2}, 3/2, \sqrt{5/2}.$$

Stage	r	m	n
1	$d/2 < r \leq d/\sqrt{2}$	1	7
2	$d/\sqrt{2} < r \leq d$	4	33
3	$d < r < d\sqrt{5}/2$	6	59
4	$r = d\sqrt{5}/2$	5	52
5	$d\sqrt{5}/2 < r < 5d\sqrt{2}/6$	10	129
6	$r = 5d\sqrt{2}/6$	8	109
7	$5d\sqrt{2}/6 < r < 5d/4$	10	159
8	$r = 5d/4$	8	119
9	$5d/4 < r \leq d\sqrt{2}$	10	159
10	$d\sqrt{2} < r \leq 3d/2$	14	221
11	$3d/2 < r < d\sqrt{5/2}$	15	244
12	$r = d\sqrt{5/2}$	7	119

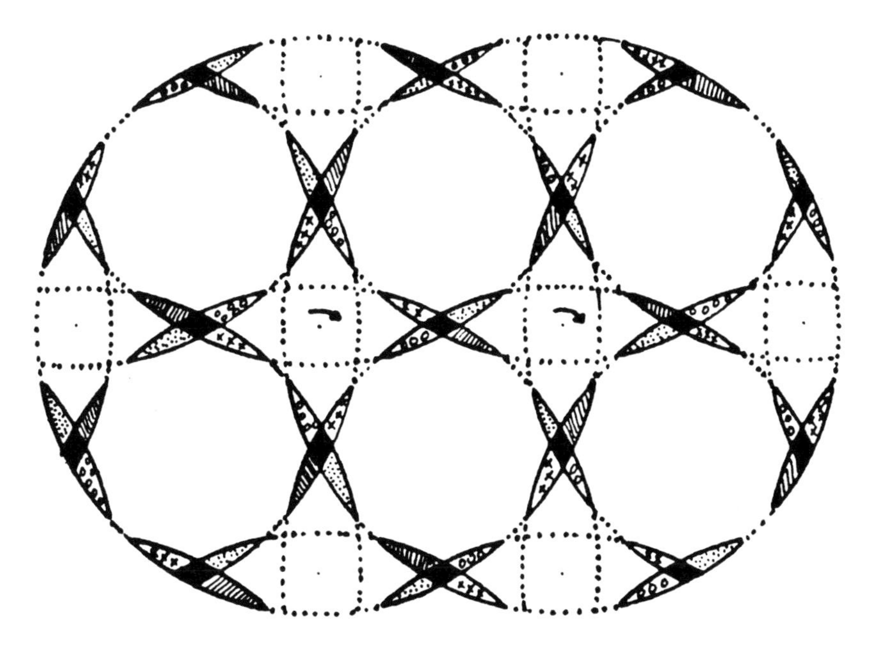

Figure 5: Butterfly puzzle.

To get an idea of the difficulties involved in obtaining answers to the above questions and solving the corresponding puzzles, consider the "butterfly puzzle" shown in Figure 5. This puzzle is obtained by restricting Stage 5 to the five new orbits formed at that stage. Note that different shadings indicate the different orbits. (If the orbits are colored with contrasting colors, e.g., red, green, yellow, blue, and black, the effect is quite pleasing. See Plate 1 on p. 9.)

For the butterfly puzzle, $p = 4$ and $q = 13$, and one may routinely verify condition (3). It can also be shown (with some effort) that 3-cycles within each orbit are attainable. Thus, taking into consideration parity constraints (cf. [**3**, Prop. 1]), one sees that there are

$$\frac{15!^4 17!}{8} > 10^{62}$$

attainable permutations of the 77 pieces! Now how about if I mix it up and you unmix it?

References

1. E.R. Berlekamp, J.H. Conway & R.K. Guy, *Winning Ways for Your Mathematical Plays*, Vol. 2, Academic Press, New York, 1982.

2. A.K. Dewdney, Computer Recreations: Bill's baffling burrs, Coffin's cornucopia, Engel's enigma, *Scientific American*, October 1985, 16–27.

3. J.A. Eidswick, Cubelike puzzles - what are they and how do you solve them?, *Amer. Math. Monthly*, **93**(1986) 157–176.

4. Douglas A. Engel, *Circle Puzzler's Manual*, General Symmetrics, Englewood, Colorado, 1986.

5. Douglas R. Hofstadter, Metamagical Themas: The Magic Cube's cubies are twiddled by cubists and solved by cubemeisters, *Scientific American*, March 1981, 20-39.

6. Martin W. Liebeck & Jan Saxl, Primitive permutation groups containing an element of large prime order, *J. London Math. Soc.*, (2) **31**(1985) 237–249.

7. C.E. Praeger, On elements of prime order in primitive permutation groups, *J. Algebra*, **60**(1979) 126–157.

8. David Singmaster, *Notes on Rubik's Magic Cube*, Enslow Publishers, Hillside NJ, 1981.

9. Helmut W. Wielandt, *Finite Permutation Groups*, Academic Press, New York, 1964.

Department of Mathematical Sciences
University of Montana
Missoula, MT 59812-1032

Metamorphoses of Polygons

Branko Grünbaum

The first three illustrations of this note show "metamorphoses" of polygons—sequences of polygons gradually changing from one regular 14-gon to another regular 14-gon. While the "star" polygons that arise at the intermediate steps can be enjoyed for their unfamiliar but attractive shapes, there is quite a lot of mathematics that can be appreciated at the same time. I should hasten to add that there is nothing difficult or deep in the mathematical aspects of these metamorphoses; in fact, most of the assertions I shall make are so obvious that any formal proofs would only obscure the situation.

First, a brief explanation of the diagrams in Figures 1, 2 and 3. Each begins and ends with a regular polygon, that is, with a polygon in which all vertices are alike, and all edges (sides) are alike. The intermediate polygons are "regular" to a lesser degree—only the vertices of each are alike, while the edges are of two kinds. In each of the three sequences, a finite number of polygons is shown; however, they are only instances that happen to have been selected from among families of polygons that change in shape continuously, reaching through gradual change from one of the two extreme specimens to the other.

Now, the first mathematical point to be made is that all the polygons shown in the diagrams are 14-gons, despite the seemingly obvious presence of heptagons. Clearly, this calls for some explanation, and it brings up errors made almost two centuries ago and perpetuated ever since. We'd better start with some definitions.

Given an integer $n > 1$, an n-gon P is any collection of n points $A_1, A_2, \ldots, A_{n-1}, A_n$, called **vertices** of P, together with the n segments $A_1A_2, A_2A_3, \ldots, A_{n-1}A_n, A_nA_1$, called the **edges** of P. In general, the vertices can be points in any setting in which it makes sense to talk about segments; in the present note we shall restrict ourselves, without exception, to the traditional Euclidean plane. The edges are straight line segments; however, since the coincidence of two consecutive vertices has not been excluded, some of the segments that form edges can be reduced to single points. The definition of "polygon" does not preclude this from happening; neither does it preclude many other kinds of coincidence or overlap. In fact, one could even admit the case in which all vertices coincide, though it should be granted that such a polygon is not of great interest; in order to avoid repetitious exceptions, we shall exclude such polygons from all further considerations.

Although some earlier writers made hints in the direction of such a general definition of polygons, the first explicit statement appears in Meister [7]—an intriguing work, with which posterity has dealt very poorly (as will be explained below). The same definition reappeared in Poinsot's often-quoted paper [9], and has since become standard—except that some authors balk at admitting consecutive vertices that coincide, while other workers forbid the coincidence of any vertices, or even the placement of a vertex at any point of an edge (other than its two endpoints). One hint

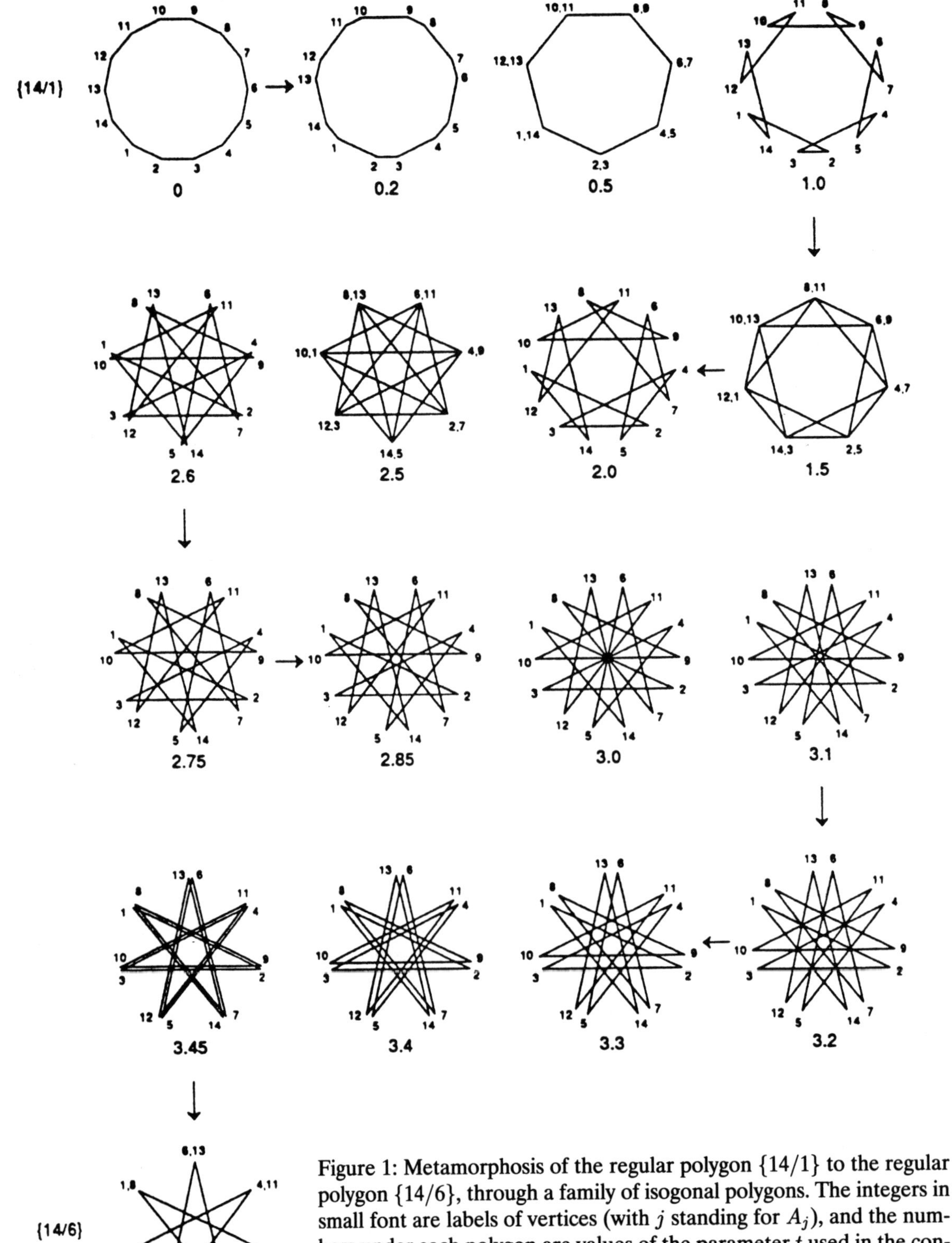

Figure 1: Metamorphosis of the regular polygon $\{14/1\}$ to the regular polygon $\{14/6\}$, through a family of isogonal polygons. The integers in small font are labels of vertices (with j standing for A_j), and the numbers under each polygon are values of the parameter t used in the construction, as discussed in the text. The different parts of the figure show polygons of distinct types (in the sense defined in the text).

why such a negative attitude is unjustified appears in our diagrams: if polygons with coinciding vertices were banned from the discourse, our continuous families of polygons would shatter into many fragments (which would then be unrelated), and some of the extreme polygons would also be ruled out of existence. Clearly, the universe becomes more orderly if vertices may coincide.

Next, we should consider what is meant by "regular" polygon. The idea, from time immemorial, was that a polygon is regular if all its angles are congruent, and if all its edges are congruent. As can be seen by the example of the cross-shaped 12-gon that is the boundary of the union of five congruent squares, this definition has to be taken with a grain of salt: nobody would like to consider that polygon as regular. Following Möbius **[8]**, it became customary to understand the requirements just stated in a stricter sense than conveyed by the words used, namely as additionally requiring that the two edges which determine an angle be correspondingly congruent to the two edges determining any other of the angles, and analogously, that the two angles at the endpoints of each edge be correspondingly congruent and equally placed on every other edge. Clearly, this eliminates the unwanted examples. The same goal can be achieved by considering the polygon and the plane as **oriented**, so that angles can be positive as well as negative, and inserting the appropriate requirement in the definition.

However, a much more elegant way relies on symmetries, that is, isometric mappings (congruences) that may bring a polygon to coincide with itself. The above definitions (in the restricted versions) can be rephrased, equivalently, in each of the following two forms:

1. A polygon P is regular if and only if
   (a) for each pair of vertices of P there is a symmetry of P that maps the first onto the second; and
   (b) for each pair of edges of P there is a symmetry of P that maps the first onto the second.
2. A polygon P is regular if and only if for each pair of edges of P there is a symmetry of P that maps the first onto the second.

Although formally (1) seems to be more restrictive than the previous definition, and (2) more restrictive than (1), in fact all these definitions are equivalent.

By well-known general arguments, all symmetries of a regular polygon (or any other polygon) form a **group of symmetries**, in which the group operation is composition of the isometries. Condition (a) expresses the transitivity of the group of symmetries on the set of vertices, condition (b) transitivity on the set of edges. Any polygon satisfying (a) is called **isogonal**, and any polygon satisfying (b) is said to be **isotoxal.** The mathematical fact visually expressed by Figures 1, 2 and 3 is the possibility of connecting certain pairs of regular n-gons by a continuous family of isogonal n-gons.

But here again we run into the need for some explanations, since practically every text that mentions regular star-polyhedra states that each can be denoted by a symbol $\{n/d\}$, where n and d are **relatively prime,** a condition that clearly does not apply to such expressions as $\{14/12\}$. The rather dismaying story is as follows.

When starting his investigation of regular polygons, Poinsot **[9]** used definitions of polygons and regular polygons equivalent to the ones given above. He went on to say (with different words,

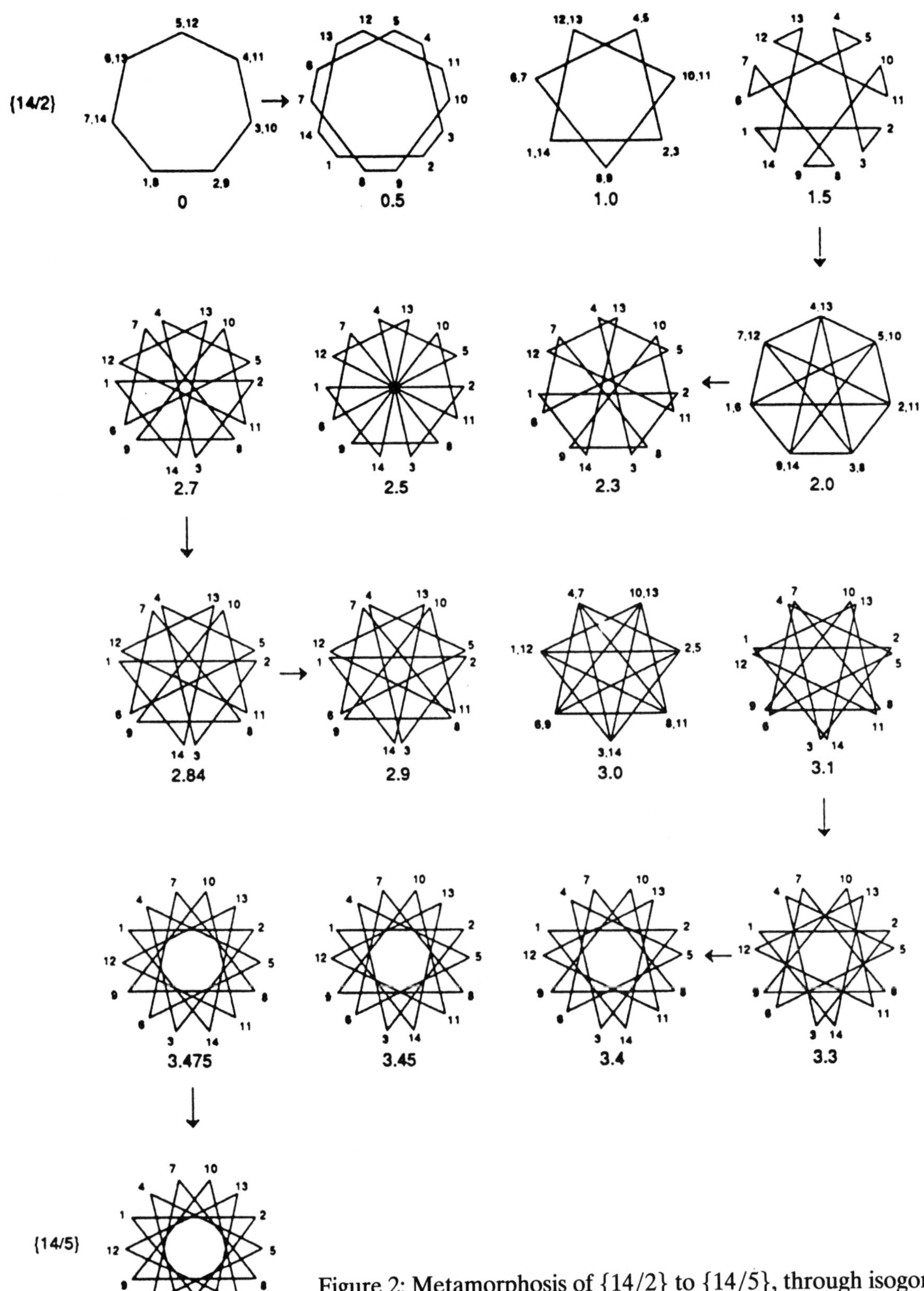

Figure 2: Metamorphosis of {14/2} to {14/5}, through isogonal polygons.

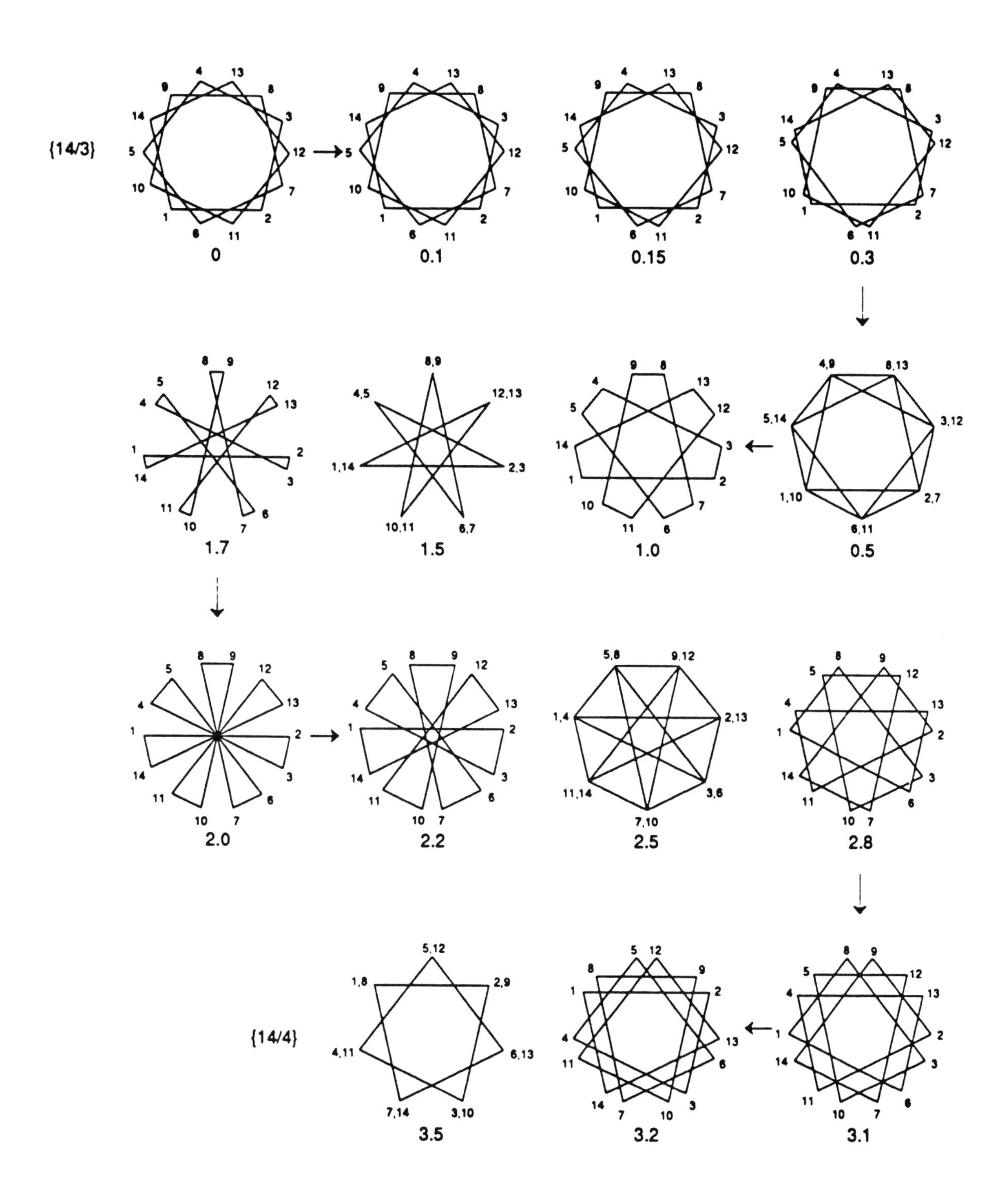

Figure 3: Metamorphosis of {14/3} to {14/4}, through isogonal polygons.

but quite correctly) that if $d \leq n/2$ is a positive integer relatively prime to n, and if n points equidistributed on a circle are connected to each other by segments each of which spans d of the arcs determined by the points, a regular polygon is obtained; this is the polygon usually denoted by $\{n/d\}$. After illustrating this construction with a few examples, Poinsot goes on to say (again, quite correctly) that if n points are equidistributed on a circle, and are connected by segments each of which spans d of the arcs, but with n and d having a common factor $k > 1$, then no regular polygon is obtained. Instead of a single polygon, one obtains a family consisting of k regular polygons of type $\{\frac{n}{k}/\frac{d}{k}\}$.

The logical error committed by Poinsot, and repeated ever since in all the texts, is the assumption that the statements of the preceding paragraph prove the nonexistence of regular polygons satisfying the requirements of the definitions, but corresponding to values of n and d that are not relatively prime. In fact, to see the fallacy of that assumption and to actually construct the polygons in all cases, all one has to do is to start with a point on the circle, connect it to a second point by a segment spanning an arc which is d/n times the length of the perimeter of the circle, connect that point to a third in the same way, and so on, until the nth step, which closes the circuit of edges of the polygon. It is obvious that the difference between n and d relatively prime or not is expressed by the fact that in the former case the nth step will be the first time the starting point is met again, while in the latter case it is the kth such encounter. Although in that case k-tuples of vertices of the polygon come to lie at the same point, they all are still distinct as vertices. Following the edges of $\{n/d\}$ one goes around the center of the polygon d times before closing the circuit—regardless of whether n and d are relatively prime or not. (If $d = n/2$, the polygon appears as a segment covered n times; since these polygons have certain special properties, in some contexts they need to be considered separately.)

We note in passing that such "unconventional" regular polygons as $\{6/2\}$, $\{8/2\}$, etc. can be used to generate "unconventional" regular polyhedra—but exploring this direction in the present paper would lead us too far afield.

It is important to understand that, for example, the regular polygon $\{20/4\}$ (in which the 20 vertices are located by fours on each of five points, which are the vertices of a regular pentagon) consists of one circuit of 20 edges winding four times around the pentagon—and not of four superimposed pentagons. The error just mentioned is one of many made by Edmund Hess (see **[5]**, page 632) in his study of isogonal polygons (which will be discussed below).

The mistakes of Poinsot and Hess are even more striking in view of the fact that Meister **[7]**, writing at a much earlier date, saw the situation correctly, and explained and illustrated it in great detail—including the diagrams of all ten regular 20-gons $\{20/1\}, \{20/2\}, \ldots, \{20/9\}, \{20/10\}$. But instead of gratitude, Meister reaped slander: it appears that the only person who read Meister's masterpiece during the first 200 or so years after its publication was the historian of mathematics Sigmund Günther; unfortunately, Günther (**[4]**, pp. 45–46) misquotes Meister's explicit statements and ascribes to Meister the same (erroneous) conclusions that were reached by Poinsot. All later writers (such as Brückner **[1]**), if they mention Meister at all, quote from Günther, thus missing the deeper understanding achieved by Meister, and helping perpetuate Poinsot's fallacy.

With these explanations, the mysteries and misgivings concerning the endpoints of each of the metamorphoses are removed, and it is time to clarify the construction of the intermediate polygons. The idea is as follows. We start, as appropriate according to the above explanations, by taking, for $j = 1, 2, \ldots, n$, on a circle C points A_j that determine with a fixed radius R of C an angle of $2\pi\frac{d}{n}\cdot j$, and connecting each A_j to A_{j+1} by a segment. This yields a regular polygon $\{n/d\}$. From now on, we shall assume that n is even since, if n is odd, it is easy to see that every isogonal n-gon must be regular, hence of no interest in the present context. The continuous families visualized in the diagrams arise by taking a parameter $t \geq 0$, and locating A_j so that the angle to the radius R is

$$\frac{2\pi}{n}\cdot(j\cdot d + (-1)^j t);$$

in other words, the vertices of the starting regular polygon are alternately moved ahead or backwards from their original position, all through the same distance. Clearly, an isogonal polygon is obtained regardless of the value of t; however, a short calculation shows that when $t = n/4$ the isogonal polygon is, in fact the regular polygon $\{n/e\}$, where $e = \frac{n}{2} - d$. We take this value of t as the end of our metamorphosis; if larger values of t are used, one obtains the same kinds of isogonal polygons as before, but in the opposite order, till one reaches the starting regular polygon $\{n/d\}$ at $t = n/2$.

In the diagrams of Figures 1, 2 and 3 the values of d are 1, 2 and 3, respectively; the value of t is indicated near each of the polygons.

The metamorphoses just described are most interesting for such values of n which are twice an odd prime—this is the reason the illustrations deal with $n = 14$. The reader may find it amusing to investigate the families that result for some other values of n that are equal to twice an odd number. It is not hard to show that the procedure just explained yields all possible isogonal polygons in these cases. However, if n is divisible by 4 another metamorphosis is possible. It is illustrated in Figure 4 for $n = 4$. We leave it to the reader to investigate what are the analogues of the sequence in Figure 4 for other values of n that are divisible by 4, and to formulate a complete description of the isogonal polygons possible in these cases.

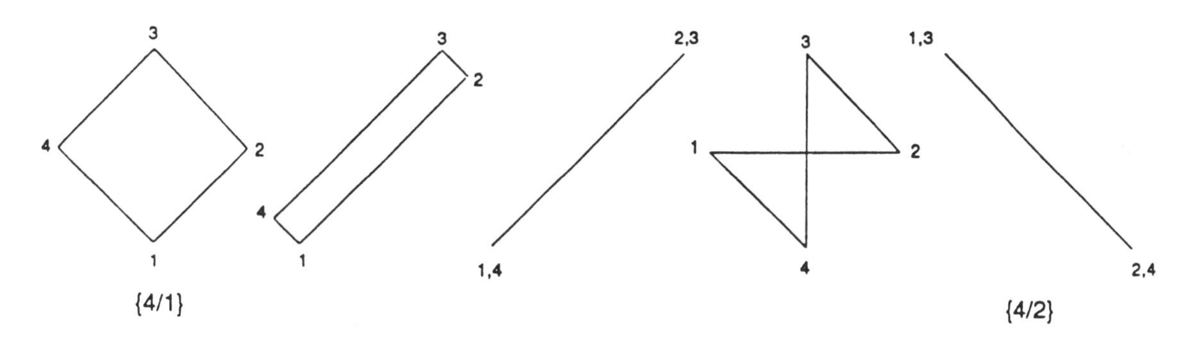

Figure 4: Metamorphosis of $\{4/1\}$ to $\{4/2\}$, through isogonal polygons.

The polygons in Figures 1, 2 and 3 are meant to show all possible **types** of isogonal 14-gons. Naturally, such a statement makes sense only if we agree on a definition of type—in other words, if we provide criteria allowing us to distinguish between polygons of types that are considered distinct. This task is less simple than it may appear at first blush, and the details are, in fact, largely a matter of convenience. The (somewhat redundant) criteria we adopted that two isogonal polygons have to satisfy in order to be considered of the same types are the following, all formulated under the assumption that a fixed correspondence has been chosen between their vertices:

(i) on the circumcircle of each, corresponding vertices have to coincide in the same sets, and following the same order;

(ii) corresponding edges meet or fail to meet in the same way in both, and, on each edge, the order of intersections by the other edges is the same as on the corresponding edge in the other polygon;

(iii) if three or more edges of one polygon meet at one point, the same is true for the corresponding edges of the other;

(iv) the symmetry groups of the two have to be isomorphic, and act on the polygons in the same way.

With this definition, it is easy to verify that Figures 1, 2 and 3 show all types of isogonal 14-gons, except the regular polygon $\{14/7\}$. Similarly, in Figure 4 are shown all types of isogonal 4-gons.

The metamorphoses in our diagrams can be used to illustrate the concepts of winding number and rotation number, and to clarify the distinction between them. For both concepts it is necessary to orient the polygon, and we shall assume that an orientation has been chosen in each case. The **rotation number** r of a polygon P can be defined as the sum of the **deflections** at the vertices of P, measured in units of the full angle; the deflection at a vertex A_j is the angle through which the extension of the incoming edge has to be turned in order to coincide with the outgoing edge, the angle being chosen to lie between $-1/2$ and $1/2$ of the full angle. (We recall that counterclockwise angles are counted as positive, clockwise ones as negative.) It follows that the rotation number is **undefined** if an edge overlaps with the following one (since there is no way to decide whether the deflection is $1/2$ or $-1/2$), or if two consecutive vertices coincide (since the single-point edge does not determine any direction). The first eventuality happens, for example, for regular polygons $\{n/d\}$ where $d = n/2$; the second occurs, in a very important role, among isogonal polygons, as we shall see shortly. Regardless of the relative primeness of n and d, the rotation number of a regular polygon $\{n/d\}$, with $1 \leq d \leq n/2$, is either d or $-d$, depending on the orientation.

While the rotation number of a polygon depends only on the polygon itself, the winding number depends on a reference point as well. If O is a point, the winding number $w(P, O)$ with respect to O of a polygon P can be computed by following a suitable ray X (or any curve, for that matter) from O to points sufficiently far, so as to reach outside a circle enclosing the whole polygon; each time the ray crosses the polygon we obtain a contribution to the winding number, the contribution being 1 if the direction of the polygon at the crossing is from right to left when looking in the direction of the ray, and -1 otherwise. For this to work, O should not lie on any line determined by the edges of P, and X should not pass through any vertex of P. It is well known that the winding number does not depend on the particular ray X chosen, that the value of $w(P, O)$ is the same as $w(P, O^*)$ if the segment OO^* meets no edge of P, and that the definition of $w(P, O)$ can be extended by continuity to all points of the plane except those on the edges of P. Additional information about the rotation and winding numbers, as well as some other functions associated with polygons, can be found in Grünbaum & Shephard [**3**].

Both the rotation number and the winding number change sign if the direction of the polygon is reversed. Hence it is in many cases convenient to assume that the orientation is such that one of these numbers is nonnegative. However, as the examples in Figures 1, 2, 3 show, it is not in all cases possible to choose the orientation so that both numbers are positive.

In connection with regular polygons, it is customary to call the absolute value of the winding number of a polygon P with respect to its center O the **density** of P. It is well known and easily shown that the density of $\{n/d\}$ is d; this holds regardless of the relative primeness of n and d, except that the density of $\{n/d\}$ is not defined if $d = n/2$. The winding numbers of isogonal polygons are also usually considered with respect to the center of the polygon; hence the winding number (with respect to the center) of an isogonal polygon is not defined if some of the edges of the polygon pass through the center.

It is of interest to follow the values of the rotation and winding numbers as we advance in each of the metamorphoses in Figures 1, 2 and 3. When t satisfies $0 \leq t < 1/2$, the rotation number has constant value $\pm d$ (the sign depending on the orientation); at $t = 1/2$ pairs of consecutive vertices coincide, and the rotation number is undefined. For t with $1/2 < t \leq n/4$ (in the present case $n = 14$) the rotation number is $\pm(d - \frac{n}{2})$. In contrast, for t with $0 \leq t < n/4 - d/2$ the winding number with respect to the center is $\pm d$, and for $n/4 - d/2 < 4 \leq n/4$ the winding number is $\pm(d - \frac{n}{2})$; for $t = n/4 - d/2$ the winding number with respect to the center is not defined. We see that in the interval $1/2 < t < n/4 - d/2$ the rotation number and the winding number have opposite signs, regardless of the orientation. It should be mentioned that although an analogous situation occurs with isogonal n-gons when n is divisible by 4, the second part of Figure 4 shows a new phenomenon: namely, the winding number is undefined, and the rotation number is 0.

Isotoxal n-gons, that is polygons in which the symmetries act transitively on the edges, behave analogously to the isogonal n-gons; they also present a variety of interesting shapes, transitional between regular polygons $\{n/d\}$ and $\{n/e\}$, where $e = \frac{n}{2} - d$. In Figures 5, 6 and 7 we show representative isotoxal 14-gons, with $d = 1, 2,$ or 3, respectively. We shall not dwell in detail on their construction or properties, but would like to point out two aspects. First the behavior regarding rotation numbers and winding numbers is simpler (and less interesting) than among isogonal polygons: the values of the two numbers coincide whenever defined, and in each of the metamorphoses, there is a single polygon for which they are not defined; at that stage, the common value changes from $\pm d$ to $\pm(d - \frac{n}{2})$.

The second aspect worth mentioning is the duality between isogonal and isotoxal polygons. In the older literature (and some of the newer as well) much is made of duality among polygons, without bothering to point out the limitations of the assertions. It is well known that in the projective plane there is a complete duality between points and lines, that is, a correspondence between points and lines that preserves incidences. Due probably in part to the widespread semantic confusion caused by the venerable tradition of using the word "line" to denote both the infinite lines (of the Euclidean or projective planes) and the finite segments of straight lines, many writers have blithely discoursed on the duality of polygons in general, and regular and other special polygons in particular. This is especially visible in the writings of Hess **[5]** and Brückner **[1]**, who have at length discussed the duality between isogonal and isotoxal polygons. We shall not repeat here the details of the objections that can be raised against duality among polygons (or polyhedra) in general, since their polyhedral formulation can be found in Grünbaum & Shephard **[2]**. It is clear that there is nothing wrong concerning the duality of convex polygons and polyhedra; also, for regular polygons

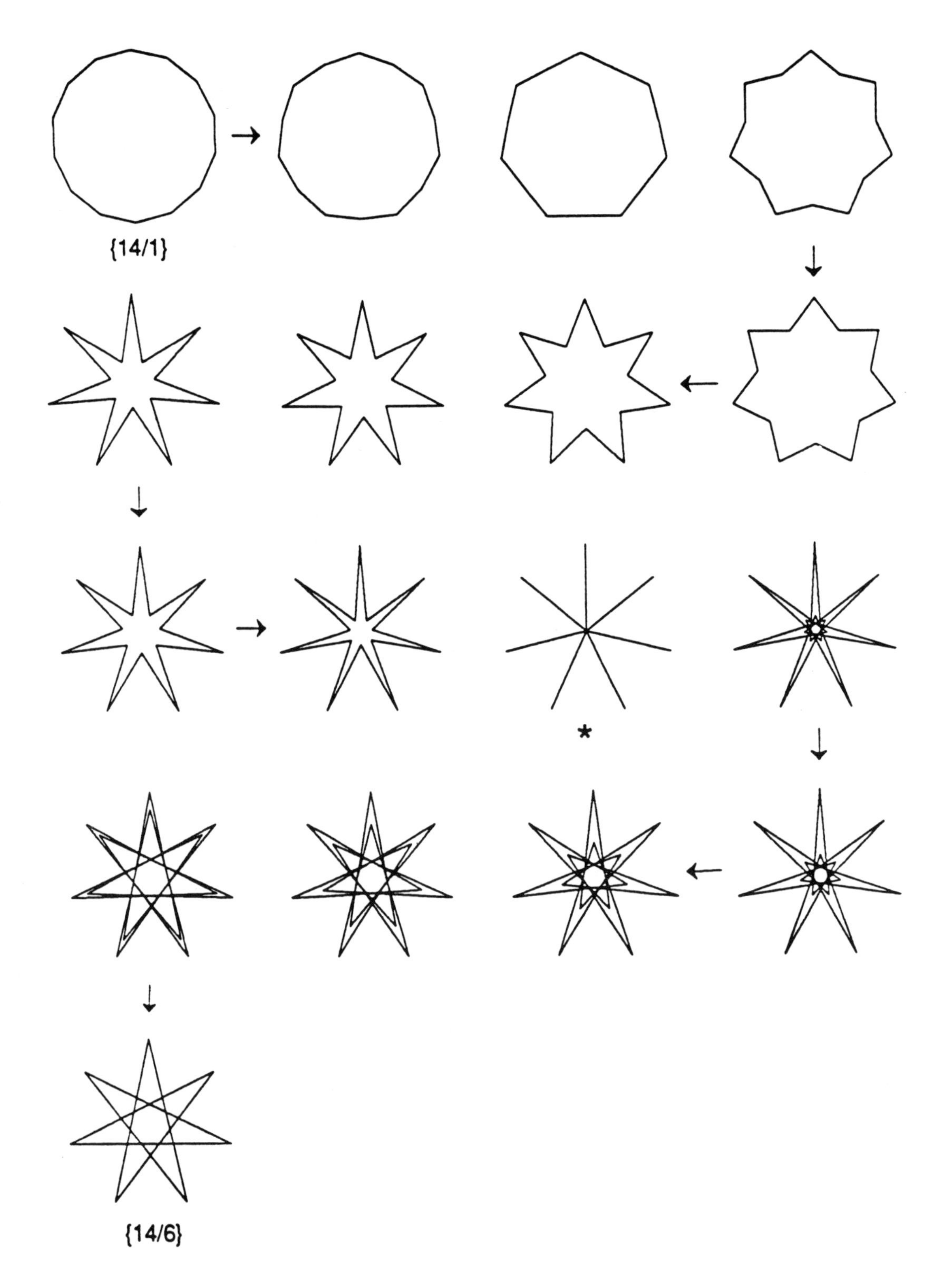

Figure 5: Metamorphosis of $\{14/1\}$ to $\{14/6\}$, through isotoxal polygons.

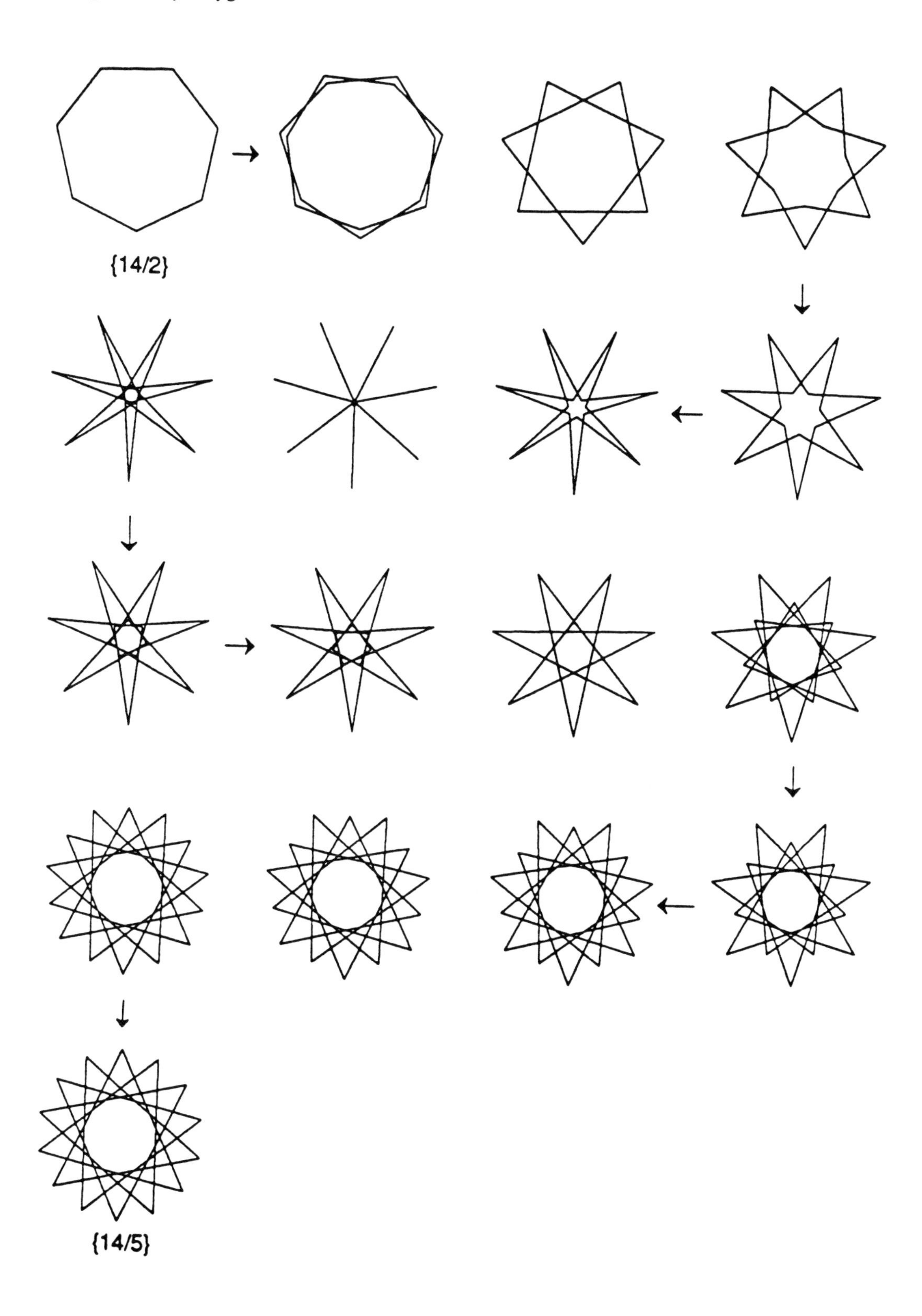

Figure 6: Metamorphosis of $\{14/2\}$ to $\{14/5\}$, through isotoxal polygons.

and polyhedra, and for many polygons that have a "center," one can give reasonably satisfactory definitions of duality. But the diagrams of the metamorphoses shown in the present paper quite clearly show what can go wrong in these dualities. The polygons in Figures 1 and 5 correspond to each other by such a duality (in fact, by reciprocation in a suitable circle) in all cases except one: the isogonal 14-gon for $t = 3$ does not admit a dual isotoxal polygon under that duality, and the isotoxal polygon marked by an asterisk does not correspond to any isogonal polygon under the same duality. Similar is the situation in the "dual" pairs of metamorphoses in Figures 2 and 6, and in Figures 3 and 7. Even more blatant is the lack of duality between the isogonal 4-gons shown in Figure 4, and the isotoxal 4-gons shown in Figure 8.

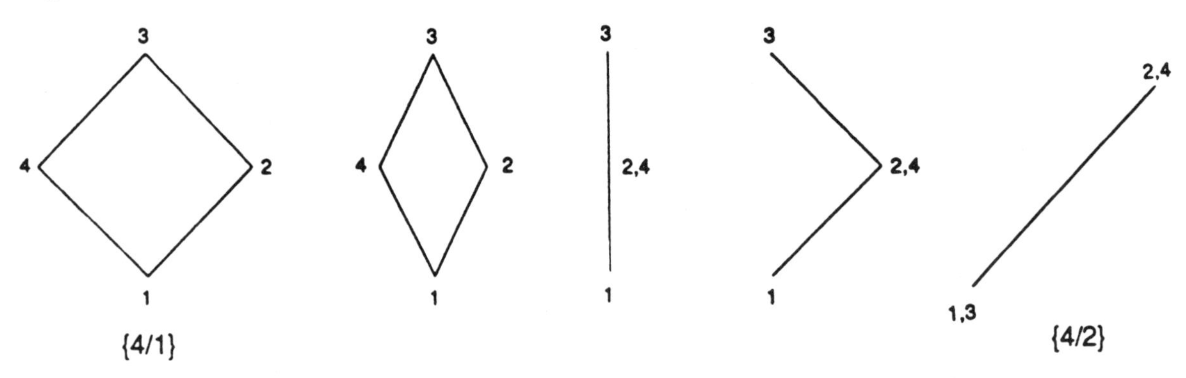

Figure 8: Metamorphosis of $\{4/1\}$ to $\{4/2\}$, through isotoxal polygons.

In conclusion, here are some historical remarks. I am not sure about the origin of the winding numbers (which play a rather prominent role in the calculation of areas), but they were certainly defined and used by Meister [7]; Steinitz [**10**], p. 4, assigns to Meister the priority of their definition. The rotation numbers were also introduced rather early, by Wiener in his frequently mentioned but apparently rarely read work [**12**]. Wiener used the word "Art" (German for "kind") instead of rotation.

Unfortunately for Wiener, and for mathematics, Hess [**5**] (mis)appropriated the term "Art" for a different concept which has not turned out to be useful; however, since Hess (and later Brückner [**1**]) used the term in the modified sense, Wiener's original concept was effectively forgotten. It was independently rediscovered only much later, by Whitney [**11**] in 1936; rotation numbers play an important role in topological considerations, as well as in the classification of polygons (see Mehlhorn & Yap [**6**]).

The work of Hess [**5**], which we have already mentioned several times, sets out to investigate isogonal and isotoxal polygons at length. Unfortunately, it is essentially devoid of any worthwhile results or insights, mainly due to his insistence of classifying the polygons by their "Art" according to his definition of this term. There seems to have been no later investigation of these interesting types of polygons, besides Brückner's [**1**] account of some of Hess's statements. It is my hope that the present paper will make the isogonal and isotoxal polygons more accessible, and will lead to their use as examples of various concepts and misconceptions. But above all I hope that there may be more appreciation given to their visually appealing qualities.

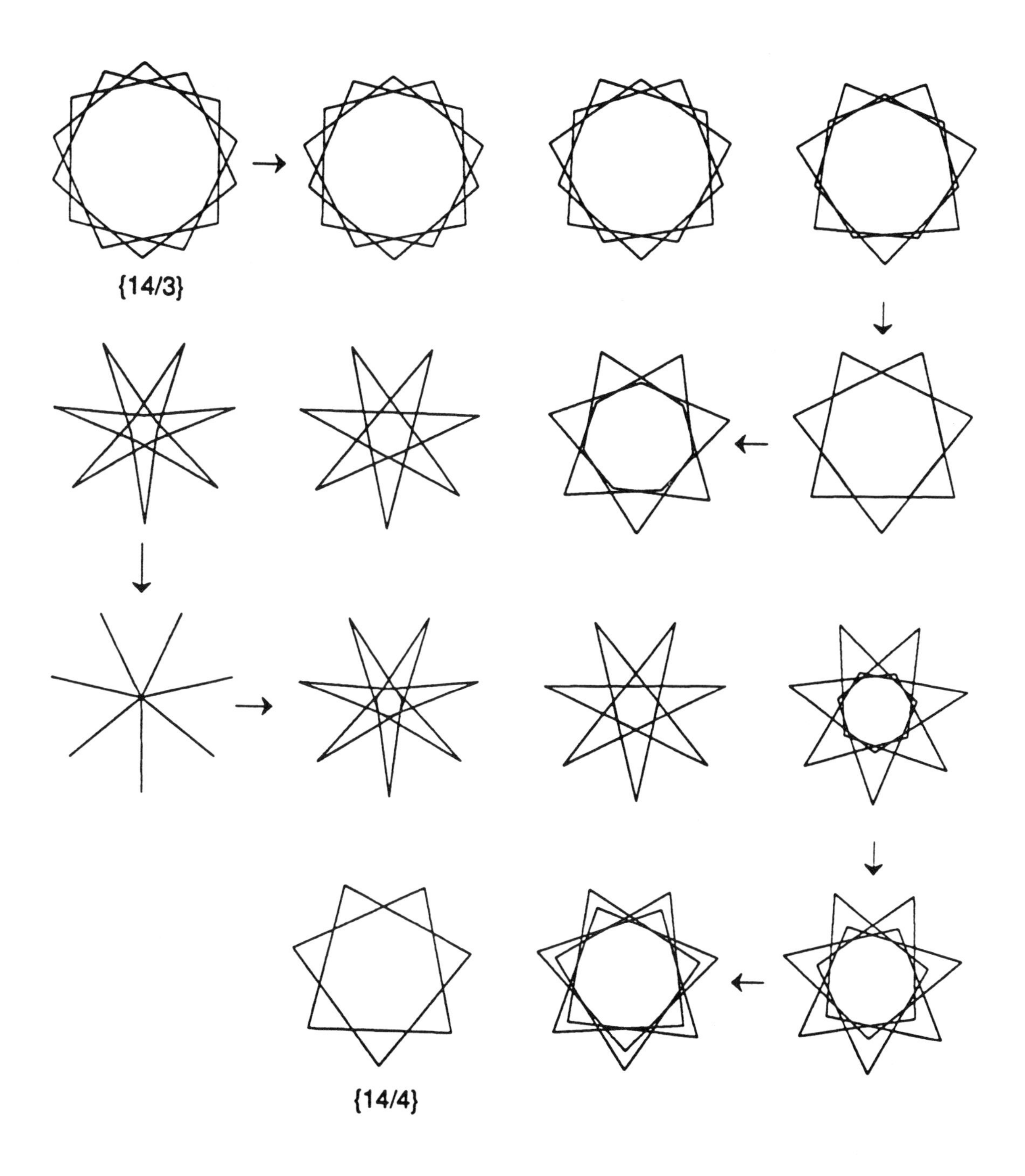

Figure 7: Metamorphosis of $\{14/3\}$ to $\{14/4\}$, through isotoxal polygons.

References

1. M. Brückner, *Vielecke und Vielflache,* Teubner, Leipzig, 1900.

2. B. Grünbaum & G. C. Shephard, Duality of Polyhedra, in *Shaping Space: A Polyhedral Approach,* M. Senechal and G. Fleck, eds., Birkhäuser, Boston, 1988, pp. 205–211.

3. B. Grünbaum & G. C. Shephard, Rotation and winding numbers for planar polygons and curves, *Trans. Amer. Math Soc.* 322 (1990), 169–187.

4. S. Günther, *Vermischte Untersuchungen zur Geschichte der mathematischen Wissenschaften,* Teubner, Leipzig, 1876.

5. E. Hess, Ueber gleicheckige und gleichkantige Polygone, *Schriften der Gesellschaft zur Beförderung der gesammten Naturwissenschaften zu Marburg,* Vol. 10, No. 2, pp. 611–743 + plates, Th. Kay, Cassel, 1874.

6. K. Mehlhorn & C.-K. Yap, Constructive Whitney–Graustein theorem: or how to untangle closed planar curves, *SIAM J. Comput.* 20 (1991), 603–621.

7. A. L. F. Meister, Generalia de genesi figurarum planarum et independentibus earum affectionibus, *Novi Comm. Soc. Reg. Scient. Gottingen* 1 (1769/70), pp. 144–180 + plates.

8. A. F. Möbius, Ueber die Bestimmung des Inhaltes eines Polyëders, *Ber. Verh. Königl. Sächs. Ges. Wiss., math.-nat. Kl.* 17 (1865), 31–68 (*Ges. Werke,* Vol. 2, pp. 473–512, Hirzel, Leipzig, 1886).

9. L. Poinsot, Mémoire sur les polygones et les polyèdres, *J. École Polytechnique* 4 (1810), 16–48 + plate. Annotated German translation in: R. Haußner, Abhandlungen über die regelmäßigen Sternkörper, Ostwald's Klassiker der exakten Wissenschaften, Nr. 151, Akademische Verlagsgesellschaft, Leipzig, 1906.

10. E. Steinitz, Polyeder und Raumeinteilungen, *Enzykl. math. Wiss.* Vol. 3 (Geometrie), 1922, Part 3AB12, pp. 1–139

11. H. Whitney, On regular closed curves in the plane, *Compositio Math.* 4 (1936), 276–284.

12. C. Wiener, *Über Vielecke und Vielflache,* Teubner, Leipzig, 1864.

C138 Padelford Hall, GN-50
University of Washington
Seattle, Washington 98195

SquaRecurves, E-Tours, Eddies, and Frenzies: Basic Families of Peano Curves on the Square Grid

Douglas M. McKenna

1 Introduction

During the last two decades there has been a dramatic rise in the use of the computer as a tool for aiding the study, teaching and illustration of mathematics. Because of its general purpose nature, the computer has been used in a great variety of mathematical activity. The renowned 100-year old Four Color Map Theorem, proved (controversially) with the indispensable aid of a computer [**1**], has quickly become the canonical example of this trend in research, but there are numerous other interesting examples as well, such as in finite groups, factorization problems and the study of prime numbers, 4-dimensional geometry, and chaotic systems and non-linear dynamics. In game analysis, the 3-D analog of Tic-Tac-Toe has been proved to be a first-player win through the judicious use of both human and computer skills, when all analytic attempts had failed [**2**].

The teaching of mathematics promises to be greatly improved with the use of computer-illustrated "MathWorlds," such as the LOGO Turtle Geometry environment [**3**] for children, or Mathematica [**4**] and similar symbolic mathematics programs for professionals. By using the computer's ability to draw accurate representations of mathematical ideas in both space and time, these interactive graphical environments help tweak the imaginations of normally curious minds, as well as the more mathematically gifted ones.

Some of the most beautiful mathematical objects ever seen belong to the study of fractal sets [**5, 6**], which has blossomed recently in large part because of the computer's ability to make high-resolution pictures whose creation by hand would have been unthinkable—if not impossible—20 years ago. In the following, I show how I have used the computer to help me discover, explore and illustrate some new families of fractal space-filling curves. I want also to display for the normally curious some fascinating mathematical patterns. To do so, we will begin with an explanation of how a certain class of geometric fractal patterns, called Koch curves, can be designed and analyzed. This will lay the groundwork for understanding how the computer has helped in investigating a small but interesting mathematical problem.

2 Koch Self-Substitution: The Whole is Some of the Whole

In the early years of this century, the mathematician von Koch devised an elegant and very simple method for constructing a continuous but non-differentiable set of points known as the Snowflake Curve. The algorithm is best illustrated by showing successive approximations in its construction

(Fig. 1). Each approximation is a refinement of the last, and the Snowflake Curve is considered to be the limiting set of these approximations as the number of refinements tends towards infinity. The limit set is now known as a fractal, and has the curious but important property of geometric self-similarity: appropriately chosen pieces of the whole are geometrically similar to the whole.

If we begin with a single line segment of unit length, which we'll call our 0th approximation, we can build the $k + 1$st approximation from the kth by dividing each of the line segments in the kth approximation into three equal segments. We then replace the central segment of the three by two segments forming part of the perimeter of an equilateral triangle. Each approximation has a length $4/3$ times the length of the previous one; thus, the length of the kth approximation would be $(4/3)^k$, which tends to infinity even though the limit set remains in a bounded area of the plane. If the three line segments of an equilateral triangle are each "expanded" outward in this way, the resulting closed curve is what is usually called the Snowflake Curve (it encloses an area equal to 8/5 the area of the original equilateral triangle).

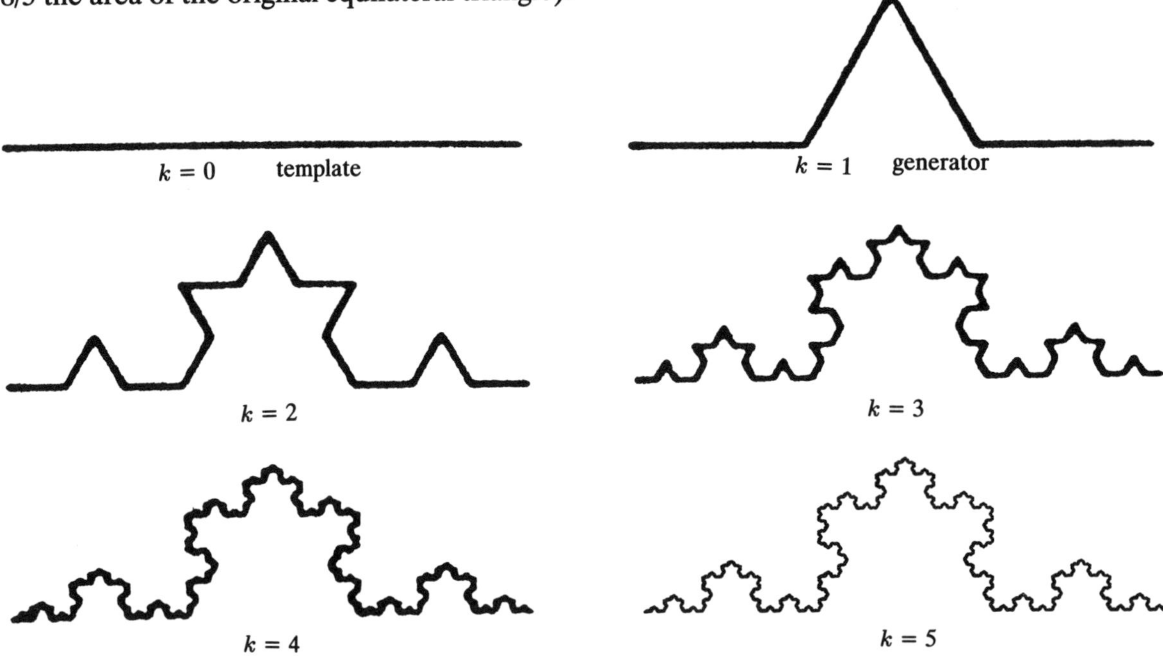

Figure 1: Approximations 0, 1, 2, 3, 4 and 5 of the open Snowflake Curve.

This process is called "self-substitution" because we are taking a given pattern (the 1st approximation) consisting of a set of line segments, which we shall call the generator, and then creating subsequent approximations by substituting scaled, oriented replicas of the original generator for each of its own constituent line segments. Every segment of the kth approximation is replaced by a copy of the generator to form the $k + 1$st approximation. The generator itself is associated with a single line segment (the 0th approximation), which we call the template segment, since it really only serves to place the eventual Koch curve somewhere in the plane. The template for the full Snowflake Curve as it is usually illustrated is the set of three connected line segments forming the closed equilateral triangle.

As we shall see, the simple recursive act of replacing elements of a set with copies of the entire set can be responsible for many interesting Koch curves and designs.

3 Fractal Dimension

The form of the Snowflake Curve is infinitely detailed yet highly regular. Fortunately, it is possible to describe and classify this form by its fractal dimension [**5**], which is a generalization of the normal Euclidean dimension. The fractal dimension of the Snowflake Curve is obtained using the Hausdorf–Besicovitch formula

$$\text{Dimension } D = \log(4)/\log(3) = 1.2618+$$

since for each substitution a set of 4 equally long line segments is being substituted for a single line segment 3 times as long as each of the 4. Note that the resultant dimension is nonintegral: greater than 1, the topological dimension of a line, but less than 2, the topological dimension of an area. It is an indication of how well the curve "fills" Euclidean space. This is actually a very intuitive idea of dimension once we've been able to see illustrations of fractal Koch curves with dimensions varying between 1.0 and 2.0.

As an example, let us look at a simple generalization of the Snowflake Curve (Fig. 2). Given an initial template line segment of length x, it is possible to construct a generator of 4 equal segments each of whose lengths varies anywhere between $x/4$ and $x/2$ ($x/3$ for the Snowflake Curve). However, instead of forming part of an equilateral triangle with the two central segments, we form an isosceles triangle such that the interior angle a varies at its base between 0 and 90° (60° for the Snowflake Curve). In this case, the fractal dimension of the limiting set will vary continuously between

$$\text{Dimension } D = \log(4)/\log(4) = 1.0$$

and

$$\text{Dimension } D = \log(4)/\log(2) = 2.0$$

as the base angle a varies from 0 to 90°. At $a = 0°$, the recursive substitutions serve only to subdivide the template segment into smaller collinear segments, so the dimension is just 1. At $a = 90°$, the isosceles triangle degenerates into two coincident line segments, perpendicular to the other two, which remain distinct with respect to the next set of substitutions. The set of points generated after a suitable number of approximations can be shown to be within any given distance of a given point in an area, and hence is topologically 2-dimensional, or space-filling, in the limit [**7**].

4 Tiles and Sub-Tiles

One of the nice properties of triangles is that not only do they tile the plane but they also tile scaled (similar) versions of themselves; that is, a triangular tile can be divided into a set of triangular sub-tiles, each similar to the first, but possibly differently oriented. A tile with this recursive property has been called both a rep-tile [**8**] and a per-tile [**9**] (for reasons of amusement and precedence, we'll use the former). Parallelograms (and thus squares and rectangles) and certain polyominoes are rep-tiles; however, the regular hexagon and the Swiss cross shape, though they can tile the plane, are not. Let us look at the Snowflake Curve with respect to the equilateral triangular rep-tiling.

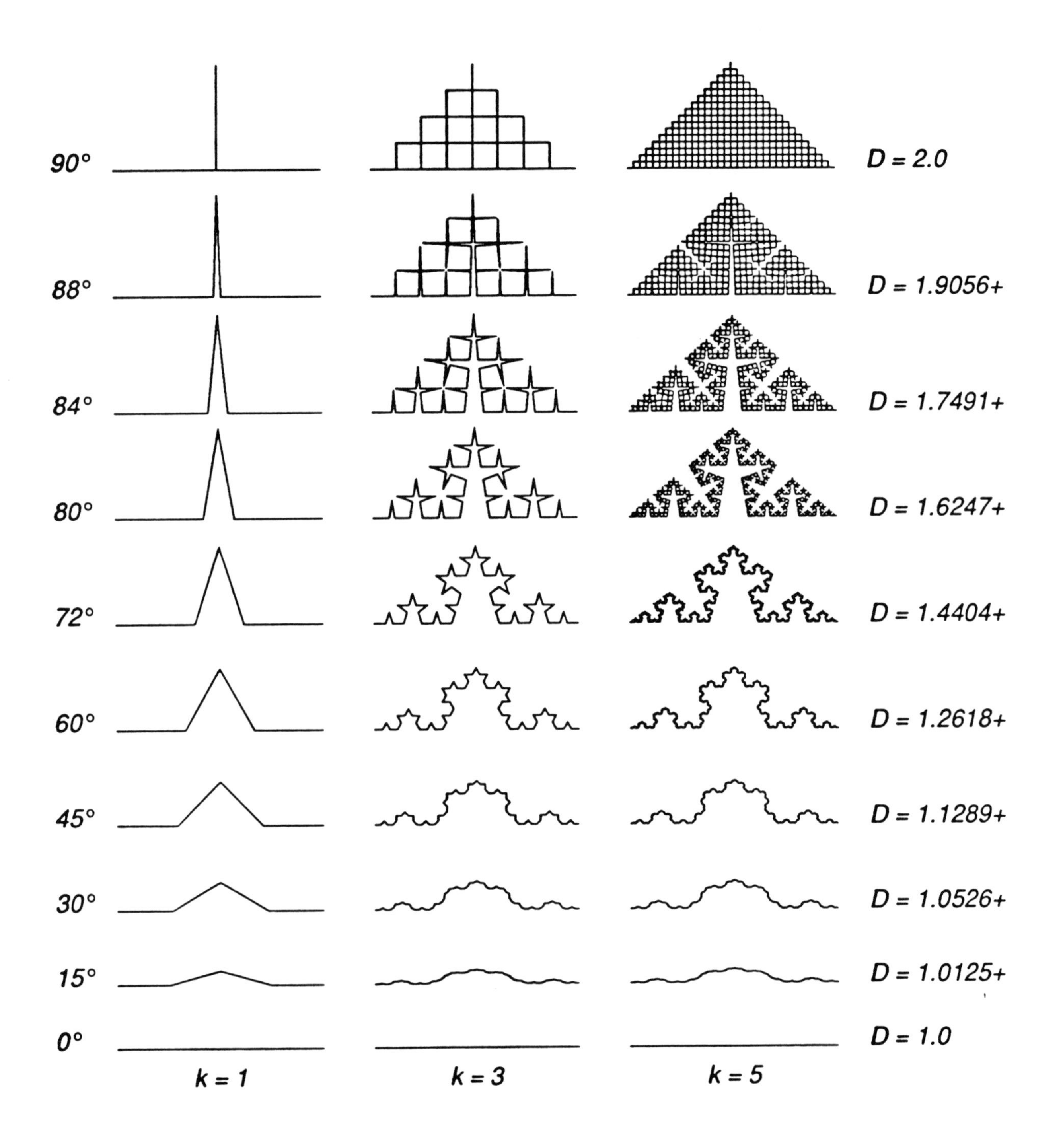

Figure 2: As base angle changes, fractal dimension varies from 1.0 (line) to 2.0 (area).

Suppose we erect an equilateral triangle with its base equal to a template segment, and subdivide it into 9 equal sub-triangles (Fig. 3a). Call this an order 3 triangular grid (I use the term grid since we will be interested in the grid lines between tiles as much as the tiles themselves). If we juxtapose the Snowflake Curve generator onto the base of the triangle, we note that each of the 4 equal segments of the generator can be taken as the base of one of the 9 sub-triangles. Because each of the generator segments has a sub-triangle associated with it, and because each sub-segment can be taken as the base of its sub-triangle, we can apply our algorithm recursively to each of the sub-triangles (Fig. 3b), defining sub-sub-triangles and so on, thus getting all subsequent approximations. From this argument, then, we see that in addition to specifying where the generator segments lie with respect to one another, we also need to say on which side of each segment we want to perform the next substitution (in the figures, this is indicated by adding a small tick mark on the appropriate side of each generator segment).

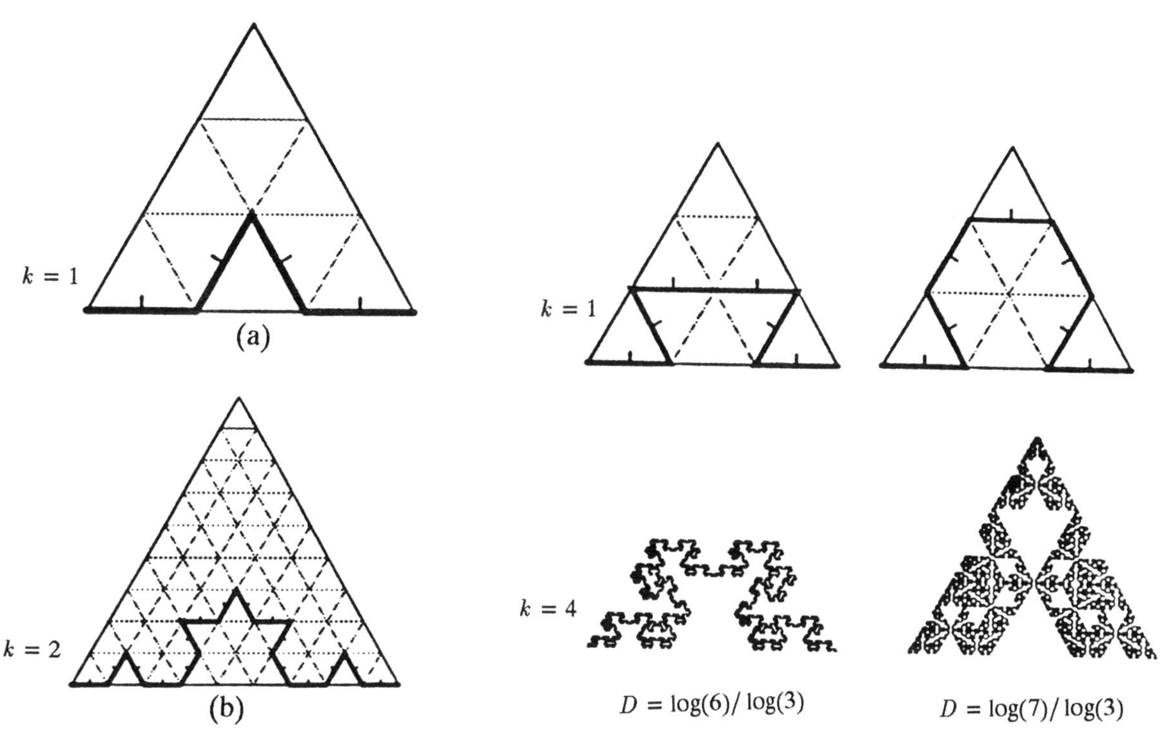

Figure 3: Order 3 equilateral triangular grid with Snowflake generator.

Figure 4: Approximations 1 and 4 for two order 3 triangular Koch curves.

Since only 4 out of the 9 possible sub-triangles in the generator are involved in the construction of the curve, another generalization of the Snowflake Curve might be to allow different numbers of generator segments and to associate them with the various combinations of the sub-triangles (Fig. 4). Notice that as the number of sub-triangles used gets closer to the maximum of 9, the resulting fractal curve tends to fill (visually at least) more and more of the original triangle's area. The fractal dimension of these curves rises also, giving a quantitative indication of this filling property. If we can find a way to include exactly one side of each of the 9 sub-triangles in some generator, we will have found a way to make a space-filling (i.e., Peano) curve, whose fractal dimension will then be exactly 2.0:

$$D = \log(9)/\log(3) = 2.0$$

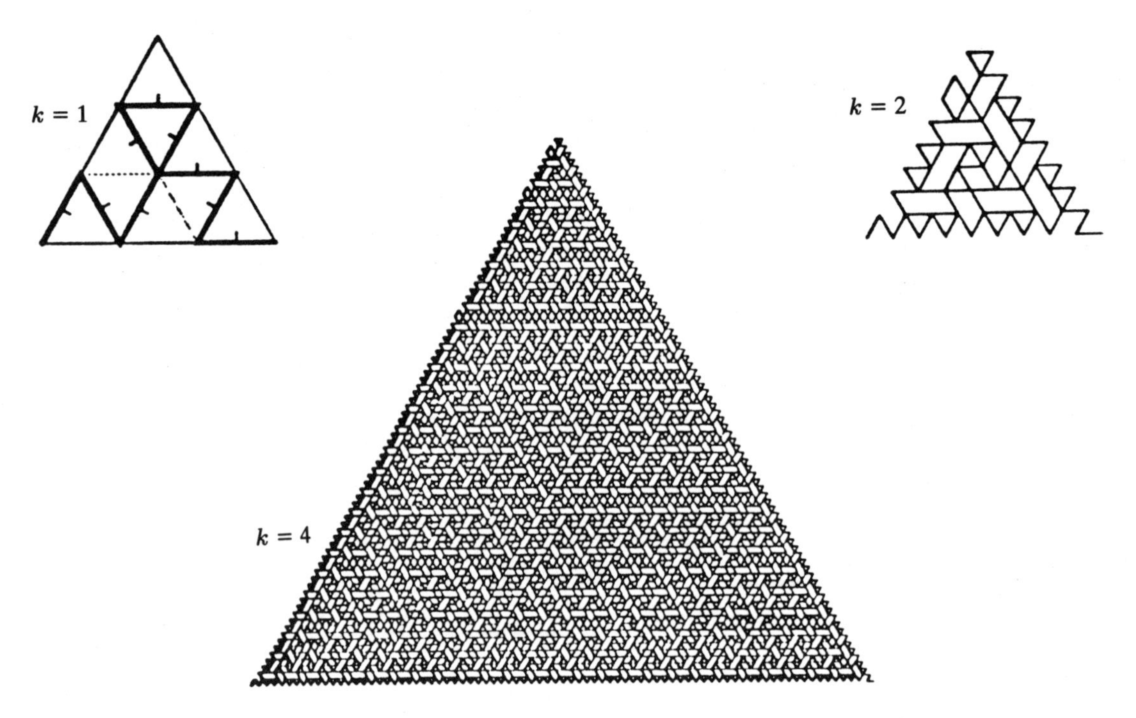

Figure 5: A self-contacting order 3 triangular space-filling Koch curve.

Figure 5 illustrates some approximations to one of the many possible Peano curves for the order 3 triangular grid. Note that the generator is self-contacting: some line segments share common endpoints. This self-contact has no bearing on the properties of the curve in the limit, since by virtue of being 2-dimensional every point will be one of self-contact, but it makes the course of approximations to the curve ambiguous and difficult to follow. It would be nice to find a generator that in all approximations builds a curve that avoids itself, but unfortunately this is impossible for all order triangular grids.

Theorem 1 *Generators on order n triangular grids always generate Peano curve approximations that are self-contacting.*

Proof. There is only one distinct generator for the order 2 triangular grid, and although it is not self-contacting on the first approximation, all subsequent approximations are, due to interactions among adjacent sub-tiles. Clearly, any self-avoiding approximation to a Peano curve must be based on a self-avoiding generator, since all approximations are nothing but a connected series of n^{k-1} suitably scaled generators. Consider the top three rows of an order n ($n \geq 3$) triangular grid (Fig. 6a), where the top 9 sub-triangles have been labeled from A to I. Since any Peano curve generator must pass all of the sub-triangles, then in particular the base of A must be passed—connected segments must lead from the start of the generator at the lower left corner of the tiling to the left corner of A and subsequent segments must lead from the right corner of A back down to the end of the generator at the lower right corner of the tiling (Fig. 6b). Exactly one of the segments leaving from or arriving at A must pass sub-triangle C which forces the other segment to pass either B or D. Since the two situations are mirror images of each other, we can choose one of them without loss of generality: B and C are passed (Fig. 6c). Once C is chosen, D must be passed by the next segment connected to C (Fig. 6d). This choice forces the next set of segments to pass F and H, respectively (Fig. 6e). These choices are forced by the fact that if a sub-triangle is not chosen to be passed at the current point in our argument, there is no way that the path can get back to passing it later and still keep from contacting itself. Having passed F and H, we are then forced to pass E and I: this leaves no possible way to pass G without self-contact at one of its corners (Fig. 6f).

QED

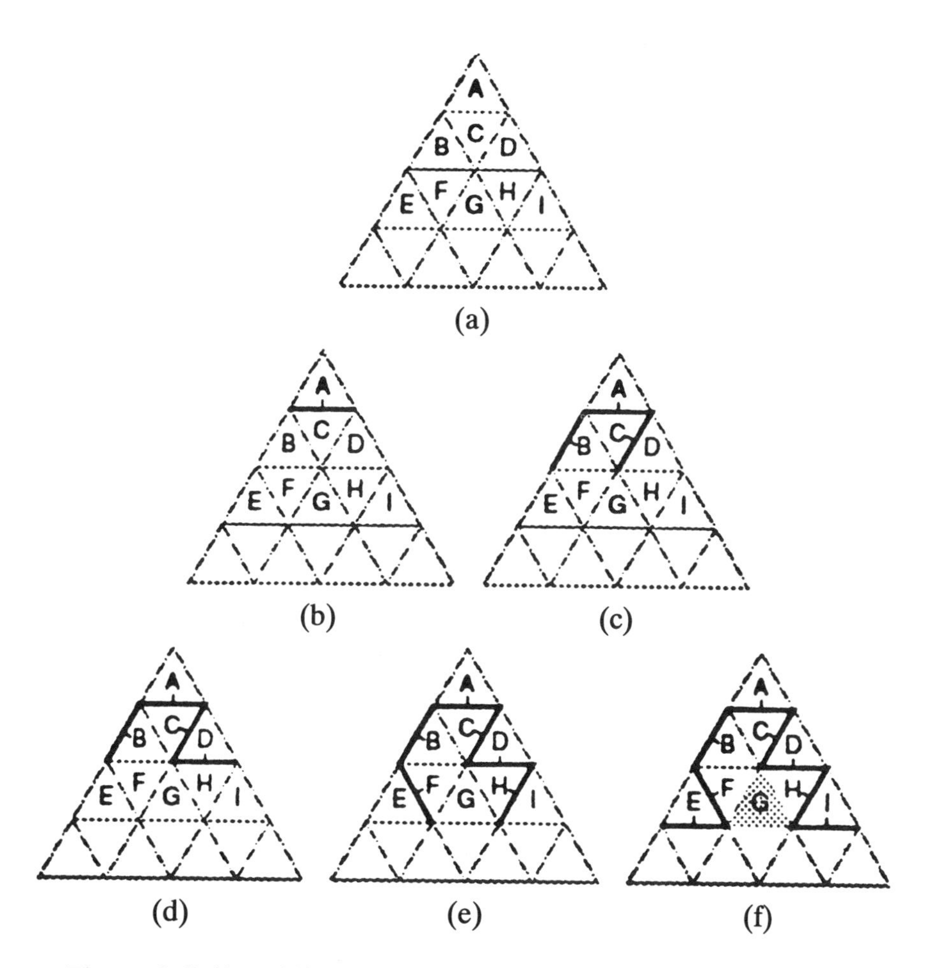

Figure 6: Self-avoiding order n triangular Peano curves do not exist.

We see then that in order for a set of line segments to generate a Peano curve (using Koch self-substitution) whose approximations consist solely of line segments along sub-tile borders, unit sub-lengths of the generator must have a 1-to-1 relationship with unit sub-tiles (or sub-areas) of the corresponding tile. The collection of sub-tiles taken in the orientations that the generator segments dictate must tile a larger similar version of themselves. Koch curves that generate space-filling curves abound on the triangular tiling, but all are self-contacting ones. This is not to say that there are no intriguing or visually pleasing space-filling Koch curves on the triangular tiling (Fig.7); however, if we are to find a generator for a space-filling Koch curve that avoids contacting itself for all its finite approximations, we must look to other tilings.

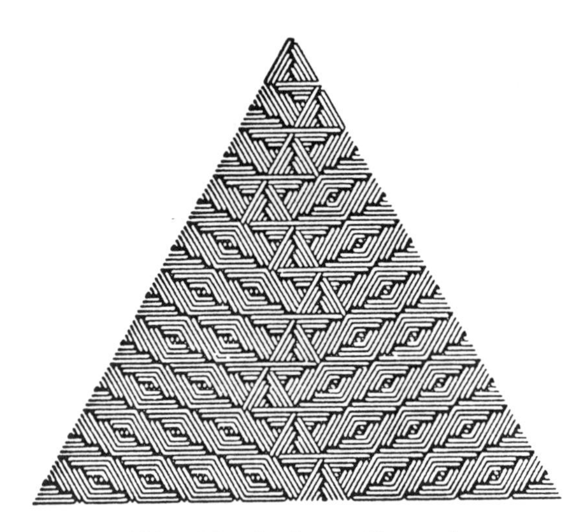

Figure 7: An order 10 space-filling Herringbone Curve (generator left as an exercise).

5 Generators on Fixed Square Grids

Squares are also rep-tiles, so it is natural to look for Peano curves based on square tilings. Many have been devised, such as Peano's original space-filler, or the Hilbert Curve (Fig. 8, right). Typically, these curves are drawn as self-avoiding space-filling paths. To do so, however, requires modifying each's underlying Koch geometry to make a given approximation self-avoiding. The generator for Peano's curve is already self-contacting; the curve is usually illustrated by rounding all self-contacting corners **[10]**. The Hilbert Curve can be drawn by connecting lines between the centers of adjacent sub-tiles in a Koch curve that sweeps the grid (this transformation can convert any self-contacting, but not self-intersecting, Koch space-filling curve into a self-avoiding one).

Consider the unit square, and subdivide it into a grid of n by n sub-squares. We call this a fixed square grid of order n since each sub-square in the grid remains parallel to the original square (each is a scaled and translated copy of the original, with no rotation). When we begin looking for space-filling generators on the first few orders, it is easy to find ones that manage to travel among all sub-squares without self-contact, thereby saving us the trouble of proving none exist, as in the case of triangular grids. However, this property of self-avoidance is necessary but not sufficient to generate an entire curve approximation free of self-contact, because it is a global property of the curve as well as a local one for the generator. In general, adjacent instances of the generator pattern will interfere with one another. But so far we haven't shown that there is no pattern of n^2 connected line segments that generates a curve free of self-contact for any approximation; all we know is that if they exist they don't readily come to mind.

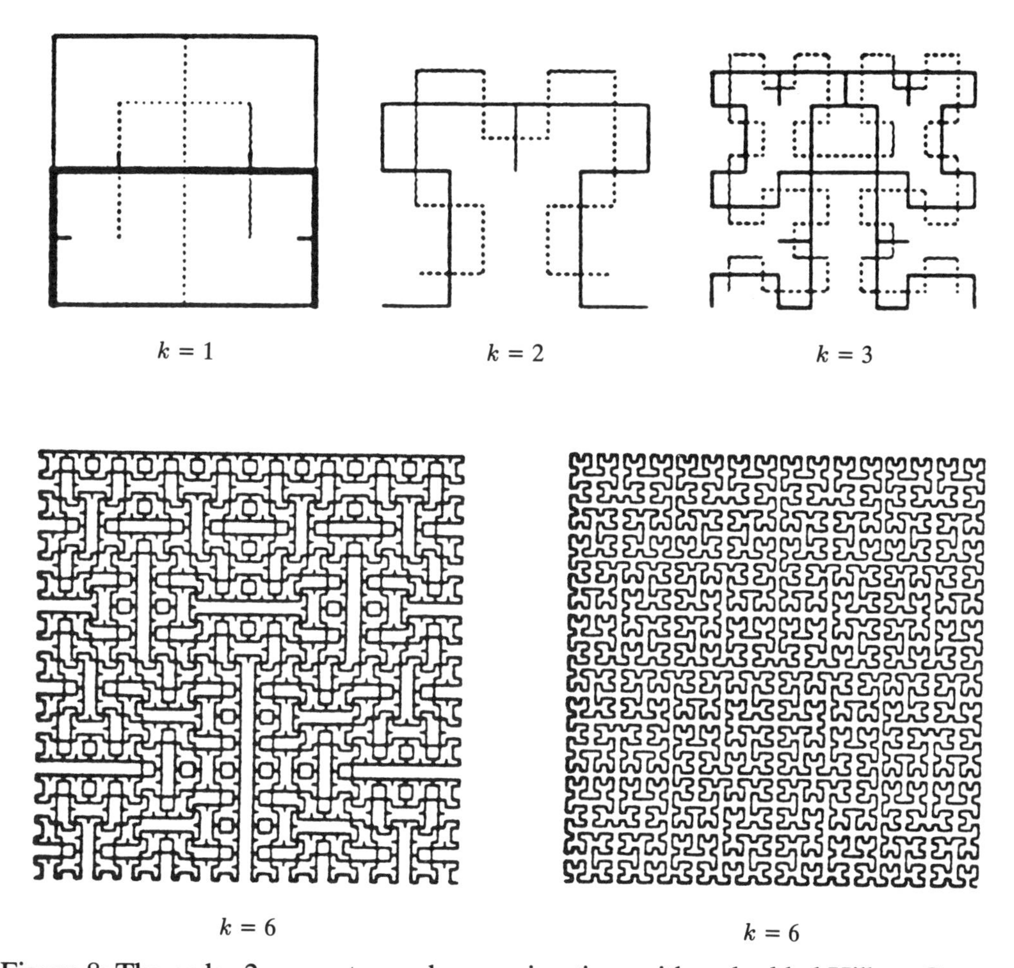

Figure 8: The order 2 generator and approximations with embedded Hilbert Curve.

There is only one possible generator curve for an order 2 fixed square grid. When it is expanded, the result is a beautiful pattern with lots of coincident line segments (Fig. 8, left). By drawing lines between the centers of adjacent sub-squares as they are passed by this curve, one can trace out the Hilbert Curve (Fig. 8, right).

On the order 3 grid, it is not hard to go through all the possibilities by hand to find that there are two ways that a generator pattern can pass each of the 9 sub-squares in a 1-to-1 fashion. The two are mirror images of one another, so there is only one distinct generator for the 3 by 3 fixed square grid. When this generator is expanded, the approximations are self-contacting.

It is possible to find by hand a number of generators on the order 4 square grid, but there is enough freedom to move around among the 16 sub-squares to make it hard to convince ourselves that all the possibilities have been found. It is at this point that we enlist the aid of a computer programmed to find all the possibilities in a systematic way.

6 An Exhaustive Search Yields a Discovery

The search we need to do can be illustrated in terms of the following recreational problem:

> **The Cabbie's Dilemma.** One December day, a tourist in some large city finds himself on the corner of 7th Ave. and 41st St., where streets run perpendicular to avenues in the typical city grid (unlike most cities, however, all roads are two-way here). Each city block has one famous skyscraper on it, and each intersection has a one-of-a-kind holiday decoration draped across it. The tourist hails a cab and asks the driver to proceed to 7th Ave. and 36th St. (just 5 blocks away), but to do it while touring all buildings in the 5 by 5 block area between 7th and 2nd Avenues, and between 41st and 36th Streets. The tourist can look out of only one side of the cab at a time, although at intersections after viewing the decorations he can move to either side as is necessary to view the next new building. He asks the driver to take a route that allows him to see each of the 25 buildings exactly once, and although the tourist is not interested in seeing all the intersection decorations, if he does see one, he doesn't want to see it again (who could blame him?). U-turns are illegal. How many distinct tours are there to his destination without leaving the 5-by-5 block area?

To solve this, we can write a program that takes as input the order n of a fixed square grid (5 in the above problem), and which outputs the total set of distinct, self-avoiding generators (tours) for that grid. The number of these generators for various order grids and the actual generators for the first 5 orders are shown in Figure 9. The program uses standard backtracking techniques to traverse an undirected graph in which the nodes are the vertices (intersections) of the square sub-tiles, the arcs of the graph are the sides (roads) of the sub-tiles, and the faces bordered by arcs are the sub-tiles themselves (buildings). Beginning at the lower left corner node, partial generators are built by finding all legal arcs from the current node to neighboring nodes. Each candidate arc is added to the generator being built, and the search algorithm is called recursively to continue with the new node on the other end of the added arc. As each node is added to the path, it is marked as "visited" so that it will be excluded from consideration at later recursive stages. This ensures that only self-avoiding tours will be found. In addition, faces of the graph are "bound" to the consecutive arcs passed so that they too will not be available during later stages. This ensures that there will always be a 1-to-1 relationship between arcs (generator segments) and faces (square sub-tiles). Whenever a partial path gets to the lower right corner node, and is of length n^2, it is a generator and we save it in an output list. When all the possibilities have been found, we cull the distinct generators from the list, which will have a size twice the final count (see Lemma 1 below), since for each generator found, the search will also find its mirror image.

When we look at the 7 distinct self-avoiding generators on the order 4 tiling, we can quickly see that expanding any one of them even once creates an approximation that is self-contacting. However, by carefully checking each of the 138 distinct self-avoiding generators for the order 5 fixed square grid, we find that exactly one of them can generate a Peano curve that at all approximations remains self-avoiding. Since none of the generators on any of the smaller order grids have this property, this curve—which, because of its resemblance to the letter E, I call either the **E Curve Recurve** or the **E-Tour Detour** ("the longest distance between two points")—is the simplest

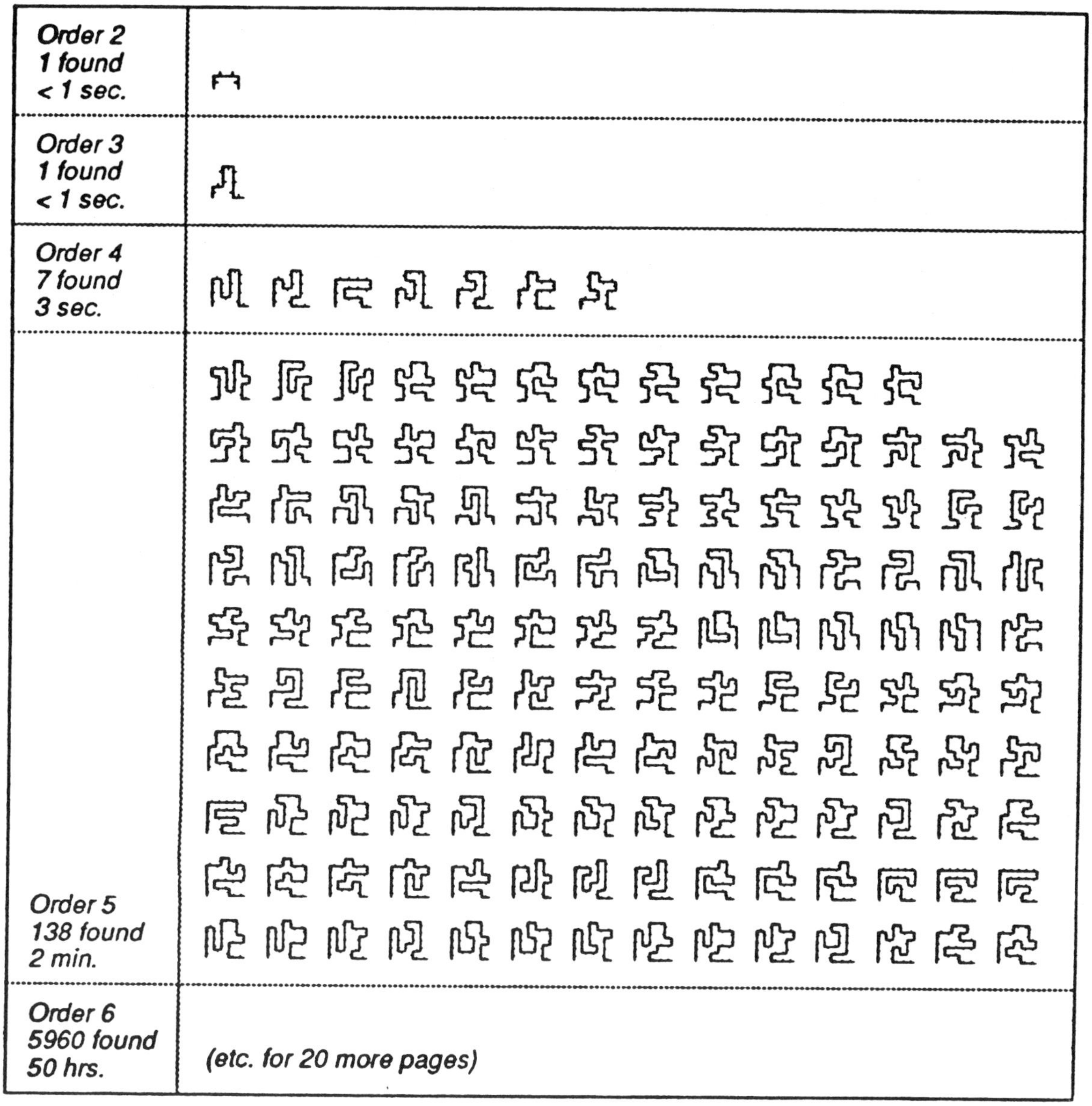

Order 2 *1 found* *< 1 sec.*	
Order 3 *1 found* *< 1 sec.*	
Order 4 *7 found* *3 sec.*	
Order 5 *138 found* *2 min.*	
Order 6 *5960 found* *50 hrs.*	*(etc. for 20 more pages)*

Figure 9: All distinct Koch square space-filling generators or orders 2 through 5.

Koch substitution Peano curve that fills a square without self-contact at any finite approximation (Fig. 10) [**11**].

Unlike the generators for the Hilbert Curve or Peano's curve, the E Curve generator seems curiously asymmetric. In fact, all 138 self-avoiding generators are asymmetric, which leads us to the following lemma.

Lemma 1. *There are no bilaterally symmetric self-avoiding Peano curve generators on order* n *fixed square grids when* n *is odd.*

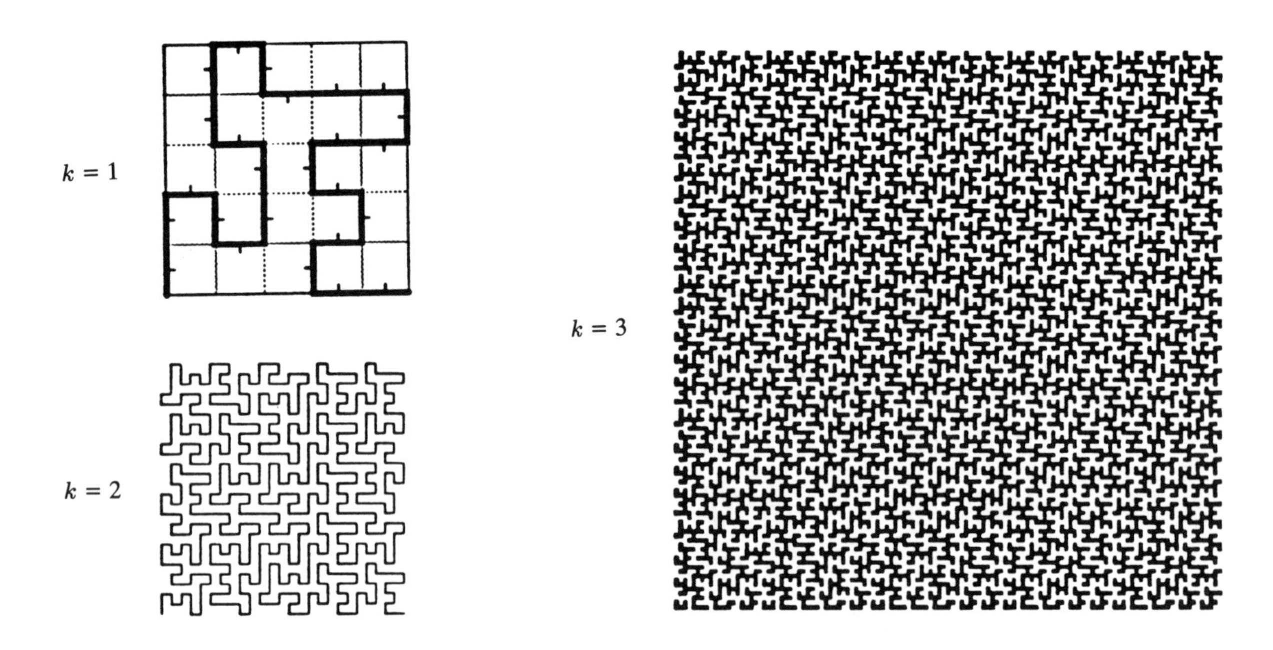

Figure 10: The E Curve generator with approximations 2 and 3 (filled).

Proof. Assume a bilaterally symmetric generator exists. On an odd order square grid (Fig. 11a), there will always be a central column of sub-squares. Since the self-avoiding generator path consists of a connected series of line segments leading from the lower left corner $(0,0)$ of the square tiling to the lower right corner $(n,0)$, by a parity argument it must pass an odd number of sub-squares in the central column in order to get to the right side (Fig. 11b). However, any odd number of crossings greater than one will lead to the path having an S-like configuration with respect to the central column, and such a path cannot have bilateral symmetry. Thus there can only be one crossing of the central column. Exactly one sub-square, which we label A, will be associated with the segment crossing the central column. Consider the sub-square at either the top or bottom of the central column, whichever is furthest from A, and call it B (Fig. 11c). Since every sub-square must be associated with exactly one line segment of the generator, then B must be associated with a segment passing either its right or left edge (a segment along its top or bottom edge would be a second segment crossing the central column). Whichever side of B we choose to associate with this segment (Fig. 11d), if the candidate generator is symmetric, then another symmetric segment must pass the other side of B and be associated with B's neighbor, C (Fig. 11e). But since C is next to the perimeter of the grid, its associated segment cannot continue to be connected to other segments leading to the nearest corner of the grid, since the only possible candidate sub-square is C, which has already been used (Fig. 11f). Hence the assumption of a bilaterally symmetric generator is contradicted. QED

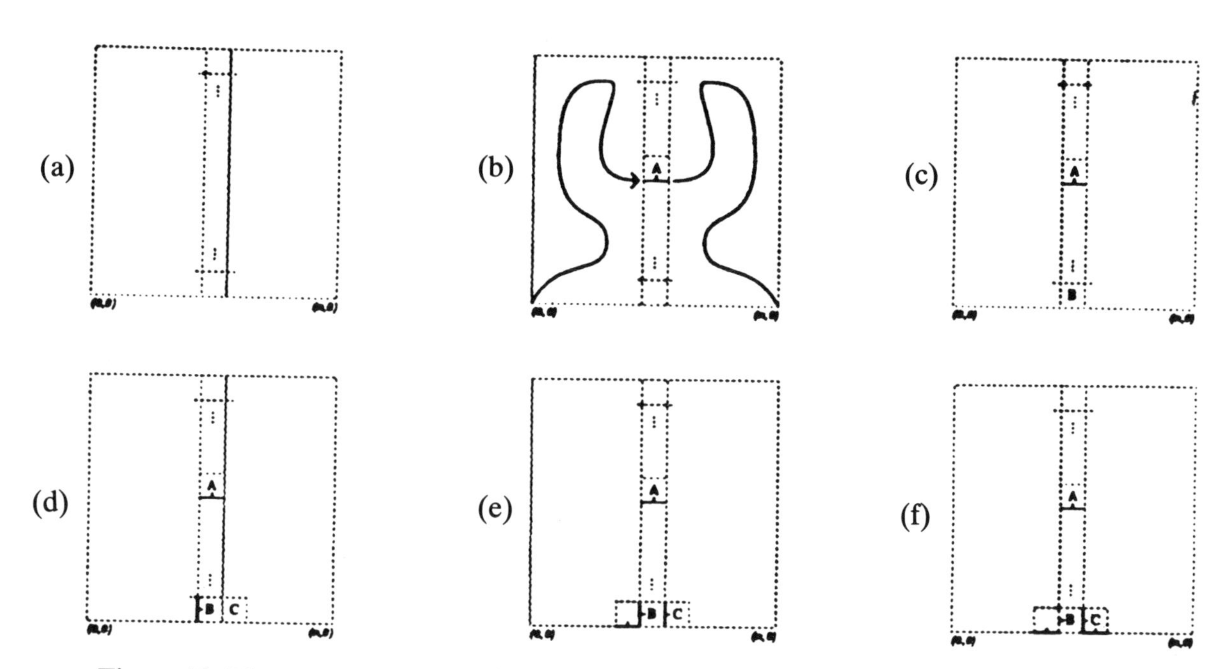

Figure 11: There are no symmetric self-avoiding generators on odd order square grids.

Bilateral symmetry on even-order grids is a different matter. By the same argument as above, there can be only one crossing of the central separation of the two central columns of an even-order grid, and it must consist of two horizontal line segments that pass two adjacent sub-squares in the columns, associated either both above or both below the two segments. However, the rest of the argument for odd n does not apply, and although direct inspection of all 5690 possible generators for the order 6 grid shows that none is bilaterally symmetric (but some are tantalizingly close), we can construct a bilaterally symmetric even order generator for a higher order even grid (Fig. 12).

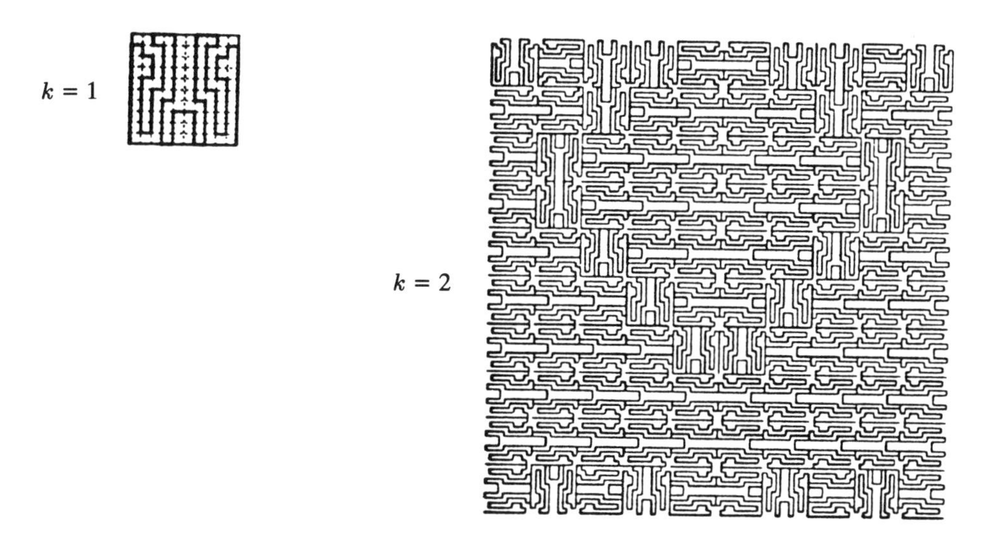

Figure 12: Self-contacting, symmetric, even order 10 Peano curve.

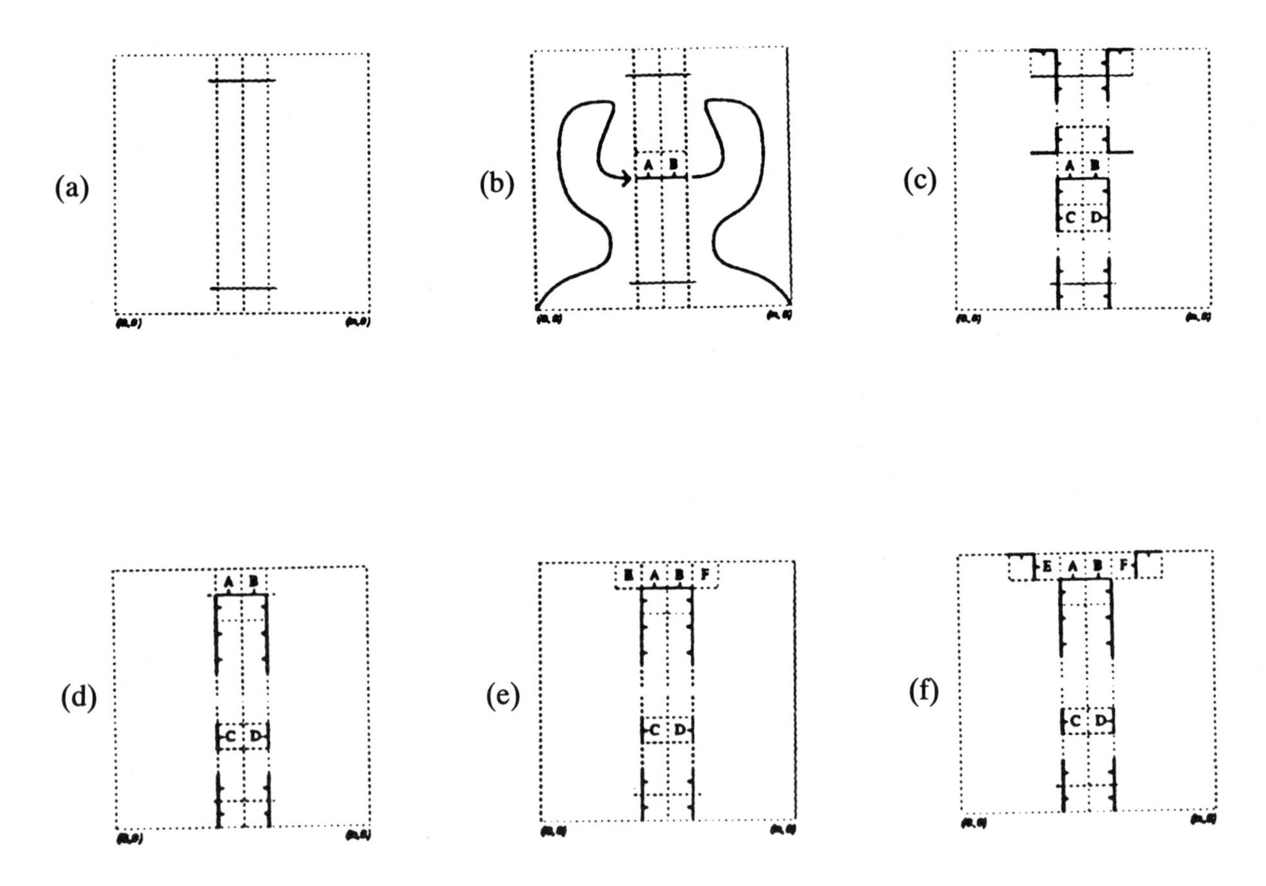

Figure 13: There are no symmetric self-avoiding Koch/Peano curves on even order square grids.

Again we note that even when a generator, consisting of n^2 connected line segments, is self-avoiding, this does not necessarily mean that approximations to the Peano curve it generates are self-avoiding. In fact, we cannot build bilaterally symmetric self-avoiding approximations for any order n.

Theorem 2 *Bilaterally symmetric Peano curve generators on order n fixed square grids can never generate self-avoiding approximations, for $k > 1$.*

Proof. By Lemma 1 there are no bilaterally symmetric generators for odd order grids, so we only need to look at even order grids.

The unique order 2 generator expands into self-contacting 2nd and higher approximations. By enumeration, we can see that all order 4 generators are asymmetric. For orders 6 and higher, we consider the central two columns of sub-squares (Fig. 13a). As in the odd case, a symmetric path can only cross these central two columns in one place, and symmetry dictates that the two adjacent sub-squares, A and B, that are associated with the two unique horizontal crossing segments must have the same orientation (Fig 13b). All other squares in the left and right columns are forced to be associated with line segments on their left or right sides (Fig. 13c), respectively, in such a way that the orientations of the generator pattern at the next recursive iteration will be 180° opposed to one another, as in the sub-tiles C and D. Thus, if the generator pattern is bilaterally symmetric, it suffices to show that the generator must somewhere touch the top of its square grid. If A and B are not in the top row of the two central columns, then the forced vertical segments above them must touch the top edge. If A and B are in the top row (Fig. 13d), then the segments associated with A and B must pass along their bases rather than tops in order to avoid the top edge of the grid. However, in this case sub-squares E and F must then have segments up their outside vertical sides (Fig. 13e, f), which lead up to the grid's top edge. Thus there will always be self-contact along the central division of the even grid under the constraint of bilateral symmetry. QED

7 The E Curve and the SquaRecurve Family

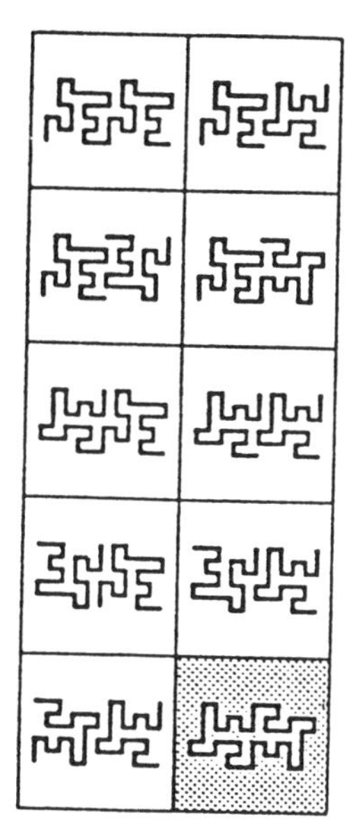

Figure 14

The reason the E Curve works as it does is because the generator has a kind of rotational (as opposed to bilateral) symmetry that allows it to mesh in a self-avoiding manner with any adjacent copy of itself regardless of the orientations of the two underlying adjacent sub-squares (Fig. 14). Note that although all of the possible combinations remain self-avoiding, the one that forms an immediate loop would never occur in any approximation since it implies that there would have to have been two coincident line segments in the previous approximation.

Since we have found a relatively simple generator for a self-avoiding Koch space-filling curve on the order 5 square grid, it is reasonable to wonder what generators of self-avoiding curves can be found for higher orders. Initial attempts to create by hand a generator for order 6 having the same general shape as the E Curve's do not prove fruitful, so again we resort to our full enumeration of 5690 distinct candidate generators. One can check this collection by hand or by computer to ascertain that among them finite approximations to their respective Peano curves are all self-contacting. The reason for this almost certainly lies in the fact that n here is even and/or that order 6 is too small, both qualities being a constraint on the ability to be rotationally symmetric; however, the full enumeration makes no use of this fact.

As Figure 9 illustrates, both the number of generators found and the CPU time it takes the program to find them grow exponentially with the size of the tiling. This combinatorial explosion arises because each time the algorithm chooses a candidate line segment along an edge of some sub-square, in the worst case it has 5 possible choices of next line segment to make. Choices near the perimeter are more constrained; however, as the order n gets larger, the number of perimeter choices becomes negligible compared to interior choices, so the number of choices in the average case gets closer to the number of choices in the worst case, which is on the order of 5^{n^2}.

A full enumeration of the order 7 square rep-tiling clearly will require too many resources unless a considerably more intelligent algorithm can be found for "pruning" the backtracking search tree. The algorithm used here to search the first six orders already employs a number of special purpose pruning tests, such as never taking the inappropriate turn when choosing a next segment along the perimeter, analyzing the current position for forced choices, etc. However, if we attempt to construct by hand a generator for the 7 by 7 grid that has the same type of structure as the E Curve, we will quickly be rewarded (Fig. 15). This curve is self-avoiding at all finite approximations because it has the same type of rotational meshing symmetry as the E Curve.

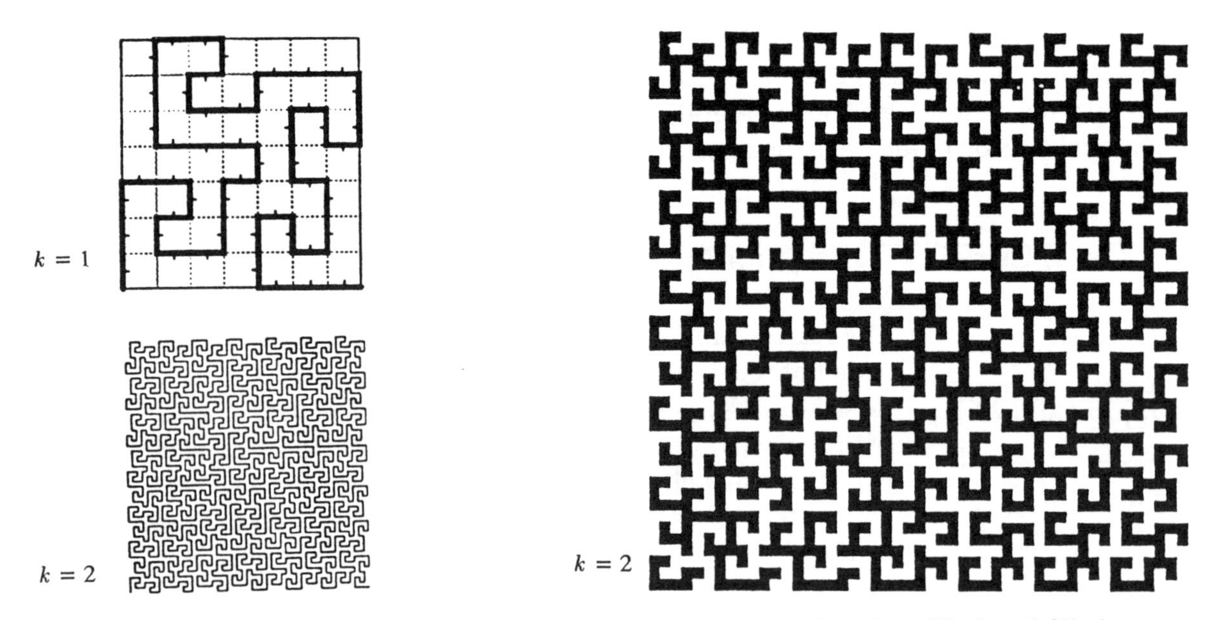

Figure 15: Order 7 SquaRecurve generator and approximation 2, unfilled and filled.

For each odd order n, we can construct such a related Peano generator (Fig. 16). I call this series of Peano curves "SquaRecurves" (short for Square Recursive Curves)[12]. SquaRecurves are characterized by four double spirals that "grow" out of the four "buds" of the E Curve. Each successive SquaRecurve generator has spirals that are wound a quarter of a turn more tightly.

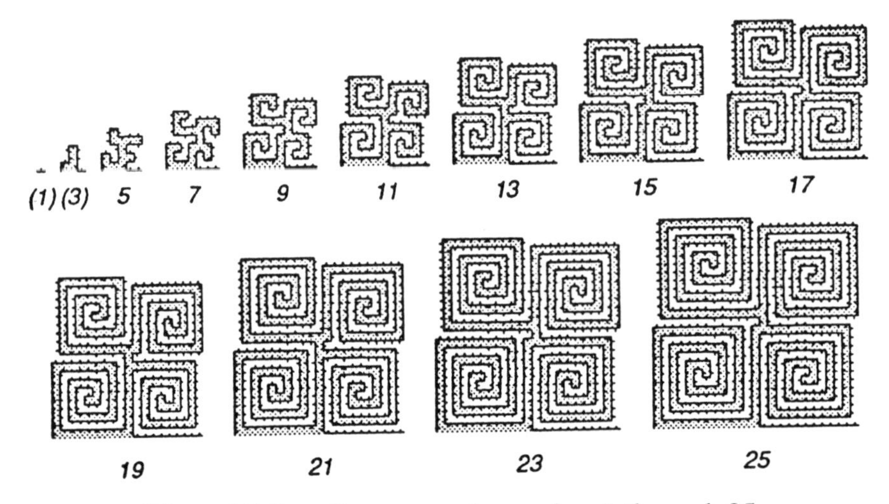

Figure 16: SquaRecurve series, orders 1 through 25.

The four double spirals connect with each other near the central sub-square of the odd n by n tiling. Each generator has a length of n^2 line segments, and each of the component segments is associated on a 1-to-1 basis with a component sub-square past which the segment travels.

If we want to include the order **1** and **3** grids in the SquaRecurve series, then it seems appropriate to call the generator for the order **1** grid trivial, since it is just a single line segment; and the generator for the order **3** tiling degenerate. There simply is not enough freedom to allow a self-avoiding curve in the order **3** grid under the combined constraints of maintaining the 1-to-1 mapping between segments and sub-squares while also maintaining the rotational symmetry.

8 New Irk: Unique?

Although there are clearly tens of thousands of generators for self-contacting Peano curves on the order 7 grid, we have been able to home in on one that will form a self-avoiding curve for all approximations by using a rotationally symmetric criteria in its construction. Since there appears to be at least one self-avoiding Peano curve for each odd order grid greater than 3, is it possible to say that the SquaRecurve for a given odd order tiling is unique, as is true of the E Curve?

In general, the answer to this is no, as can easily be seen by the following counterexample. Consider the order 25 grid, for which there is a SquaRecurve generator (Fig. 16, lower right). However, if we look at the second approximation of the E Curve (Fig. 10, upper right), we notice that it is composed of 25 copies of the 5 by 5 grid, all arranged (precisely because of the E Curve's structure) to tile the original square. If we disregard the origin of how these segments were placed on this 25 by 25 grid, we can see that it is itself a generator for a self-avoiding Peano curve. Its kth approximation is the E Curve's $2k$th approximation.

Instead of considering this generator as approximation 2 of the E Curve, we can think of it as a composition of one order **5** generator with another order **5** generator, both of which happen to be the same. To compose two generators, though, does not necessarily mean that they be the same, or even that the grid order be the same: we can compose an order n generator with an order m generator to create an order $n * m$ generator. For instance, the composition of the E Curve generator with the order **7** SquaRecurve generator yields a generator for the order **35** grid that is not the SquaRecurve for that order but which will have self-avoiding approximations (Fig. 17). The process of composition does not affect the rotational symmetry property of the product,

Figure 17: Order 35 generators: composition is not commutative.

since the structure of the perimeter of the composite generator depends only on the second of the two composed generators. Note that composing two generators is a commutative operation with respect to the order of the composite generator (since it's just a multiplication), but that it is not with respect to the composite generator itself. The order $\mathbf{5} * \mathbf{7}$ generator is clearly distinct from the order $\mathbf{7} * \mathbf{5}$ generator (although the limit sets, remember, are not).

With the above in mind, we can think of creating the kth approximation of a given generator G as an exponentiation G^k, with the order **1** generator (just the unit straight line) being the identity generator. The question of uniqueness now should be reformulated: Are the SquaRecurves analogous to prime numbers in the space of generators on odd-order grids?

Again, the answer to this is no, with simple proof by counterexample, shown to the author by Scott Kim in 1983. The E Curve can be generalized into a family of self-avoiding space-filling generators (E Tours) for every other odd order grid n, $n = 4i + 1, i = 0, 1, 2, \ldots$ (Fig. 18).

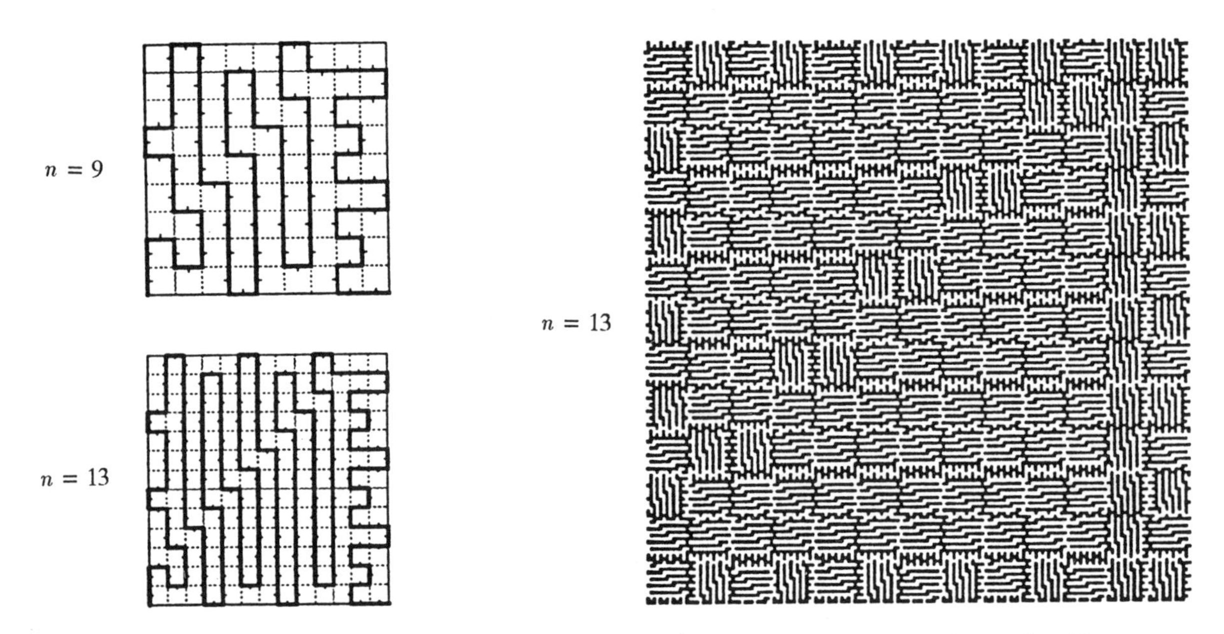

Figure 18: Generators for orders 9 and 13 E Tours, with filled $k = 2$ for order 13.

9 Perturbing the Fixed Square Grid

We know from tiling theory that if we perturb all four sides of a regular square tile in a rotationally symmetric way around the center of each side, the resulting figure remains a tile. Although we lose a degree of freedom—since in general the perturbed tile becomes untilable with its mirror image—the resultant tiling's underlying geometry is still intimately related to the original.

Suppose we rearrange the perimeter sub-squares of the order 5 square grid, creating both a hole and an accompanying peninsula on each side (Fig. 19). The resulting grid is no longer a square, and no longer a strict rep-tile, but has the same area, shares the same starting and ending corner as the original square grid and the same tiling characteristics as the original square. If a

Figure 19: Approximation 3 of unique generator for perturbed order 5 square grid.

path traveling from one of its corners to the adjacent corner manages to pass each of the same 25 sub-squares in a 1-to-1 fashion as described previously, then it qualifies as being a Peano curve generator, although not necessarily a generator for a self-avoiding curve.

However, the perimeter is now more convoluted than in the unperturbed case, and so we should expect the set of possible generators to be smaller than those of the normal square grid.

Thus, if we try by hand to find a generator for this grid based on the E Curve's shape, we quickly find one (Fig. 19, upper right). Furthermore, when we do a full enumeration of all possible generators embedded in this perturbed order 5 square grid, we discover that there are no other possibilities, even for generators of self-contacting Koch/Peano curves. And all finite exponentiations of this unique generator are self-avoiding. Clearly something crystalline is going on here, and a closer look reveals the underlying structure. The generator is made up of 5 exact copies of an open-ended 5-segment pattern, connected end to end. Each segment of one of these patterns is associated on a 1-to-1 basis with exactly one square out of the 5 squares that together form a Swiss cross tile. These 5 tiles exactly cover the perturbed order **5** grid. And most importantly, the placement and orientations of the 5 patterns are determined by a larger mirror image of the same pattern. Thus, this unique generator for the perturbed grid turns out to be the composition of a much simpler 5-segment generator with its own mirror image **[13]**.

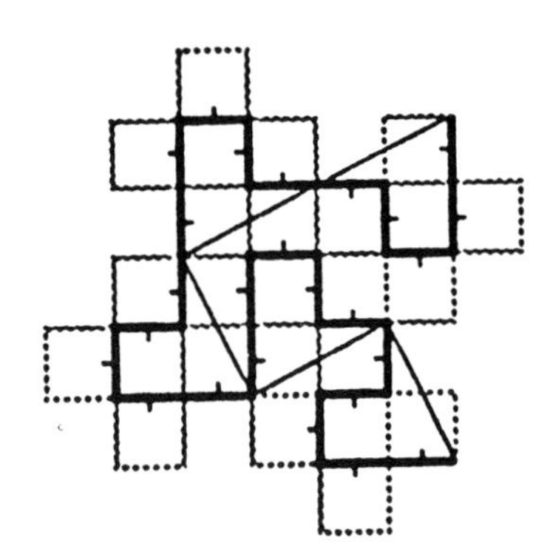

10 Variable Square Grids

If we orient the 5-segment pattern so that its ends coincide with the ends of the unit line segment, we see that its segments are embedded in a square grid that is rotated at an angle of

$$\theta = \arccos 1/\sqrt{5}$$

Let us call this generator $\mathbf{g}_{1,2}$. When it is exponentiated, each successive iteration is on a grid rotated by θ more radians than the last, with the sub-square sizes decreasing by a factor of $\sqrt{5}$. I call this a variable square grid, whereas the fixed square grid does not allow this kind of rotation. Note that when we use the mirror image of the Swiss cross generator for every even Koch substitution, then the rotation will be by $-\theta$, reorienting the grid back to an angle of 0, which is the unperturbed fixed square grid.

On the fixed square grid, we needed only one number, the order n, to specify the grid. On variable square grids we need two numbers, n and m, to specify the grid. A variable square grid, $\mathbf{G}_{m,n}$, will rotate by an angle

$$\theta = \arccos m/\sqrt{n^2 + m^2}$$

on each Koch substitution. Each $\mathbf{G}_{m,n}$ has associated with it a perturbed square tile composed of $n^2 + m^2$ square sub-tiles, called $\mathbf{T}_{m,n}$. Generators for self-avoiding Koch/Peano curves travel among these sub-tiles in the same way as generators on the fixed square grid do: their lengths must be equal to the number of sub-tiles, and unit segments of the generator must be associated on a 1-to-1 basis with the $n^2 + m^2$ sub-tiles of $\mathbf{T}_{m,n}$. In fact, if m is 0, then the variable square grid becomes just the fixed square grid of order n, with $\theta = 0$.

The simplest variable square grid is $\mathbf{G}_{1,2}$, which we have seen is associated with the Swiss cross tile, and which has a unique Koch/Peano curve generator embedded in it. This generator is actually easily generalized into two different infinite families of space-filling curves, which have a very similar spiral structure.

The first family, which I call Eddy Curves, is found on $\mathbf{G}_{1,n}$, where n is even (Fig. 20). For each $\mathbf{G}_{1,n}$ a generator $\mathbf{g}_{1,n}$ can be constructed from $\mathbf{g}_{1,n-1}$ by simply adding line segments around the perimeter sub-tiles of $\mathbf{T}_{1,n-1}$ in a spiraling manner. Thus each successive generator $\mathbf{g}_{1,n+1}$ contains the previous generator $\mathbf{g}_{1,n}$ embedded within it, and as such is shaped like a double-spiral that progressively winds tighter and tighter as n rises. I like to think of these curves as the two-dimensional equivalents of the familiar Greek key design found in architecture since ancient times.

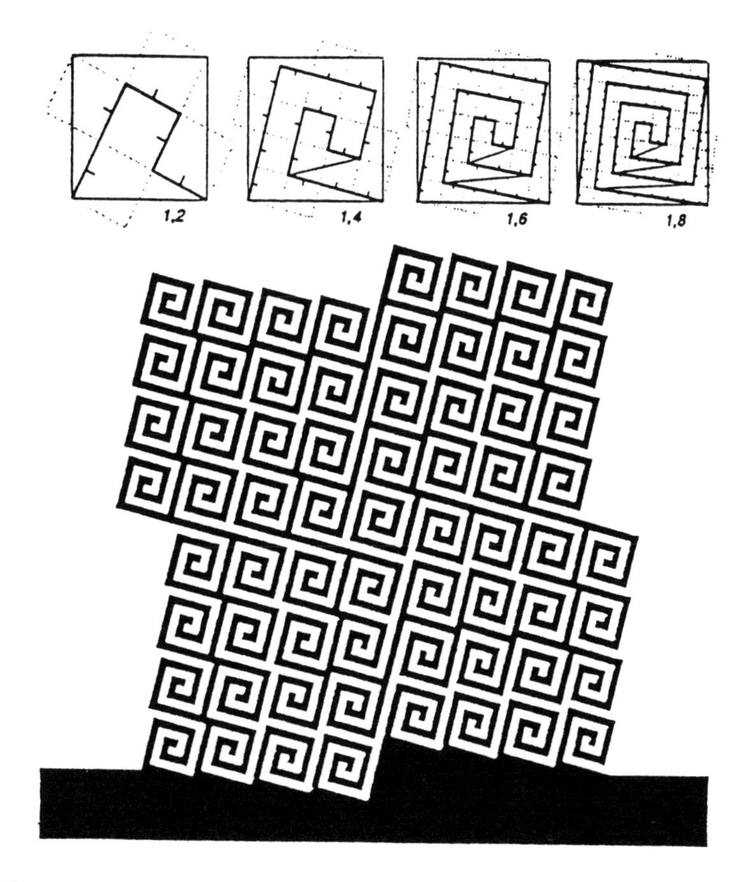

Figure 20: Eddy Curves and filled approximation 2 for Eddy 1, 8.

The second general family has the same recursive spiraling structure as the Eddy Curves, although they look quite different. I call them Frenzy Curves due to obvious visual energy they display (Fig. 21). Frenzy Curves are found on $\mathbf{G}_{n,n+1}$ for all $n > 0$. Like the Eddy Curves, each successive generator $\mathbf{g}_{n,n+1}$ contains the previous generator $\mathbf{g}_{n-1,n}$ embedded within it in a double-spiraling manner. Instead of smooth series of collinear segments along the sides of the square spiral, like the Eddy Curves have, the sides of the spirals in the Frenzy curves are zig-zags.

Both of these families share the same first member, $\mathbf{g}_{1,2}$, and generalize from it in slightly different ways using a spiral structure. A similar spiral generalization can also be applied to Gosper's FlowSnake curve [**14**] on a variable hexagonal grid.

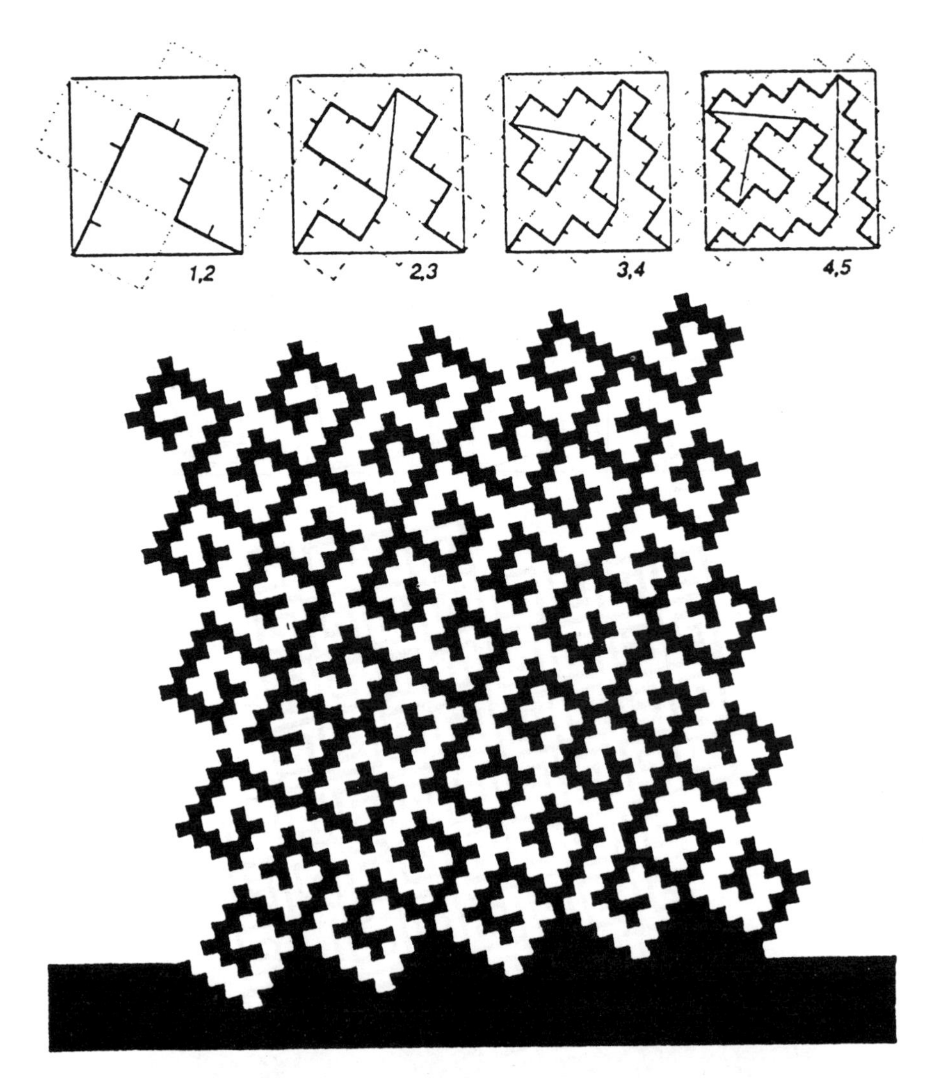

Figure 21: Frenzy Curves and filled approximation 2 for Frenzy 4, 5.

Like the SquaRecurves on the fixed square grid, there are other self-avoiding curves besides the ones in these families. Although $\mathbf{g}_{1,2}$ is unique on $\mathbf{G}_{1,2}$, $\mathbf{g}_{1,4}$ is not unique on $\mathbf{G}_{1,4}$—there is one other possible Peano curve generator on $\mathbf{G}_{1,4}$ and it is self-avoiding (Fig 22). Both $\mathbf{g}_{2,3}$ and $\mathbf{g}_{3,4}$ are unique on their respective grids, which fact can be proved as before using machine enumeration. The numbers for higher order variable square grids are not known.

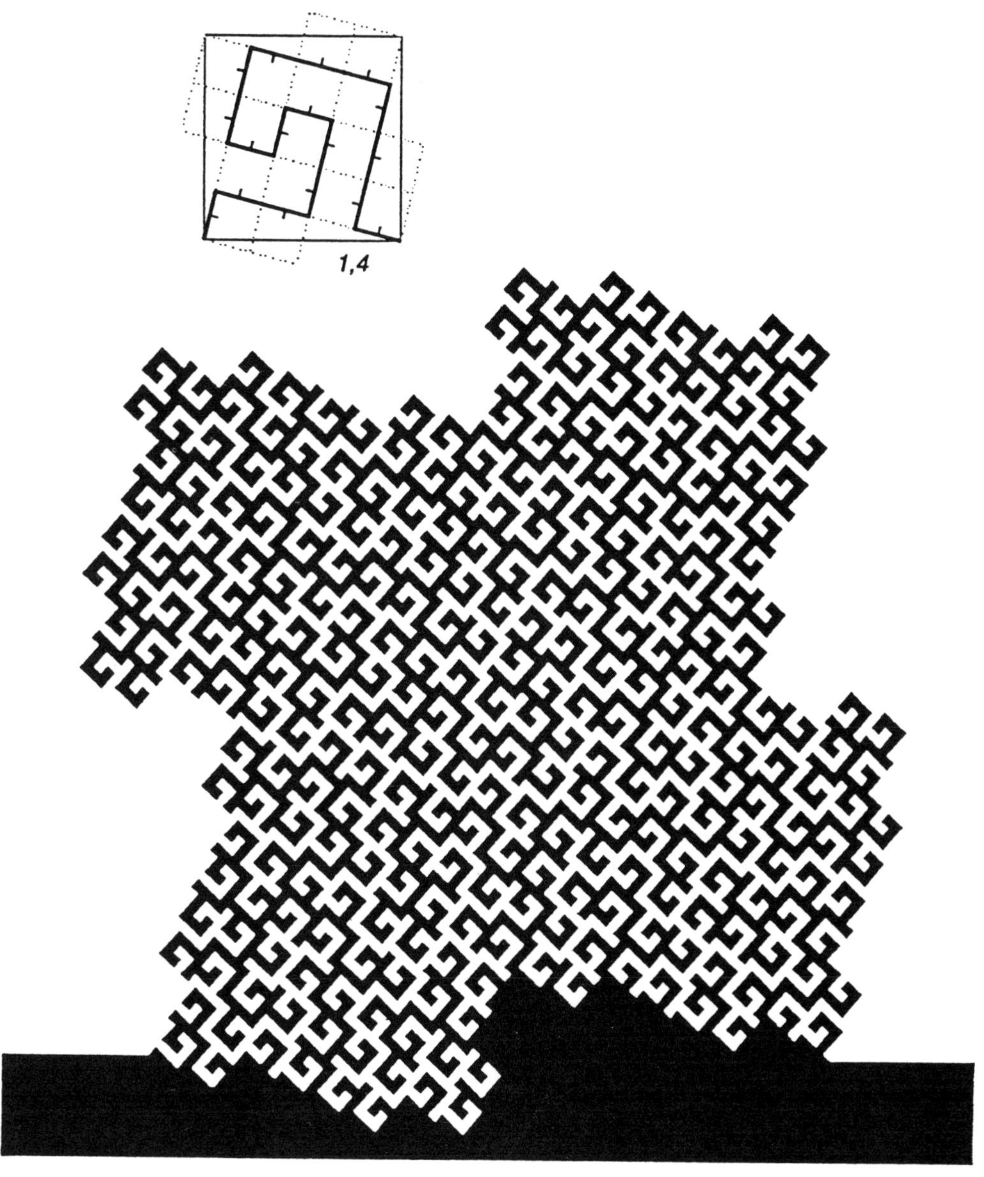

Figure 22: Eddy (1,4)'s only cousin, approximation $k = 3$, filled.

11 Conclusion

Self-avoiding, space-filling paths embedded in square grids in such a way that they can be multiplied under the composition operator of Koch substitution are subject to a variety of constraints that limit their form. These constraints subside somewhat as the grid order rises. On the fixed square grid, the corner tiles are the most limited, and other perimeter tiles are less so, with the interior the least. Thus, there are no self-avoiding curves at all on the analogous triangular grid, and none on the square grid until we get to order **5**, where the E Curve is unique. However, as the square grid order becomes larger, we find at least three kinds of generators for self-avoiding Koch/Peano curves for various odd order fixed square grids: SquaRecurves for all odd orders, E-Tours for every other odd order, and compositions of any number of these for composite odd orders.

Fixed square grids are special cases of variable square grids, in which many space-filling patterns can be found. Two related families, the Eddy Curves and the Frenzy Curves, are both analogous to Gosper's FlowSnake curve and its generalizations on variable hexagonal grids. The constraints on self-avoiding Koch/Peano curves embedded in variable square grids are less than those of the fixed square grids, because the underlying recursive tiling geometry helps keep the generator curves away from the perimeter.

Programming a computer to enumerate all possibilities is useful both as a method of discovery and as a method of proof of uniqueness or non-existence. In this instance, the computer's aid lets us look just far enough to see the beginning of an interesting geometric realm of constraint and freedom in the simple square tilings. The space of self-avoiding Peano curves with strict Koch geometry is not yet fully explored. Until a full enumeration of just the next few orders of the simple square grid is available, it will be difficult to see the patterns of form and variation within the tiling imposed by the constraints of the problem.

The full search of the order **5** fixed square grid has been performed over the years using four different implementations in three different computer languages on 5 different computers, all yielding the same figure of 138 generators with the unique E Curve among them. This helps build confidence in it as a proof (although all of the above was performed solely by the author). The order 6 search, which took 50 hours of CPU time on an IBM 3330 mainframe computer in 1981, has been performed only once, albeit using one of the same implementations mentioned above. The limitations of using the computer should also be apparent. Only the smallest order grids are amenable to full-scale enumeration, and although those full enumerations that we are able to create help us explore and visualize, they don't tell us much about any underlying mathematics. That is still up to us humans.

12 Acknowledgements

I'm grateful to the University of Montana computer center for allowing me the free use of their DEC-20 computer during 1978–79. Drawing fractals at 300 baud over the phone line at that time would have otherwise been far too expensive. Thanks are also due to Dr. V. Alan Norton of IBM Research for his unwavering and optimistic support during difficult times.

The illustrations in this paper were designed and implemented as PostScript programs that print directly on many laser printers and phototypesetters.

Endnotes & References

1. K. Appel & W. Haken, Every planar map is four colorable, *Illinois J. of Math.* 21 (1977) 429–567.

2. Oren Patashnik, Qubic: 4 x 4 x 4 Tic-Tac-Toe, *Math. Mag.* 53 (Sept. 1980) 202–216.

3. Seymour Papert, *Mindstorms,* Basic Books, 1980.

4. Stephen Wolfram, *Mathematica,* Addison-Wesley, 1988.

5. B. B. Mandelbrot, *The Fractal Geometry of Nature,* W. H. Freeman and Co., 1982 (illustrated by numerous people, including this author).

6. H. O. Peitgen and P. H. Richter, *The Beauty of Fractals,* Springer-Verlag, 1986.

7. c.f. "Cesaro Sweep," Fractal Geometry of Nature, p. 64.

8. S. Golomb, Replicating Figures in the Plane, *Math. Gaz.* 48 (1964) 403–412.

9. Mandelbrot, *Fractal Geometry of Nature,* p. 46.

10. Original Peano curve, *Fractal Geometry of Nature,* p. 63.

11. I discovered the E Curve in 1978. It was independently discovered a few years later (Dekker, Recurrent Sets, *Advances in Math.* 44 (1982) 78–120) (reference from SIGGRAPH '88 Course Notes "Fractals Part II: Lindenmayer Systems, Fractals and Plants").

12. Mandelbrot once proposed calling these curves Peano–McKenna Curves (private communication, 1979), along the same lines as his renaming of Gosper's FlowSnake Curve to "Peano–Gosper Curve"; however they did not make it into his book's taxonomy of geometric fractals.

13. This generator appears to have been discovered independently by many different people, including Gosper, Mandelbrot and myself. See plate 49, *Fractal Geometry of Nature.*

14. William Gosper (from Mandelbrot, *Fractals: Form, Chance and Dimension,* Freeman, 1977).

93 Wolcott Road
Brookline, MA 02167-3108

Fun With Tessellations*

John F. Rigby

This article falls into two parts. The first part (Section 1) contains examples showing how tessellations can be used to prove some theorems in elementary geometry that appear at first sight to have no connexion with tessellations. The second part (Sections 2–6) stems from my delight in making perfect colorings of tessellations, mainly in the hyperbolic plane; in it we shall meet strange shapes that I call regular convex dendroids, but these may already exist in some hyperbolic arboretum under another name.

1 Tessellations and elementary geometry

Since this volume contains an article on the art of bellringing, it seems appropriate to mention that my story begins when I was four or five years old and my father returned one Saturday from a ringers' outing bringing home for me a present: a box of small cardboard squares each with a letter of the alphabet printed on it. The box was labelled "Word Making and Word Taking" and the letters were clearly intended for some word game, though I do not recall ever seeing the rules of the game. But the letters did have their uses when I was learning to spell: by one of those tricks of childhood memory I clearly recollect producing the word TOUNGE, since when I have never had any trouble in spelling "tongue".

The backs of these cardboard squares had (and still have) a rich red glossy surface, and I am told that from the very day on which I received the gift I used to turn them all over and make patterns with them. One favorite pattern that I can remember constructing a few years later is shown in Figure 1, where black squares have been laid on a white surface; no one told me at that time that I had constructed one of the eight semiregular tessellations of the plane, but the incident shows that my interest in tessellations began at an early age.

If at the same age I had been given cardboard equilateral triangles of three different sizes, I wonder whether I should have succeeded in making the tessellation shown in Figure 2 (where white triangles have been laid on a black surface). During a week's Mathematics Camp for bright 13-year-olds from schools in mid-Glamorgan in 1986, I supplied the pupils with equilateral triangles in three sizes, and also with scalene triangles, and most of them were able to fit them together to produce Figure 2. This is more than just a nice pattern, as we shall now see. The centre of each small equilateral triangle is equidistant from the centres of the six neighboring small equilateral triangles; hence the centres of all the small equilateral triangles form an equilateral triangular

*Sections 2–6 of this article are adapted and reprinted by permission from a longer article in the journal *Leonardo* [**8**].

Figure 1: One of the eight semiregular tessellations of the plane.

lattice, as shown in Figure 3. The centres of the remaining equilateral triangles of the tessellation clearly lie at the centres of the triangles of the lattice, and so the centres of *all* the equilateral triangles form a smaller equilateral triangular lattice (Figure 3). Thus in Figure 4, which forms a small part of the tessellation, the centres of the three equilateral triangles form an equilateral triangle; this is **Napoleon's theorem** [2], and it is true whatever the shape of the scalene triangle since the tessellation of Figure 2 can be constructed starting with any scalene triangle.

Figure 2: A tessellation using three sizes of equilateral triangle.

When I was an undergraduate at Cambridge there was a story told about an eminent physicist who remarked during a lecture that a certain result could be proved "without any mathematics at all, just by using group theory". I feel that in the same way we have proved Napoleon's theorem without any geometry at all, just by using tessellations.

Napoleon's theorem is also valid if the triangles are erected inwards on the sides of a triangle,

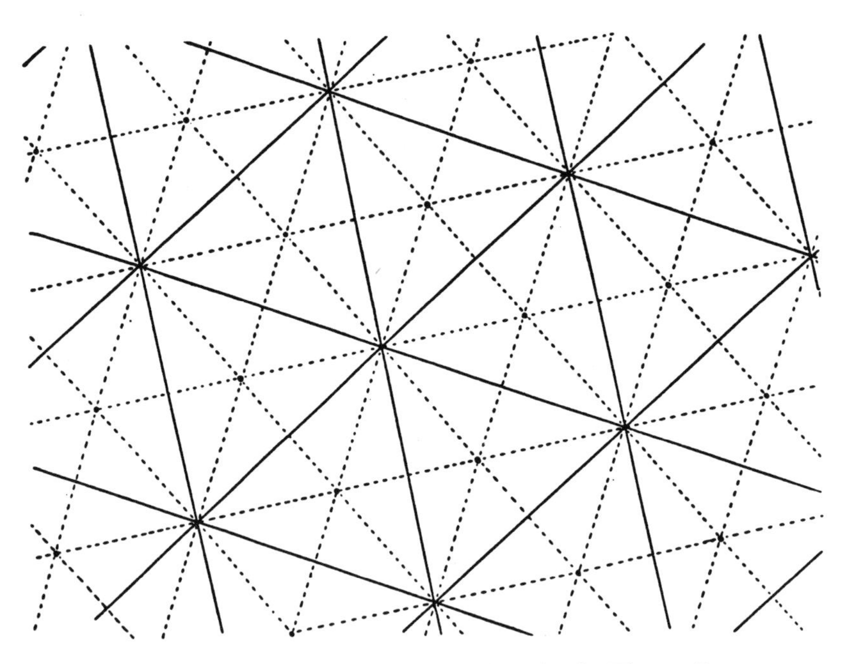

Figure 3: The centres of the triangles in Figure 2.

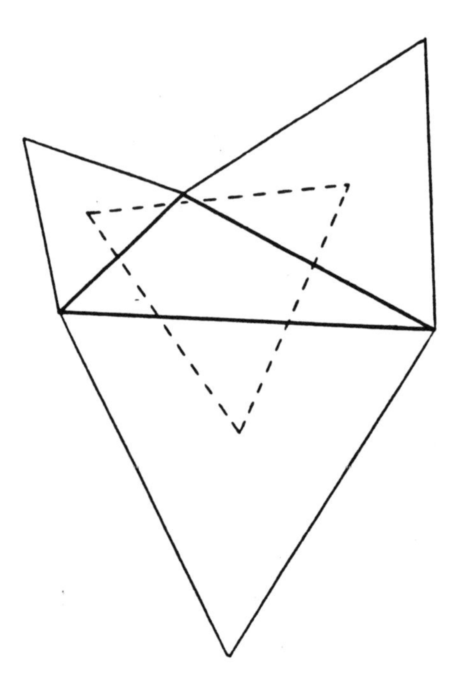

Figure 4: Napoleon's theorem.

and a similar proof can be given in this case also, as long as we are prepared to extend our idea of a tessellation; but there are generalizations of the theorem that cannot be proved using tessellations [**6**].

In a similar way the tessellation of Figure 5 can be used to prove that the centres of squares erected on the sides of a parallelogram, as in Figure 6, form a square.

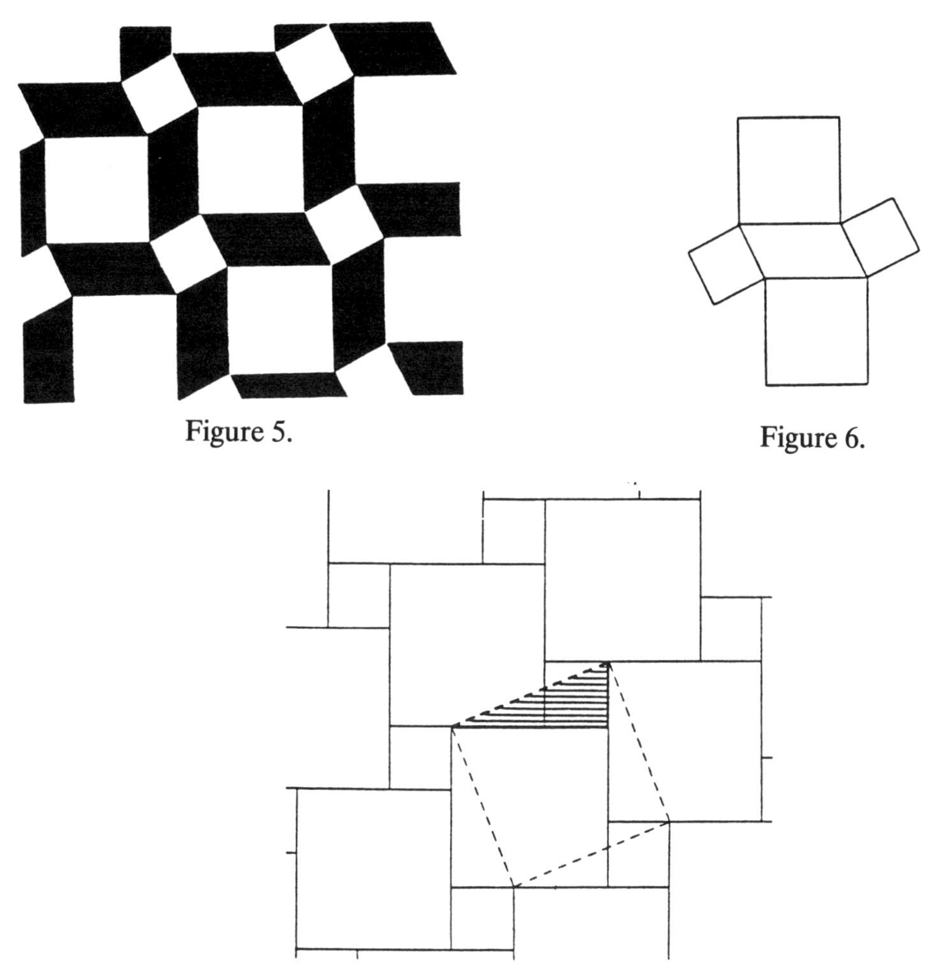

Figure 5.

Figure 6.

Figure 7: A tessellation proves Pythagoras's theorem.

Figure 7 shows a tessellation formed from squares of two sizes. The centres of the small squares form a square lattice, and by translating this lattice diagonally up to the right we see that the top right-hand corners of the small squares form a square lattice. Thus the four broken line-segments in the figure form a square, divided up into five pieces; three of these pieces can clearly be re-assembled to form a large square of the lattice, and the remaining two pieces form a small square. Thus if we consider the shaded triangle in the figure we see that the square on its hypotenuse is equal to the sum of the squares on the other two sides. This proof of Pythagoras's theorem by dissecting the square on the hypotenuse is not new—I remember it in one of my school geometry text books—but the use of the tessellation shows clearly how the dissection is carried out and why it works [**5**].

2 Compound tessellations and perfect colorings

The rest of this article is concerned with an aspect of **visual mathematics**: the delight of regular colorings of patterns and the way in which shapes (some of them strange) fit together.

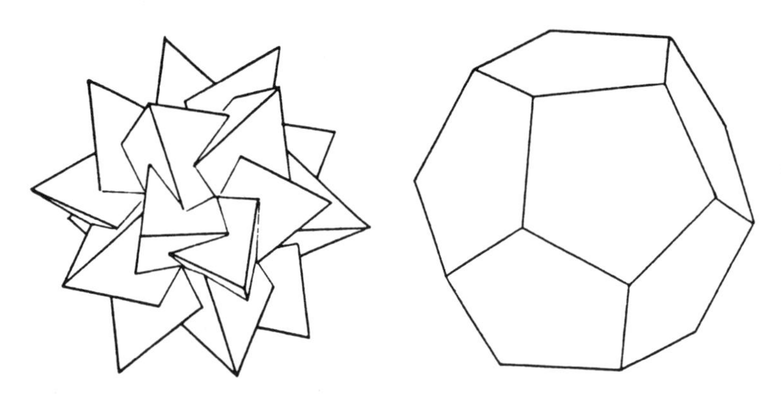

Figure 8: Five tetrahedra inscribed in a dodecahedron.

The two best-known regular compound polyhedra are Kepler's **stella octangula** (two tetrahedra with their eight vertices at the vertices of a cube) and five tetrahedra inscribed in a regular dodecahedron (Figure 8). Now the tetrahedron, cube and dodecahedron can be "blown up" like balloons to form regular tessellations on a sphere (Figure 9; this figure is adapted from Figure 7/1 in **[4]**); the compound polyhedra just mentioned then produce **compound tessellations** on a sphere, compounds of two tetrahedral and five tetrahedral tessellations respectively. This idea can be extended to plane tessellations; for example in Figure 10a the vertices of a regular tessellation of hexagons have been divided into four sets, and the vertices in each set form a regular hexagonal tessellation.

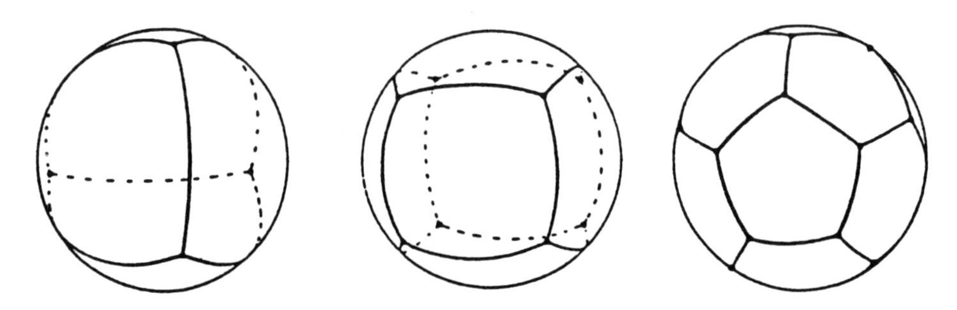

Figure 9: Regular tessellations on a sphere.

This last figure, even if we draw the edges in four different colors, lacks the visual appeal of the compound polyhedra (the five tetrahedra inspired the artist Escher to create the carving "Polyhedron with Flowers"; see **[3]** for example), but there is a way of remedying this situation. The original hexagonal tessellation in Figure 10a has a **dual** tessellation of triangles (Figure 11): each vertex of the hexagonal tessellation lies at the centre of a triangle and each vertex of the triangular

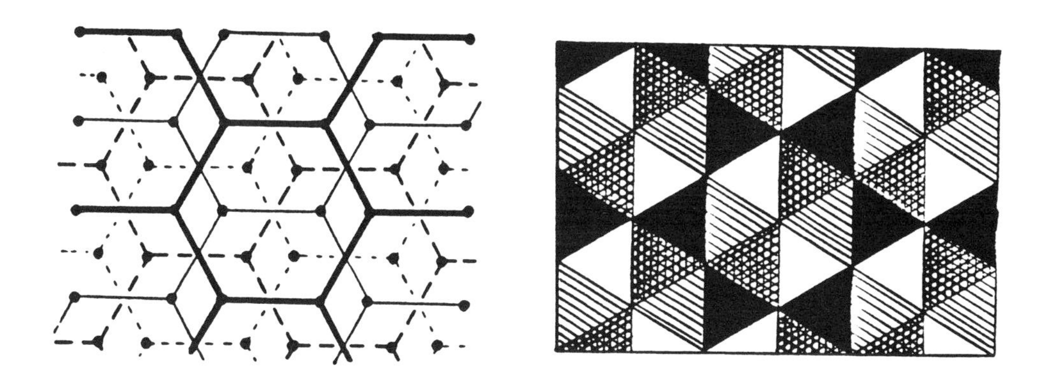

Figure 10: A compound of four hexagonal tessellations and the colored dual.

tessellation lies at the centre of a hexagon. Imagine each vertex of the hexagonal tessellation to be colored with one of four colors, according to which component of the compound tessellation it belongs to, then color each face of the dual triangular tessellation with the color of the vertex at its centre (Figure 10b). In this way the original compound constructed from the vertices of the hexagonal tessellation gives rise to a coloring of the dual triangular tessellation. This coloring is clearly "regular", in a sense that we shall describe in detail later; it is called a **perfect** coloring of the triangular tessellation.

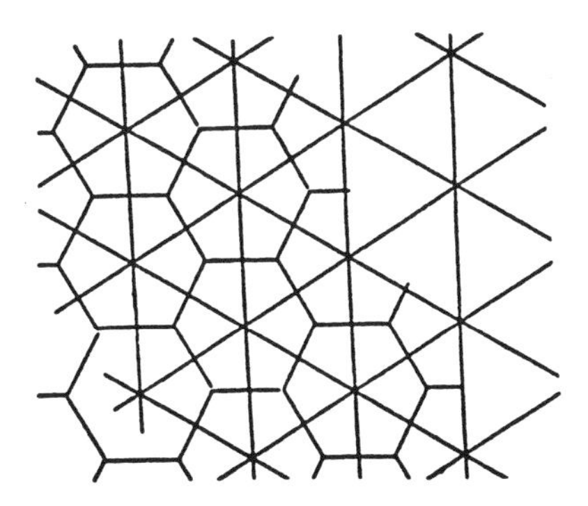

Figure 11: Dual hexagonal and triangular tessellations.

In a similar way, the *stella octangula* inscribed in a cube gives rise to a perfect coloring, in two colors, of the dual octahedral tessellation on a sphere, and the five tetrahedra inscribed in a dodecahedron give rise to a perfect coloring, in five colors, of the dual icosahedral tessellation (Figure 12).

In this way, regular compound tessellations lead us to perfect colorings of regular tessellations. Is the converse true: do perfect colorings always lead us back to regular compounds? The investigation of this question in the hyperbolic plane has interesting consequences.

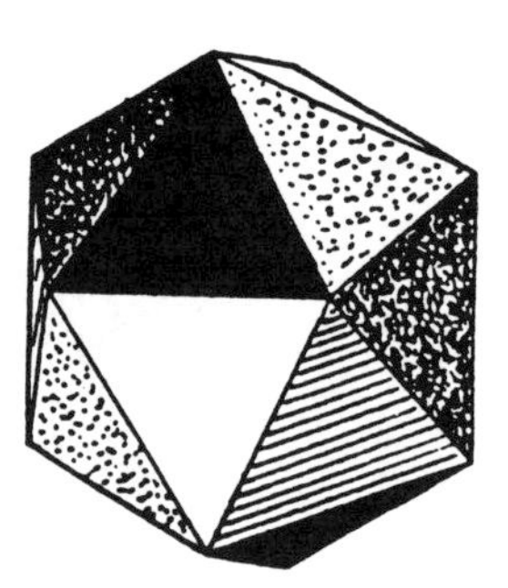

Figure 12: A perfect 5-coloring of the icosahedron.

3 Regular compound tessellations

The regular tessellation of regular p-gons, meeting q at each vertex, is denoted by the **Schläfli symbol** $\{p, q\}$; thus the dodecahedral tessellation on a sphere is $\{5,3\}$ and the triangular tessellation in a Euclidean plane is $\{3,6\}$. A **symmetry** of a tessellation is an isometry that transforms the tessellation to itself. The complete symmetry group of $\{p, q\}$ is denoted by $[p, q]$, and its "direct" subgroup, generated by rotations, is denoted by $[p, q]+$.

There are various different definitions of the term "regular compound tessellation" [7]; we shall use the definition given by Coxeter in [1], with some extra terminology. It may happen that all the vertices of a $\{p, q\}$ occur among the vertices of an $\{m, n\}$. Applying $[m, n]+$, we derive a set of several $\{p, q\}$'s forming a **vertex-regular compound**. If this compound is not symmetrical by a reflexion, it is said to be **chiral** and it then exists in **laevo** and **dextro** (i.e., left-handed and right-handed) varieties, which can be superposed to form a new **dichiral** compound with twice as many components. If there are d components (on which either $[m, n]$ or $[m, n]+$ is transitive), and if each vertex of the $\{m, n\}$ belongs to c of them, the vertex-regular compound is denoted by

$$c\{m, n\}[d\{p, q\}].$$

Thus the *stella octangula*, the compound of five tetrahedra, the compound of ten tetrahedra formed by combining the *laevo* and *dextro* varieties of the previous compound, the compound of five cubes inscribed in a dodecahedron, and the hexagonal compound in Figure 10a, are denoted respectively by

$$\{4,3\}[2\{3,3\}],\ \{5,3\}[5\{3,3\}],\ 2\{5,3\}[10\{3,3\}],\ 2\{5,3\}[5\{4,3\}],\ \{6,3\}[4\{6,3\}].$$

Many vertex-regular compounds are also **face-regular**, and the notation can be extended to indicate this fact, but we shall not need the extra notation. We shall be concerned mainly with compounds in which $c = 1$, i.e. in which each vertex of the underlying tessellation occurs as a vertex of only one component of the compound; these will be called **plain** regular compounds. The dichiral compound obtained by superposing the *laevo* and *dextro* varieties of a chiral compound is not a plain compound.

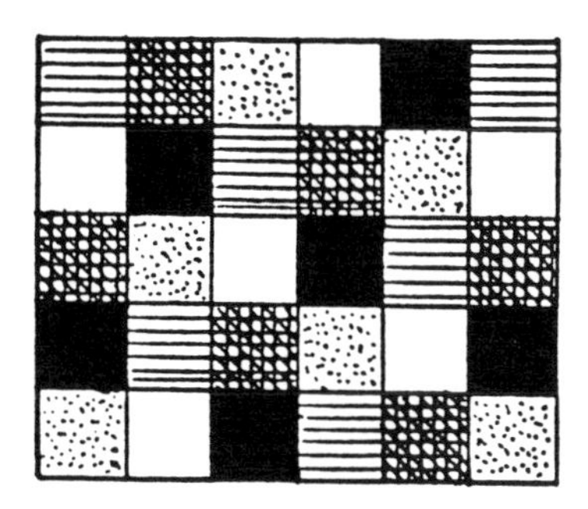

Figure 13: A chirally perfect coloring.

4 Perfect colorings

Suppose that a finite number of colors is used to color a tessellation, each tile of the tessellation being colored in one color. If every symmetry of the tessellation permutes the colors, the coloring is a **fully perfect** coloring; if every rotational symmetry permutes the colors but reflexional symmetries jumble the colors up, the coloring is **chirally perfect**. The word "perfect" will be used for both cases. For instance the coloring in Figure 13, using five colors, is chirally perfect but not fully perfect, because a reflexional symmetry of the square tessellation maps some black squares to black, some to white, etc., whereas every rotational symmetry permutes the colors; likewise the coloring in Figure 12 is chirally perfect. The coloring in Figure 10b is fully perfect.

It is obvious that a plain regular compound derived from a regular tessellation gives rise to a fully perfect (or a chirally perfect) coloring of the dual tessellation, in the manner described in Section 2, because every symmetry (or every rotational symmetry) of the tessellation permutes the components of the compound amongst themselves.

We now consider the converse process. Suppose we are given a perfectly colored regular tessellation. In order to attempt to convert it back into a regular compound tessellation we join the centre of each black tile, for instance, to all the black centres nearest to it, thus creating a "black tessellation"; we then do the same for every other color. Do the resulting colored tessellations form a regular compound?

Consider first perfectly colored tessellations on a sphere. The two cases mentioned in Section 2 give rise to the *stella octangula* and the regular compound of five tetrahedra. The only other perfect colorings are the colorings of the cube, octahedron, dodecahedron and icosahedron using three, four, six and ten colors respectively (in which opposite faces have the same color) and the colorings of all five regular polyhedra in which different faces always have different colors. These colorings do not give rise to regular compounds in the normal sense; they can be regarded as trivial cases, but the colorings of $2n$-hedra in n colors are connected with regular compounds in which the components are **hosohedra**, having just two diametrically opposite vertices like a beach ball [**9**].

In the Euclidean plane, every perfectly colored regular tessellation gives rise to a regular compound. These regular compounds have been listed in [**1**], but it is worth mentioning that some are more regular than others [**7**]. For instance, one of the simplest perfect colorings, shown in Figure 14a, gives rise to the compound shown in Figure 14b. This is vertex-regular, but the triangular faces are of two different types: some faces have a vertex of the other component at their centres,

Figure 14: A perfect coloring leading to a compound which is not face-regular.

but other faces do not. Here we have a simple example of a compound that is not face-regular.

5 Perfect colorings in the hyperbolic plane

It must be assumed that the reader has some familiarity with the hyperbolic plane. We shall use the **Poincaré model**, in which the entire infinite plane is represented inside a bounding circle. In the regular hyperbolic tessellations illustrated, all the tiles are regular and congruent to each other in "hyperbolic reality", but they are perforce distorted in the illustrations (rather as images are distorted in a convex mirror; but the distortion of physical reality in a convex mirror is geometrically different from the distortion of "hyperbolic reality" in the Poincaré model).

Let us consider various simple perfect colorings, to find out whether they lead back to regular compounds; very often they do not.

(a) Figure 15 shows the unique perfect coloring of $\{8,3\}$ in three colors. If we join the centre of each black octagon to the eight nearest black centres, we obtain the tessellation $\{4,8\}$. The two other colors yield two more $\{4,8\}$'s, giving a regular compound tessellation

$$\{3,8\}[3\{4,8\}].$$

We can similarly use the perfect coloring of $\{2n,3\}$ in three colors to obtain the regular compound tessellation

$$\{3,2n\}[3\{n,2n\}].$$

(b) Figure 16a shows the unique perfect coloring of $\{3,8\}$ in two colors. If we join the centre of each black triangle to the six nearest black centres, we obtain the tessellation shown in Figure 16b. This is a quasi-regular tessellation. Since the six faces meeting at a vertex are alternately triangles and quadrangles, we can denote the tessellation by 3.4.3.4.3.4 or $(3.4)^3$; it was used by Escher as the basis of his woodcut *Circle Limit III* (see for instance **[3]**). The white triangles in Figure 16a give another tessellation of the same type, and the two together give the non-regular compound

$$\{8,3\}\left[2(3.4)^3\right].$$

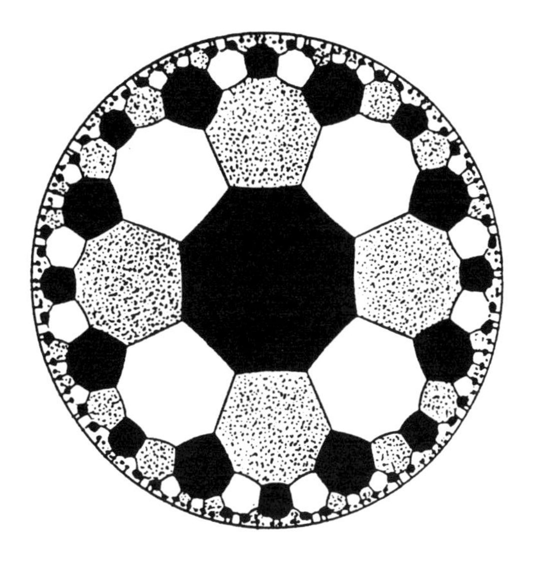

Figure 15: The perfect 3-coloring of {8,3}.

Figure 16: The perfect 2-coloring of {3,8}
and one of the two induced quasi-regular tessellations.

We can similarly use the perfect coloring of $\{k, 2n\}$ in two colors to obtain the compound

$$\{2n, k\}\left[2(k.n)^k\right].$$

If $k = n$ this compound is vertex regular but not face regular (like the compound in Figure 14b), and the notation for it then becomes

$$\{2n, n\}[2\{n, 2n\}].$$

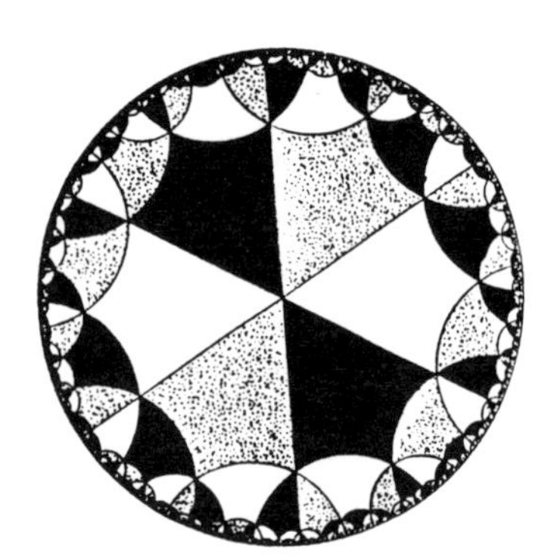
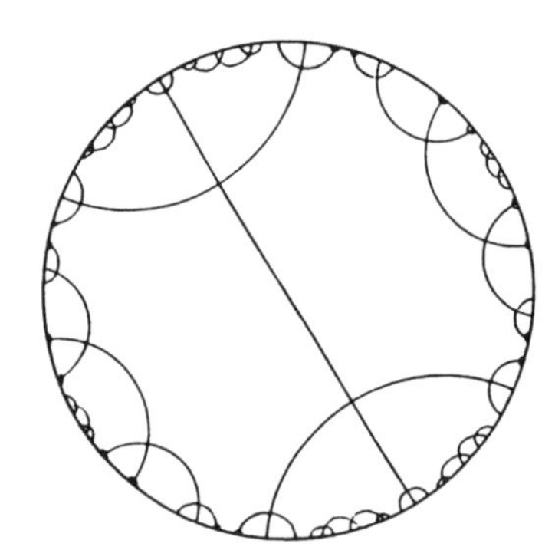
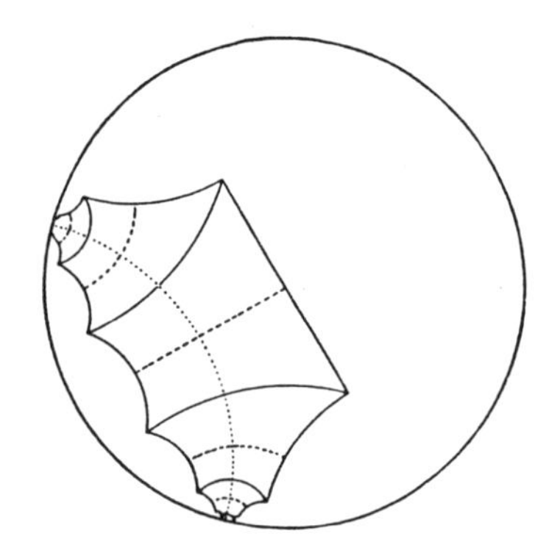

Figure 17: The perfect 3-coloring of {4,6} induces tessellations with infinite regions.

(c) Figure 17a shows the unique perfect coloring of {4,6} in three colors. If we join the centre of each black square (i.e., regular quadrangle) to the four nearest black centres, we obtain the situation shown in Figure 17b; the regions bounded by the edges are infinite regions, not polygons, but they are still convex. One such region is analyzed in Figure 17c; its vertices lie on the two branches of an equidistant curve or **hypercycle**, and the axis of the hypercycle is a line of symmetry of the region running down its entire length. There are other lines of symmetry also, of two types, indicated by broken and unbroken lines and all perpendicular to the main axis. The region can be regarded as an infinite version of a regular polygon; it is regular because its symmetry group is transitive on its flags (a flag consists of a vertex, an edge incident with the vertex, and the region itself). We shall call it an **infinite backbone**; the main axis is its **spinal cord**, and the broken or the unbroken lines divide it up into **vertebrae**.

Since four backbones meet at each vertex of the tessellation, we shall denote the tessellation by $\{2\omega, 4\}$; the reason for "2ω" rather than "ω" will be explained later. (The notation $\{\infty, 4\}$ is already in standard use to denote the tessellation of right-angled infinite polygons whose vertices lie on horocycles [**4**, p.97]). This is only a partial notation, since a backbone with a right angle at each vertex is not unique. The compound tessellation obtained from Figure 17a is

$$\{6, 4\}[3\{2\omega, 4\}].$$

One way in which a tessellation of backbones is less regular than a tessellation of polygons or a tessellation such as $\{\infty, 4\}$ is that it does not have a dual.

During the actual coloring of a tessellation, rules are necessary for progressing from one black tile to another. In figures such as 15a, 16a, and 20a, which give rise to compound tessellations of polygons, only one rule is needed, although it may have to be used in both left-handed and right-handed versions. In figures such as 17a, where the compound tessellations are formed from infinite shapes whose boundaries are not connected, two rules may be needed, one for proceeding from a vertex of the infinite shape to an adjacent vertex and one for "crossing over" to another portion of the edge. The rules of procedure should be obvious in all the figures.

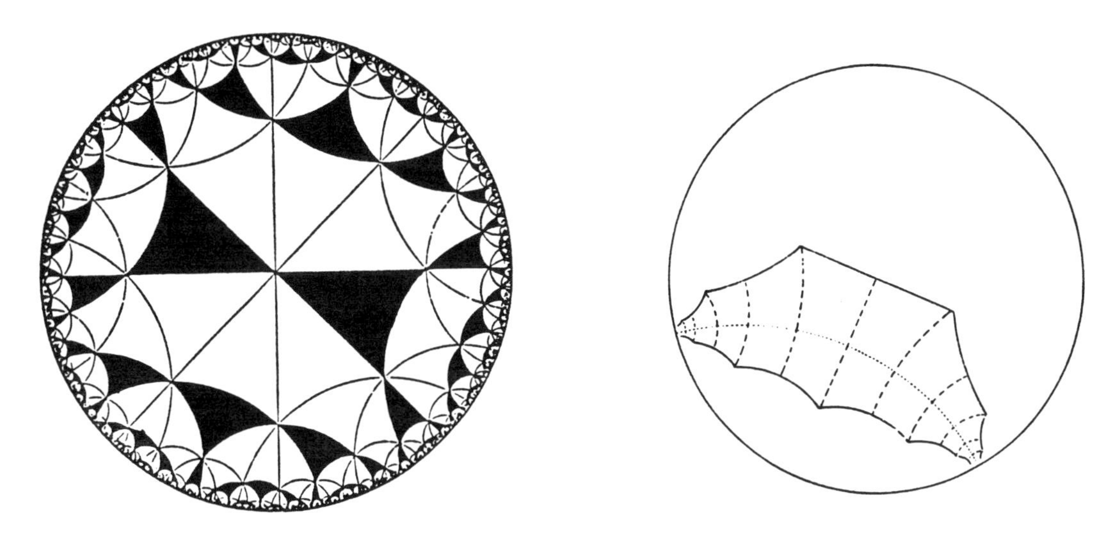

Figure 18: One of four colors in the perfect 4-coloring of $\{3,8\}$.

(d) Figure 18a indicates the unique perfect coloring of $\{3,8\}$ in four colors. As the number of colors increases it becomes less helpful to use various types of black-and-white shading; since the tiles of any color can be obtained from the black tiles by a suitable rotation of the tessellation, the complete coloring can be deduced from Figure 18a, and the reader may like to complete the coloring of this figure and the subsequent ones. The same procedure as before produces a tessellation formed from backbones like that shown in Figure 18b. This differs from Figure 17c in that the main axis is no longer a line of symmetry, but is an axis of glide reflexion; also all the lines of symmetry perpendicular to the main axis are of the same type, joining a vertex to the midpoint of the opposite side. In Figure 17c these lines of symmetry are of two types, either joining two vertices or joining two midpoints. This is precisely the distinction that occurs between regular polygons with an odd number of vertices and those with an even number of vertices; so we can think of the backbone in Figure 17c as having an even number of vertices, which is why we call it $\{2\omega, 4\}$, and the backbone in Figure 18b as having an odd number of vertices, so we shall call it $\{2\omega + 1, 3\}$. On the axis in Figure 18b, midway between adjacent lines of symmetry, are centres of (twofold rotational) symmetry. The compound tessellation obtained from Figure 18a is

$$\{8,3\}[4\{2\omega + 1,3\}].$$

(e) Figure 19a indicates one of the perfect colorings of $\{4,6\}$ in six colors. From the centres of the black tiles we obtain a tessellation of shapes like that shown in Figure 19b. This shape also is convex; instead of a single main axis it has a spinal cord (shown by dotted lines) that is branched like a tree, so we shall mix our metaphors and call it a convex **dendroid**. Each section of the spinal

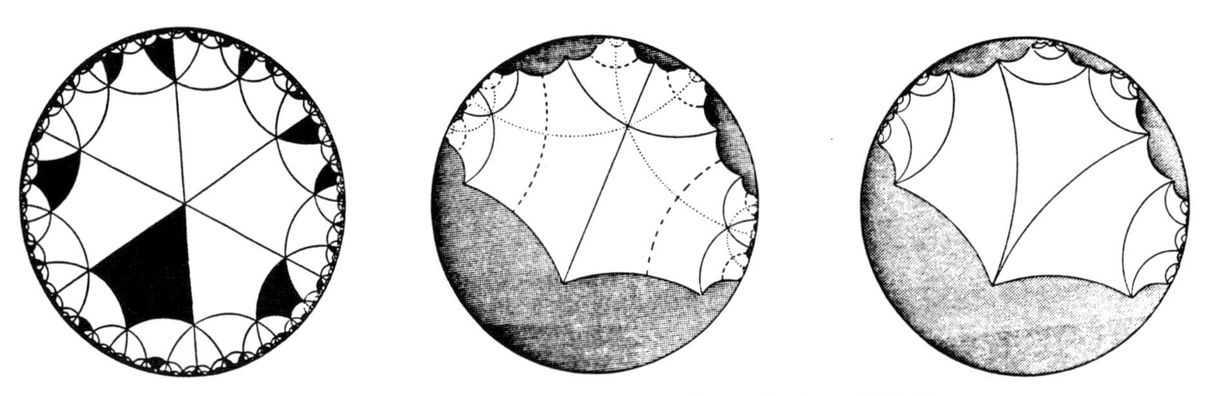

Figure 19: One of six colors in a perfect coloring of {4,6}.

cord is a line of symmetry. The broken lines of symmetry divide the dendroid into vertebrae, each of which is joined to four others. There are other lines of symmetry, indicated by unbroken lines in Figure 19b. Figure 19c shows another way of piecing together the dendroid: squares with side-length b and quadrangles with side-lengths b and c alternately, where $b > c$, joined along the sides of length b. Although it has lines of symmetry in many directions, the dendroid can still be regarded as an infinite version of a regular polygon; let us denote it simply but incompletely by $\{\omega, 4\}$. The compound tessellation obtained from Figure 19a is

$$\{6,4\}[6\{\omega,4\}].$$

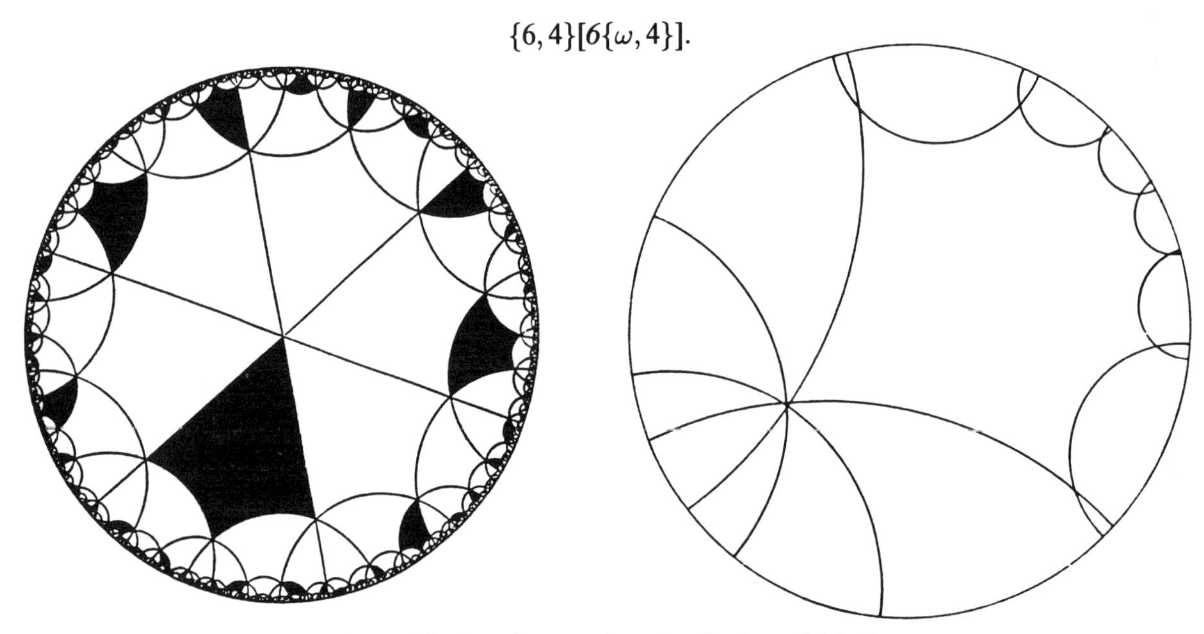

Figure 20: Another perfect 6-coloring of {4,6}.

(f) Figure 20a indicates another perfect coloring of {4,6} in six colors. Calculations show that there are eight black centres nearest to any given black centre, and we obtain the tessellation of octagons indicated in Figure 20b. These octagons are not regular; they are equilateral but their angles are alternately of two different sizes. Also the angles at each vertex of the tessellation are alternately of the two different sizes. The symmetry group of the tessellation is transitive on vertices, edges and faces, so we can call it a **transitive** tessellation. If we wish to compare this with a more familiar example, the only (non-regular) transitive tessellation in the Euclidean plane is the diamond tessellation of Figure 21a; a simpler example of a transitive tessellation in the hyperbolic plane is shown in Figure 21b, where the tiles are equilateral hexagons with angles of 45° and 135° alternately. We denote the tessellation in Figure 20b by trans{8,8} although such a tessellation of octagons meeting eight at a vertex is not unique. The compound tessellation obtained from Figure 20a is

$$\{6,4\}[6\ \mathrm{trans}\{8,8\}].$$

The tessellations of Figures 16b and 20b are similar in that in both cases angles of two different sizes occur alternately at each vertex. It is the way in which the vertices are joined up that produces a quasi-regular tessellation in the one case and a transitive tessellation in the other.

Figure 21: Transitive tessellations in the Euclidean and hyperbolic planes.

(g) If we relax the condition that says we must join each vertex to the nearest vertices of the same color, but still require that two joined vertices must be at a fixed distance apart, we can obtain variants of some of the above tessellations. For instance, suppose in example (d) we join each vertex of each backbone (Figure 18b) to the two nearest vertices on the opposite side of the backbone, and then delete the original sides of the backbones, each component of the compound becomes a transitive tessellation of diamonds, trans{4,6}.

(h) The centres of the tiles of {6,8} are the vertices of the dual tessellation {8,6}. The perfect coloring of {6,8} in two colors gives rise to a compound tessellation with two components. To obtain one of the components, choose alternate vertices of {8,6} and join each chosen vertex to its twelve nearest chosen neighbors; this gives the quasi-regular tessellation $(4.6)^6$, which is transitive on its vertices. If we color alternate vertices of each of these two components in two different colors, this is equivalent to a perfect coloring of {6,8} in four colors. A component of the corresponding compound is obtained by choosing alternate vertices of $(4.6)^6$ and joining each to its nearest chosen neighbors. A drawing of this in the Poincaré model is not helpful; each chosen vertex is joined by edges of length c to the opposite vertices of the six quadrangles to which it belongs, but is not joined to the two other chosen vertices of the hexagons to which it belongs because these are not nearest neighbors. Each (infinite) tile of the component tessellation can be constructed by joining dodecagons and equilateral triangles; the dodecagons have edges of lengths b and c alternately, where $b > c$, and the triangles have edges of length b; each edge of length b in a dodecagon is joined to an edge of a triangle, and conversely; the edges of the infinite tile have length c. The tiles are convex and regular, but are more complicated than the previous infinite tiles (cf. Figure 19c).

The reader may like to perform a similar operation starting with {8,12}, and investigate the four components of the corresponding compound. Each component is a quasi-regular tessellation of triangles and dendroids. It should not be difficult to find perfect colorings leading to quasi-regular tessellations of two different types of dendroid, and to transitive tessellations of non-regular dendroids, but I have not yet carried this particular investigation any further.

6 Further investigations

In a longer version of this article **[8]**, I discuss perfect colorings of {7,3} and {3,7}; these give rise to the regular maps $\{7,3\}_8$ and $\{3,7\}_8$, and the chiral coloring of {3,7} in seven colors (Plate 2a on p. 10) provides a simple way of deriving the symmetry group of the maps. I am investigating the use of perfect colorings to derive the symmetry groups of other regular maps.

To derive a regular map from a perfect coloring, we need the concept of **equivalent tiles** in the coloring: two tiles of the same color are **equivalent** if they are surrounded in the same way by tiles of the other colors, or if the way in which one tile is surrounded is the mirror image of the way in which the other is surrounded. We obtain the regular map associated with a perfect coloring by "identifying equivalent tiles".

The symmetries of a tessellation induce a group of permutations of the colors used in any perfect coloring; this group is connected with the symmetry group of the corresponding regular map and with the number of inequivalent tiles. During an investigation into such permutation groups I produced the chirally perfect coloring of {3,8} in ten colors indicated in Figure 22. Having completed this coloring in ten bright colors (Plate 2b), I looked for equivalent tiles of the coloring; but it turns out that the induced permutation group of the colors is A_{10}, the alternating group of order 10!/2.

The number of inequivalent tiles is therefore 604800. Since we are supposed to be considering aspects of the subject that can be appreciated visually, this seems a good point to come to an end, except to mention a simpler and more obvious fully perfect coloring of {3,8} in ten colors that I found later (Figure 23). This gives rise to a new regular compound not listed in **[1]**:

$$\{8,3\}[\mathit{10}\{8,6\}];$$

(see also **[7]**). Four other perfect colorings for the reader to analyze are shown in Plates 3 and 4 on pages 11 and 12.

Figure 22: A chirally perfect 10-coloring of $\{3,8\}$.

Figure 23: A fully perfect 10-coloring of $\{3,8\}$.

References

1. H.S.M. Coxeter, Regular compound tessellations of the hyperbolic plane, *Proc. Roy. Soc. London Ser. A* **278**(1964) 147–167.

2. H.S.M. Coxeter & S.L. Greitzer, Geometry Revisited, Mathematical Association of America, Washington DC, 1967.

3. B. Ernst, The Magic Mirror of M.C. Escher, Random House, New York, 1976.

4. L. Fejes Tóth, Regular Figures, Pergamon Press 1964.

5. K.O. Friedrichs, From Pythagoras to Einstein, Random House, New Mathematical Library **16**, New York, 1965.

6. J.F. Rigby, Napoleon Revisited, *J. Geom.* **33**(1988) 129–146.

7. J.F. Rigby, Some new regular compound tessellations, *Proc. Roy. Soc. London Ser. A* **422**(1989) 311–318.

8. J.F. Rigby, Compound tilings and perfect colorings, *Leonardo* **24**(1991).

9. J.F. Rigby, Regular hosohedral compound tessellations, (unpublished manuscript, 1988).

University of Wales College of Cardiff,
School of Mathematics,
Senghennydd Road,
Cardiff CF2 4AG,
Wales, United Kingdom.

Escher: A Mathematician in Spite of Himself*

D. Schattschneider

Introduction

This article first appeared in *Structural Topology*, **15**(1988) 9–22, along with a French translation "Escher: Mathematicien Malgre Lui." It is reprinted with permission. That issue of *Structural Topology*, as well as volume 16, is devoted to the work of M.C. Escher.

The book *Visions of Symmetry: Notebooks, Periodic Drawings, and Related Work of M. C. Escher*, W. H. Freeman & Co., 1990, by Doris Schattschneider, contains a full discussion of all of Escher's work on regular division of the plane, as well as color reproductions of his 1941–1942 notebooks and his 150 periodic drawings. The research for and preparation of the book were funded in part by the National Endowment for the Humanities.

Abstract

A picture is worth a thousand words . . . and so the graphic work of the Dutch artist, M. C. Escher (1898–1972), has been captured by mathematicians for its clever visual expression of abstract concepts. However, Escher's finished graphic works show only a fraction of his mathematical ability. Escher hotly denied understanding any mathematics, although he acknowledged more affinity with mathematicians than other artists.

No doubt his idea of mathematics was manipulating symbols and creating formulas, or producing or deciphering "hocus pocus" texts. However, Escher was a mathematician, a true researcher, exploring many mathematical questions arising from colored tilings of the plane. He created for himself a set of categories, invented a notation of classification, and ignoring the "accepted" system and restrictions imposed by mathematicians and crystallographers, explored more deeply than any of these 'professionals,' questions of colored periodic tilings of the plane. His notebooks of 1941–1942 reveal the extent of the explorations of this closet mathematician.

The graphic work of the Dutch artist M.C. Escher has always held a fascination for mathematicians. At a glance, it is evident that Escher **used** mathematics in the creation of many of his works. (Indeed, a more careful look at his preliminary sketches shows his painstaking attention to mathematically determined detail.) Far more striking is Escher's success in creating visual realizations of many difficult-to-explain abstract mathematical concepts: duality, recursion (self-reference), infinity, topological change, dimension.

---

*Reprinted with permission from *Structural Topology* **15**(1988) 9–22.

Escher's notes reveal that he studied mathematical papers on plane tilings by G. Pólya and F. Haag, and used M. Brückner's classic work *Vielecke und Vielflache* as a reference. He also corresponded with G. Pólya, H. S. M. Coxeter, R. Penrose, and other mathematicians and scientists. In a lecture in 1953 he admitted, "... I have often felt closer to people who work scientifically (though I certainly do not do so myself) than to my fellow artists." [1, p.71] Despite his frequent and vigorous denials of having any mathematical understanding or of 'working scientifically,' Escher did possess a deep mathematical curiosity, tenacity, and, ultimately, understanding of a large body of mathematical knowledge. One of the subjects of Escher's research is today called "color symmetry." Only recently, more than forty years after his inquiries, has the field begun to receive serious attention from mathematicians and other scientists.

Escher was a secret mathematician. He had no formal mathematical credentials, and, in fact, seemed as mystified as any layman by the "hocus pocus text" of mathematicians' explanations. But his self-professed mania for producing interlocking creature-shaped tiles to fill the plane led him to systematically explore the possibilities and constraints of such tiles and tilings, as well as the related question of coloring them. He alludes to this in a paragraph from his book *Regelmatige vlakverdeling* (*Regular Division of the Plane*) published in 1958:

> At first I had no idea at all of the possibility of systematically building up my figures. I did not know as a result of my study of the literature on the subject, as far as this was possible for someone untrained in mathematics, and especially as a result of my putting forward my own layman's theory, which forced me to think through the possibilities. It remains an extremely absorbing activity, a real mania to which I have become addicted, and from which I sometimes find it hard to tear myself away. [1, p.164]

The very fact that Escher considered himself not a mathematician allowed him to pursue his investigations in his own way, unhampered by advice and criticism from "experts." He suggests in another paragraph from *Regelmatige vlakverdeling* the professionals' narrowness of view:

> In mathematical quarters, the regular division of the plane has been considered theoretically, since it forms part of crystallography. Does this mean that it is an exclusively mathematical question? In my opinion, it does not. Crystallographers have put forward a definition of the idea, they have ascertained which and how many systems or ways there are of dividing a plane in a regular manner. In doing so, they have opened the gate leading to an extensive domain, but they have not entered this domain themselves. By their very nature they are more interested in the way in which the gate is opened than in the garden lying behind it. [1, p. 156]

What Escher did, he did for himself, for his own understanding and use, not for publication or explanation to scientists. His questions were not theirs, nor was his mode of inquiry. He was sparked from *within* to create tilings of the plane.

In the beginning (1926–27) he achieved only very modest success, and in frustration, sought out sources for study. In 1936 he visited the Alhambra in Granada, Spain, as well as the mosque, La Mezquita, at Cordoba, and made careful colored drawings of the geometric tilings he found there. Also, during this period, his brother B. G. Escher, a geologist at the University of Leiden, brought to his attention the paper by Pólya [8], which he carefully copied in its entirety. Pólya's paper was important to Escher not because of its words (Pólya discussed the plane symmetry **groups**, a concept which Escher never mentions), but because of the tilings it displayed, illustrating each of the seventeen plane symmetry groups. Escher's understanding was an intuitive, visual understanding—from the illustrations he clearly grasped the role that isometries play in the creation of periodic tilings. His reliance on illustration alone is further evidenced in his study of the 1924 paper by F. Haag [5]. Escher's workbook carefully records the author, title, journal, and pages, but then all that is copied from the paper are Haag's illustrations of tilings by quadrilaterals, pentagons, and hexagons.

After absorbing the visual information in these sources, Escher set out to mold his own theory of the creation, characterization, and classification of periodic tilings. His goal was to create interlocking tilings by "creature" tiles, colored so that each tile was clearly visible and not bordered by another tile of the same color (that is, the tiling must be "map colored") and using a minimum number of colors. He wanted to know how to create tiles, in what different ways tiles could be interlocked with adjacent congruent tiles, and how they could be colored according to his constraints. As his study progressed, the coloring question led him to explore how tilings requiring only two colors were related to tilings requiring three colors, and also how to choose colorings compatible with the symmetries of the tiling. Escher's investigations were guided by the goal of a tangible end result. His tilings would be rendered, either as a block print, a fragment of a graphic work or produced as ceramic tiles for a wall mural or cylindrical column. Thus the practical aspect of production influenced his theory of classification.

The fact that Escher undertook a systematic study of these many aspects of tiling, rather than be satisfied with a few interesting original examples, may seem out of character. Yet this was not so strange. He once wrote in a letter, "My affinity with the exact approach to natural phenomena is probably related to the milieu in which I grew up as a boy: my father and three of my brothers were all trained in the exact sciences or engineering, and I have always had an enormous respect for these things." [6, p. 30]

As he sought to understand how to create tiles, Escher invented his own symbolic notation to describe how portions of the edges of a tile related to each other and to edges of adjacent tiles.

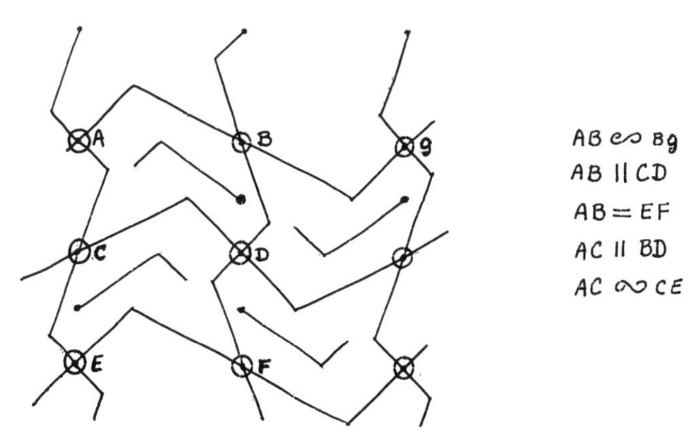

Figure 1: Escher's symbolic notation succinctly describes how edges of tiles are related. The first three lines (starting with the top line) are interpreted as follows: AB half-turns into BG; AB glide-reflects into CD; AB translates into EF.

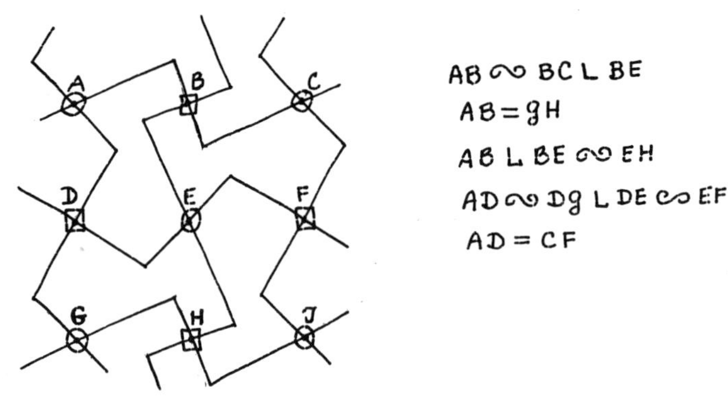

Figure 2: More of Escher's notation. The top line reads: AB half-turns into BC which quarter turns into BE.

Although some of his symbols are those used by mathematicians, Escher gave them new meaning. The sketches in Figures 1 and 2 are copied from an early workbook, and illustrate his notation. The notation, although unconventional, is excellent—it is succinct and symbolically suggestive of the isometries which relate edge pieces of tiles.

In this early workbook, we see Escher methodically considering, sorting, labeling and relabeling the various possibilities for periodic tilings by a single asymmetric tile, using translations, glide-reflections, 2- and 4-fold rotations. He used other symbolic notation which visually suggests how tiles are related to adjacent tiles: 2-fold rotation centers were marked with small circles and 4-fold centers with small squares. In addition, each tile in a tiling was given an asymmetric mark (usually a "hook"), and the positioning of these marks clearly identifies which isometry carries a given tile into an adjacent tile. (See Figures 1 and 2.) The use of all of the symbolic notation renders additional explanation unnecessary—and indeed, Escher gives very few words of explanation in his notes. The interplay of Escher's "pure research" with his artistic goal of highly recognizable interlocked creatures is shown by his frequent emphatic self-reminder that appears next to the schematic and symbolic representations of tilings: "voorbeeld maken!" (make an example!).

In the years 1936–41, Escher produced forty-three colored periodic drawings which cover a wide variety of symmetry types. His research gave him mastery of the geometric constraints and so liberated his witty artistic genius. Over the course of his lifetime, he produced over one hundred fifty such periodic drawings; more than forty of these are reproduced in [7]. In 1941–42 he set down a final version of his investigations in a notebook, and used his classification scheme to categorize each of his creature tilings. His notebook begins with a schematic chart (Figure 3) which gives a complete visual description of each of twenty-four types of tiling, including lattice type (A: parallelogram, B: rhombus, C: rectangle, D: square, E: isosceles right triangle), location of rotation centers, relation of tiles to adjacent tiles by isometries, and coloring of tiles. Combining the visual information, we can characterize for each type which isometries preserve and which reverse color.

Taking Escher's advice, we will illustrate the content of his symbolism with an example. His chart shows for type IXD a square lattice, with two 4-fold centers of rotation (diametrically opposite) and two 2-fold centers of rotation (diametrically opposite) at the corners of each square. The 4-fold rotations reverse the colors of the tiles and the 2-fold rotations preserve them (Figure 4a). Escher's drawing no. 13 of dragonflies (Figure 4b) bears his classification IXD∗. The under-

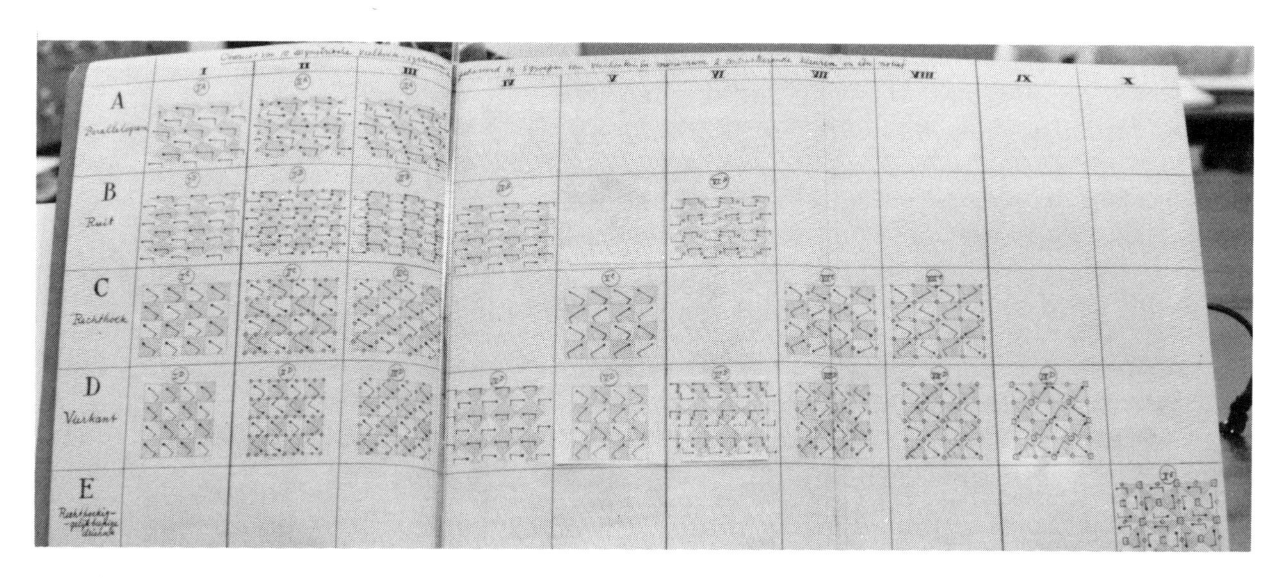

Figure 3: Escher's overview chart which classifies 24 different types of 2-colored tilings.

lying square lattice can be seen in the drawing; indeed, the edges of each dragonfly "tile" meet at the four corners of a square. We can observe that the 2- and 4-fold rotations which carry a given dragonfly into a neighboring one are located at these corners, and the 4-fold rotations carry each dark (blue) dragonfly into a white one and vice-versa, while the 2-fold rotations carry each dragonfly into one of the same color. The $*$ in the classification symbol of this drawing indicates that the motif (dragonfly) has reflection symmetry. In his classification system, Escher did not allow reflections to carry tiles into adjacent tiles; however he made note of the fact when a motif was bilaterally symmetric (using the $*$), and often (but not always) this induced reflection symmetry into the whole tiling.

Unlike crystallographers, Escher noted the exact shape of a lattice tile, whether or not all possible symmetries of that shape occur in the tiling. For example, many of his tilings have his D (square) lattice classification even though the tiling may have no 2- or 4-fold rotation symmetry. Here the practical (execution) aspect of his considerations is apparent. Square tiles are most easily produced by standard tile works. In addition, his "lattice polygon" in each of the five categories A, B, C, D, E, corresponds to a single tile (what mathematicians call a fundamental domain), rather than to a translation unit (what crystallographers call a unit cell).

In Escher's view, tilings which required three colors (in order to be map-colored) were closely linked to those which were 2-colorable, and were derived by a dynamic process of **transition**. This process is best understood by viewing Escher's own schematic explanation. Figure 5 shows how, beginning with a 2-colored tiling of Escher's type IA, a segment of the boundary (with endpoints A, B) is pivoted about the point A to create a new boundary segment (with endpoints A, C), and then all tiles of the original (top) tiling are so altered (using the translation isometries which carry the tiles into adjacent tiles) to create a new tiling (middle)which requires three colors. Escher labels this new tiling IA-IA. The process can be continued by pivoting (about A) the boundary segment with endpoints A, C to the new position A, D. Altering all tiles of the (middle) tiling in this manner results in a new tiling of type IA (bottom) which can be 2-colored. Escher carefully applies this

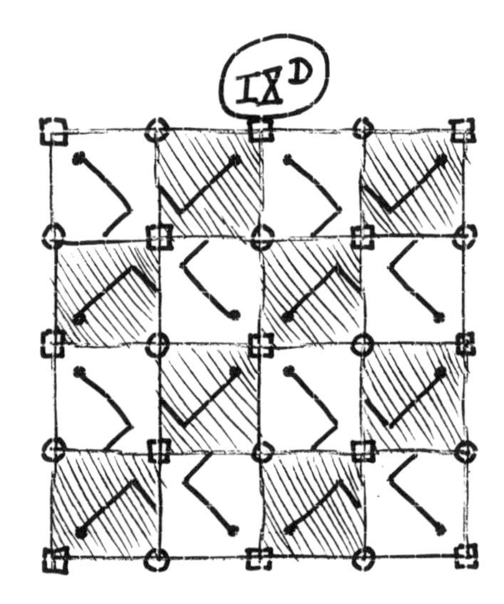

Figure 4a: Escher's type IXD from the chart in Figure 3.

Figure 4b: Escher's drawing no. 13 of dragonflies (winter 1937–38) is classified as type IXD∗.

process to all of his two-colorable tilings (in Figure 3), exploring the many variations possible, and the possible linkages of 2-color types using the twice-applied process described above.

His notebook shows a summary (a directed graph!) of these linkages of "Escher types" using the transition process, with five clearly separated categories (Figure 6). These categories are, in fact, the five different plane symmetry groups **p1**, **p2**, **pg**, **pgg**, and **p4**.

Thus, although he did not use the crystallographer's notation of symmetry groups, he made a fundamental discovery: the transition process changed the shape of the tiles (often even changing the isohedral type of a tiling), and changed the number of colors required, but did not change the symmetry group of the tiling.

Map-coloring a tiling is fairly easy to achieve, and can be done in a variety of ways. However, from the beginning, Escher seems to have linked the process of coloring a tiling with the process of symmetry operations. Inevitably this led to "perfect" colorings, that is, colorings in which all symmetries of the uncolored tiling are compatible with the coloring. What this means is that if a

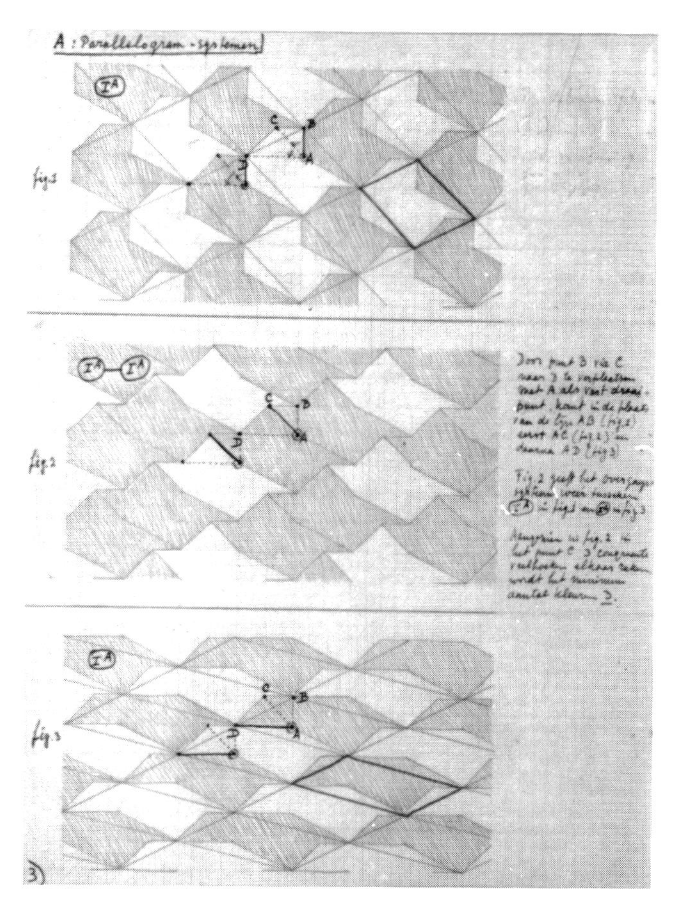

Figure 5: Escher's notebook page that describes the process of transition from a type IA tiling to IA-IA to a new IA tiling.

tiling of congruent tiles has some tiles colored red and others colored white, then every symmetry that sends one red tile to a white tile must send every red tile to a white tile. In other words, each symmetry of the uncolored tiling must induce a permutation on the set of colors of the tiling. The idea of having the symmetries of a tiling "drive" the colors was not considered by mathematicians and crystallographers until the 1950's.

Escher was fascinated with the concept of duality (*Day and Night, Sky and Water, Heaven and Hell,* ...), and his use of two contrasting colors or two interlocked motifs served to emphasize such duality. His notebook includes a chapter specifically investigating the creation of tilings by two different motifs. Beginning with a tiling by a single tile shape (and requiring either two or three colors), he split the tile into two distinct (differently shaped) tiles and then used the symmetries of the tiling to split all of the tiles in the same way. He was mainly interested in creating tilings by two motifs which could be colored in just two colors, giving further emphasis to the implied duality. Thus some of his tilings of this type consist of two motifs, with all copies of one motif colored the same color and all copies of the other motif a single contrasting color (so that either motif can play the role of figure while the other is the ground). But other tilings created by the splitting process

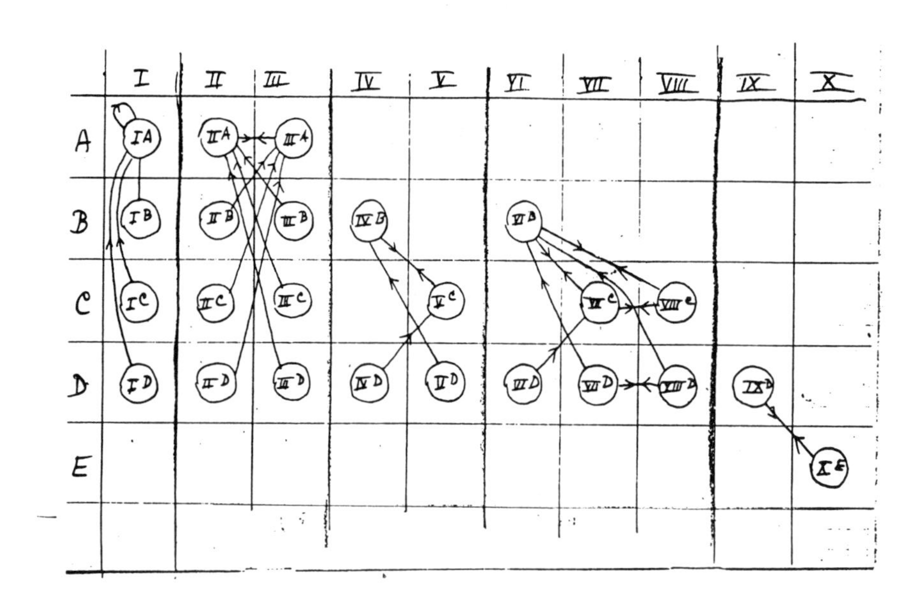

Figure 6: Escher's digraph showing which of his types are related by the transition process.

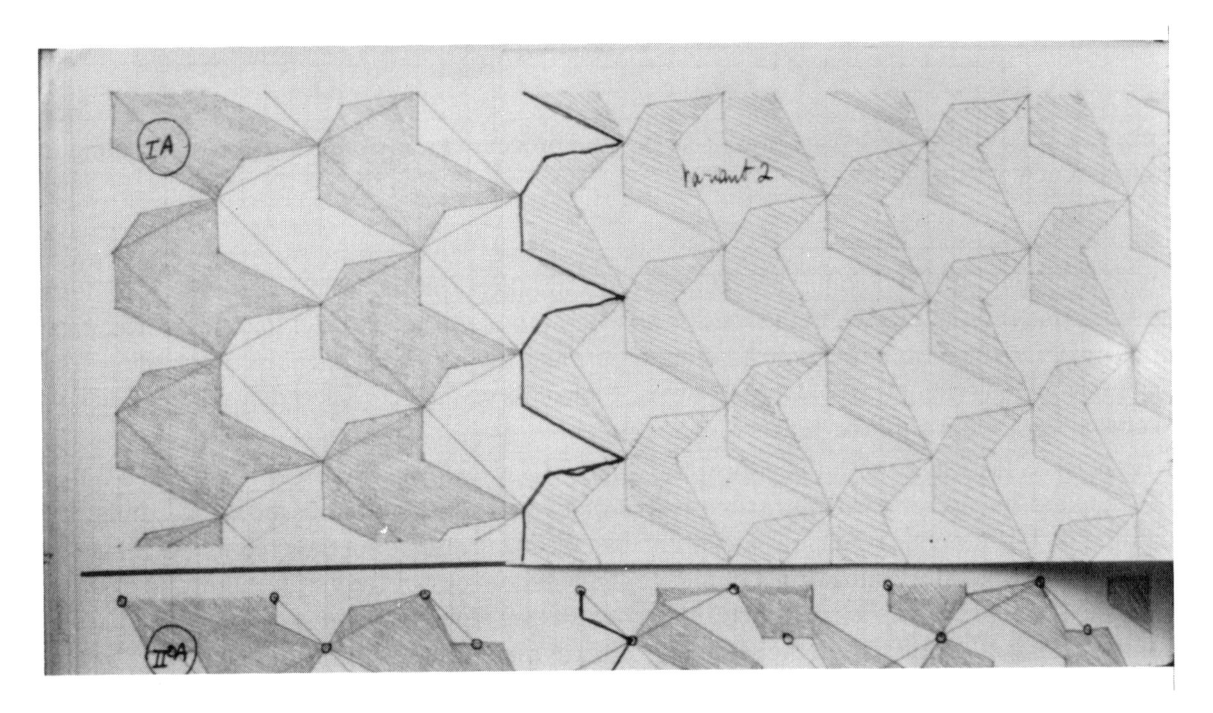

Figure 7: The splitting of a type IA tiling into a tiling by two motifs (tiles), each motif always colored the same color (Escher's variant 2).

show two motifs, in which some copies of each motif are one color, and other copies of each motif are a contrasting color (however, the coloring of these tilings is always "perfect," compatible with the symmetries of the tiling). For a few of his types, he notes distinct ways in which the tiles can be split. Figure 7 shows one of his two schematic illustrations of this process applied to his type IA; variant 2 shown is the second of two types of splittings. Figure 8 shows a finished rendering of a tiling of this type, the embossed cover of his 1958 book *Regelmatige vlakverdeling.* Only in the last few years have such tilings been the subject of mathematical investigation (they are termed 2-isohedral tilings by Grünbaum & Shephard [4]).

Figure 8: Escher's design for the cover of *Regelmatige vlakverdeling.*

Escher devotes the last section of his notebook to a very compact description of the perfectly colored tilings based on what he terms a "triangle system." These are the tilings which have 3-fold or 6-fold rotation centers. He notes that three colors are almost always necessary for such tilings, and invents a notation which records for a given tiling the following information: triangle system, number of distinct motifs, 3-fold (only) or 6-fold rotations, number of colors. Some of his most successful colored tilings (of lizards and of butterflies, for example) show his mastery of these types.

Many who have admired Escher's work may have suspected he possessed far more mathematical talent and understanding than he was willing to profess. The evidence of Escher's private research is overwhelming. He was a (secret) mathematician.

This article is based on a talk given in July 1986 at the Eugene Strens Memorial Conference on Intuitive and Recreational Mathematics held at the University of Calgary. Partial funding for the research was received from the Faculty Development and Research Committee of Moravian College.

I wish to express my thanks to the Escher Foundation and the Haags Gemeentemuseum for the privilege of examining the Escher notebooks and periodic drawings in 1976.

References

1. F. H. Bool, J. R. Kist, J. L. Locher & F. Wierda, *M. C. Escher, His Life and Complete Graphic Work,* Harry N. Abrams, New York, 1982.

2. M. C. Escher, The authentic collection of all Escher drawings from the Gemeentemuseum The Hague on microfiche, Inter Documentation Company AG, Zug, 1980.

3. M. C. Escher, *Regelmatige vlakverdeling,* De Roos, 1958. Translated and reprinted in [1] above, pp. 155–173.

4. B. Grünbaum & G.C. Shephard, *Tilings and Patterns,* W. H. Freeman & Co., New York, 1987.

5. F. Haag, Die regelmassigen Planteilungen und Punktsysteme, *Z.Kristall.,* 58 (1923) 478–488.

6. J. L. Locher, ed., *The World of M.C. Escher,* Harry N. Abrams, New York, 1972.

7. C. H. MacGillavry, *Symmetry Aspects of M. C. Escher's Periodic Drawings,* Oosthoek, Utrecht, 1965; reprinted as *Fantasy and Symmetry, The Periodic Drawings of M. C. Escher,* Harry Abrams, New York, 1976.

8. G. Pólya, Uber die Analogie der Kristallsymmetrie in der Ebene, *Z. Kristall.,* 60 (1924) 278–282.

9. D. Schattschneider, In Black and White: How to Create Perfectly Colored Symmetric Patterns, *Comp. & Math. with Appl.,* 12B 3/4 (1986) 673–695.

10. D. Schattschneider, M.C. Escher's Classification System for his Colored Periodic Drawings, in *M.C. Escher: Art and Science,* North Holland, 1986, pp. 82–96.

11. R. L. E. Schwarzenberger, Colour Symmetry, *Bull. London Math. Soc.,* 16 (1984) 209–240.

12. M. Senechal, Point Groups and Color Symmetry, *Z. Kristall.,* 142 (1975) 1–23.

13. M. Senechal, Color Groups *Discrete Appl. Math.,* 1 (1979) 51–73.

14. T. W. Wieting, *The Mathematical Theory of Chromatic Plane Ornaments,* Marcel Dekker, New York, 1982.

Department of Mathematics,
Moravian College,
Bethlehem, PA 18018-6650

Escheresch

Athelstan Spilhaus

I call this "Escheresch" because I am going to show you a couple of intellectually fun things at the end, which are reminiscent of and, in part, connected to the work of Maurits C. Escher.

M.C. Escher was not the fairy-tale artist born in poverty, slaving in a garret and finally coming to riches and fame after he is too dead to enjoy it—posthumourously, as they say.

On the contrary, Escher was born to a well-to-do family on June 17, 1898. His father was a successful civil engineer in Holland. Even as a child, Escher loved to draw and his art teacher encouraged him to make linoleum cuts. His father, seeking to find him a "respectable profession", attuned to his love of art, sent him to study architecture at the Holland School of Architecture and Decorative Arts. But, encouraged by an artist there, Escher dropped architecture at the age of 21 and started in graphic arts, a field which he was to pursue until he died at the age of 73.

This well-heeled youngster, after finishing his architectural studies, set off for Florence, Italy. There, with ample financial resources, he met Yetta, the daughter of a silk manufacturer who had escaped with his fortunes from Russia to Switzerland during the Russian Revolution of 1917. Yetta and Escher were married in Italy on June 16, 1924, and after a short honeymoon trip to both their parents in Holland and Switzerland, settled in Rome. He made many summer excursions throughout Italy, recording impressions and photos which he transformed into his representational prints of that period in his life during the winters in Rome.

Mussolini's rise in the 30's made life uncomfortable for the Eschers, particularly when their son was forced to join the Fascist Youth Movement. Therefore, in July 1935, the family moved to Yetta's parent's home at Chateau D'Oex. One of his last representational prints of a landscape in the snow was made there.

The last of the travels that Escher made was in 1936, by sea along the Mediterranean coast to Spain. That journey marked his abrupt change from representational graphics to an obsession with pattern, from realism to the impossible, from the Moorish, interlocking geometric designs, the tessellated mosaics of the walls and floors of the Alhambra and the mosques to the substitution in the tessellated patterns of animals, plants and people in designs that could, in Escher's view, go to infinity.

He moved to Belgium and then back to the Netherlands and settled there in 1941. Not only did his travelling stop but he seldom left home. Instead of looking out at the world as he had done in his travels and representing it, he looked inward and drew on his own curious imagination.

This significant change was marked by his woodcut of 1938, "Sky and Water I". It is often said of this that the fish in the water become birds in the air. It is not true. The fish in the water become the air around the birds and the water around the fish becomes the birds.

Escher's work first attracted not artists but scientists and mathematicians who found that his intricate, eye-fooling geometry provided graphic understandable explanations of abstruse mathematical concepts. Escher himself said that he did not understand abstract mathematical thinking. Thus Caroline MacGillivray, Professor of Crystallography in Amsterdam, used Escher's prints to illustrate color symmetry in crystals. Escher did not know that there was a science called crystallography.

Similarly, Lionel S. Penrose, then Professor of Genetics, and his son Roger Penrose, now Professor at Oxford, came across Escher's work on visualizing the impossible in the mid-1950's. When they returned to England, Penrose came up with an "impossible" triangle which can exist in three parts but not as a whole. They wrote an article **[1]** about it and sent it to Escher who used it in three of his most famous sleight-of-mind prints, "The Waterfall", "Ascending and Descending" (1960) and "Up and Down". Penrose said he drew the triangle for fun and to confuse but it illustrates the concept of non-integrability. Each small part represents a three-dimensional object but, when assembled, integrated, they are inconsistent—impossible. It was Penrose who introduced Escher to the concept of the non-Euclidean conformal representation of a symmetrical pattern on a hyperbolic plane which permits tessellations, large in the center, as in "Heaven and Hell" (1960) and "Circle Limit III" (1959), but becoming smaller and smaller *ad infinitum* on the edge, or large on the edge and becoming smaller and smaller toward the center. A 1956 print is aptly named "Smaller and Smaller".

Because of his curious designs, Escher began to be recognized in the United States and articles appeared in "Time" and "Life" in the 1950's and then in Holland. But it was not until 1958, at 60 years of age, that he began to live on his own earnings. His works were much more frequently shown in science museums than in art museums. Escher was considered "too intellectual" to be an artist and he, in turn, expressed more trust in scientists than in artists.

By 1969, Escher's prints were being espoused in psychedelic circles, with the titles changed to "Dream" and "Bad Trip", and published in fluorescent inks, reflecting the worst taste of the emerging "drug culture".

Cornelius Roosevelt, grandson of President Theodore Roosevelt, collected Escher's prints beginning in the early 1950's and started a friendship with Escher that lasted almost 20 years. It was Roosevelt who sent Escher the posters and also some T-shirts made on the west coast from his prints. The interpretations which the hippies put on these is exemplified by "Reptiles" where, they thought, the glass symbolized alcohol; the lizard blowing smoke, drugs; the cactus, peyote, the source of mescaline, and the word "Job", a brand of cigarette paper, symbolized religion. This was a complete surprise and probably a shock to Escher.

Escher had the good taste to turn down a request from the "Rolling Stones", a hard rock group, to use his designs.

When he died, prints that sold in the 50's for $12 to $14 each were priced at anything from $2,000 to $20,000. The Escher family continues to hold copyrights on his designs.

When Escher's view of the world turned inward he produced his best known puzzling prints, which, art aside, were truly intellectually playful, yet he was not. His life turned inward, he cut himself off and he had few friends. He seldom discussed his wife, even with his closest friends, and several years before he died, she entered a home for the aged in Switzerland. He died after a protracted illness in 1971.

I now plan to describe some things I have done which, though they do not necessarily derive from Escher's work, are reminiscent of it and are done in a true sense of playfulness, giving joy to the creator, even if to no one else.

Escher's preoccupation with repetitive patterns was triggered by the tessellations he saw in the Alhambra. The Arabs had used tessellations for a long time in their mosaics. Thus they not only preserved and passed on the marvelous concept of zero, but they also provided a visual conception of continuing **ad infinitum** — of infinity.

Escher's contribution was to make the tessellations into his curious figures of animals, men and horses—a thing not attempted by the Mohammedans whose religion forbade the representation of animal forms ("Sky and Water").

Zero and infinity, marvelous mathematical conceptions, do not indeed exist in nature and Escher seemed to love to trick the brain by drawing apparently realistic pictures which were, however, on examination, impossible.

I produced my jigsaw maps which are essentially tessellations of the world and which are reminiscent of Escher's work (compare Escher's "Metamorphoses", I to IV and my own "Geogenesis").

In his impossible architectural drawing, Escher used well-known impossible figures but incorporated them, adding life around the figures, which further tricked the eye into thinking that they were normal until there was closer inspection ("Belvedere", "Waterfall").

Escher's preoccupation with zero and infinity led him to utilize a conformal representation on a hyperbolic plane to produce his pictures "Heaven and Hell" and "Circle Limit".

Now cartography used conformal mapping and I have produced my picture "Infinite Earth" which shows the map of the earth repeating and repeating smaller and smaller to the perimeter of the circle, Escheresch in that it is like "Heaven and Hell" and uses Coxeter's mathematical diagram. My "Infinite Earth" uses a representation of the earth in an equilateral triangle as the basic tessellation.

One of Escher's best known curious drawings is "Hands", a picture of one hand with a pencil drawing the other hand with a pencil that is in turn drawing the first hand. This, too, was derivative, although Escher may not have known it, from toys and automata in which the hand of a person was programmed (early analogue computer) to draw and even write. A simple Swiss toy came out about 1915 of a hand mechanically drawing a hand on a sheet of paper but, long before this, at the end of the 18th century, Swiss and French inventors were producing princely toys, full-sized automata, which drew pictures and wrote, and therefore could have produced the hand drawing the hand.

I have taken this one step further and built a simple machine which I will now demonstrate. Here you see that, starting with two pencils only, the hands are drawn so that ultimately each one appears to grasp the pencil. Then, erasers may be substituted for the pencils and the Escheresch machine erases the hands that hold the erasers. Escher is decomposing!

This machine caused me to ponder whether Escher had some special cross-connections between his left brain lobe and right brain lobe and right hand and left hand. It is, for instance, well-known that, while he drew with his right hand, he always wrote with his left hand. It appears that, in his picture of the hands, for both hands, drawing with his left hand, he used his right hand for the model.

There is considerable interest these days among neuroscientists in problems of dexterity and coordination which appear in certain people as dyslexia or other unique evidence of abnormal neural communication. I therefore designed a machine in which one human hand moves the paper along one rectangular coordinate while the pen is moved by the other hand along the orthogonal coordinate. This machine I call a dexsinograph. It can be used with the right hand moving the paper and the left moving the pen or *vice versa*. At present it is being investigated at a hospital for disturbed youngsters in Virginia, as a possible therapy device for children with learning disabilities.

References

1. L.S. Penrose & R. Penrose, Impossible objects, a special type of visual illusion, *Brit. J. Psych.*, **49**(1958) 31-33.

2. A. Spilhaus, New look in maps brings out patterns of plate tectonics, *Smithsonian*, Aug. 1976.

3. A. Spilhaus, World Ocean Maps: The proper places to interrupt, *Proc. Amer. Philos. Soc.*, **127**(1983) 50–60.

4. A. Spilhaus, Plate tectonics in geoforms and jigsaws, *Proc. Amer. Philos. Soc.*, **128**(1984) 257–269.

Box 1063
Middleburg, VA 22117

The Road Coloring Problem

Daniel Ullman

The tourist bureau of a certain town often gets phone calls from lost tourists who know neither where they are nor how to get where they want to go. Suppose, on behalf of the bureau, you are asked to create a scheme for reorienting lost tourists. The idea is this: you are to choose two roads emanating from each intersection in town, marking one with a red arrow and one with a blue arrow. Your goal is to be able to direct a lost tourist to, say, the town hall, by giving him a sequence of instructions each of which is either "follow the red arrow" or "follow the blue arrow", without having to ask for the tourist's whereabouts! It is not enough that the directions should lead the tourist *past* the town hall while following the instructions; the tourist must arrive at the town hall upon the completion of the last instruction. Can you complete your task?

Well, it depends. The answer is no, where there are certain obvious obstructions. For example, if it is impossible for the tourist to get to the town hall (owing to an unorthodox layout of roads), then no instructions can help him. Another obstacle which makes the assignment impossible is periodicity of the network of roads, a notion which will be defined and illustrated below. In the absence of these two obstructions, one can ask if your goal is achievable. In certain specific cases the answer is known to be yes, but what about in general? The answer is: nobody knows.

The problem is most easily described in the language of graph theory. Consider the directed graph whose vertices are the intersections in town, with a directed edge between two vertices v and w if there is a (possibly one-way) road leading from v to w. To make the problem as hard as possible, assume that there are only two roads emanating from each intersection; this is to say that the out-degree of each vertex is exactly 2. We call such a directed graph **strictly bifurcating**. In order

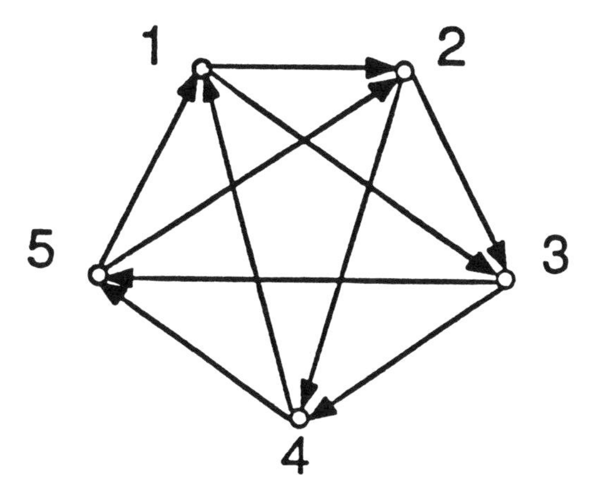

Figure 1: A strictly bifurcating directed graph.

to simplify language considerably, we use the generic term **graph** to mean a strictly bifurcating directed graph, and we denote such a graph by the letter G. Here is an example, with the five vertices numbered 1 through 5.

A **coloring** is an assignment of colors to the edges of G such that the 2 edges incident from any vertex are always colored one red and one blue. A graph together with a coloring makes a **colored graph**. Here is an example.

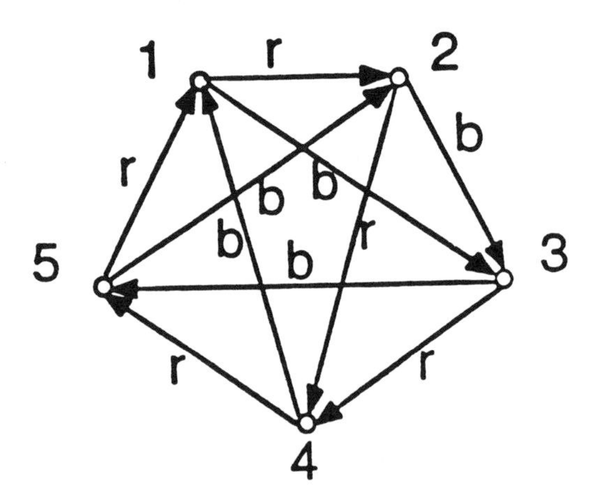

Figure 2: A colored graph.

Imagine that our lost tourist is located at some vertex of our colored graph, only we do not know which one. When we give him a sequence of instructions, which we might as well abbreviate as simply a finite sequence of the letters r and b (a "word"), he can follow our instructions by traversing the resulting path in G. Thus the words on the "alphabet" $\{r, b\}$ are in one-to-one correspondence with the paths emanating from any vertex. For example, if the tourist is at vertex 1 in Figure 2, the word *rbr* is associated with the path from vertex 1 to 2 to 3 to 4. A word w is called **resolving** if the path associated with w ends at a fixed vertex v independent of the starting vertex. The following chart shows that $w =$ *rbbbbr* is a resolving word for the colored graph above.

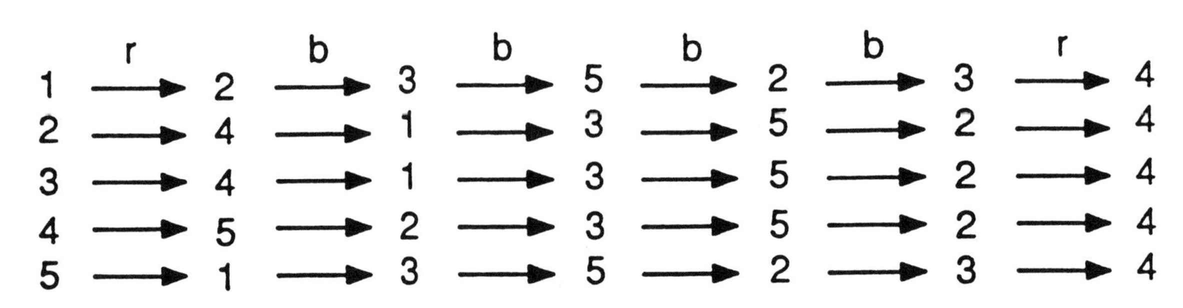

Figure 3: *rbbbbr* is a resolving word.

Do all colored graphs have resolving words? Certainly not. For example, a little experimentation shows why no resolving word exists in the colored graph of Figure 4. On this graph, after following *any* sequence of instructions, our tourist is just as lost as when he began.

One might now ask: do all graphs have colorings for which resolving words exist? Certainly

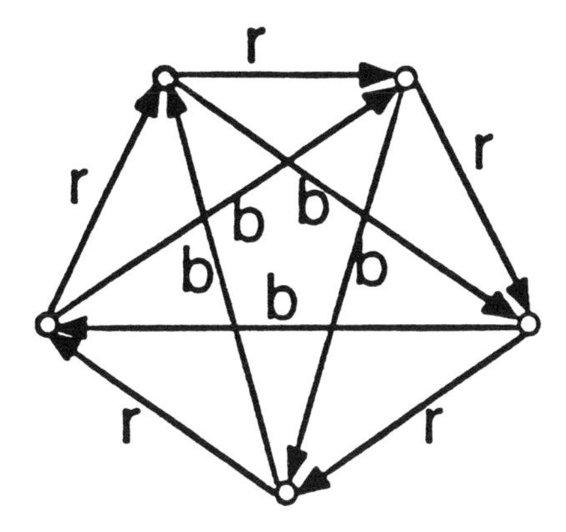

Figure 4: A colored graph with no resolving word.

some sort of connectedness is required. Let us impose an assumption of **irreducibility** (also called "strong connectedness"), which is to say that it is possible for a person to get from any intersection to any other intersection along roads. This is not strictly required according to the formulation of your task given in the first paragraph, but it is a necessary assumption if we want to be able to direct tourists to *any* intersection in town.

Even with this condition imposed, there are graphs which cannot be colored to allow a resolving word. Consider:

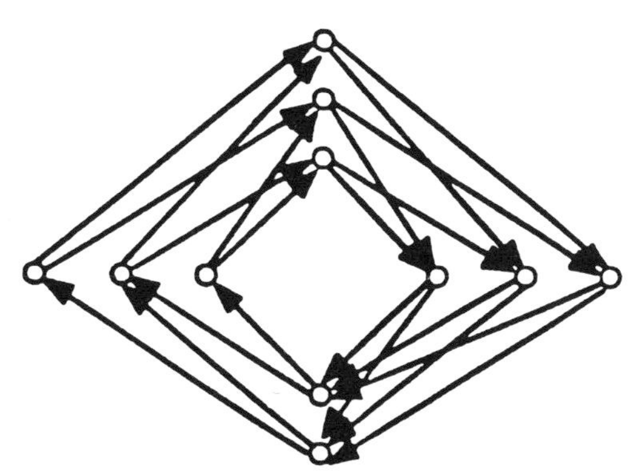

Figure 5: A graph with period 4.

The problem here is periodicity. A graph G is **periodic with period** $p \geq 2$ if its vertex set $V(G)$ can be partitioned into blocks $V_1, V_2, V_3, \ldots, V_p$, such that, if (v, w) is an edge of G and $v \in V_k$, then $w \in V_{k+1 \bmod p}$. The graph in Figure 5 is of period 4, with V_1, V_2, V_3 and V_4 the sets of vertices respectively to the north, east, south and west. Periodic graphs cannot be colored to allow a resolving word, for an obvious reason that the reader can supply for herself. A graph that is not periodic is naturally called **aperiodic**.

We are now ready to rephrase the original question.

THE ROAD COLORING PROBLEM. *Can all irreducible, aperiodic graphs be colored to allow a resolving word?*

There is another way of phrasing this question. The one-letter word r is naturally associated with a mapping from $V(G)$ to $V(G)$, the map which associates v to w if there is a road from v to w and if that road is colored red. In the same way, we associate a map to the word b. Abuse notation and let r and b stand for these functions themselves, which are elements of the semigroup of all functions from $V(G)$ to itself.

THE ROAD COLORING PROBLEM. *Does every irreducible, aperiodic graph have a coloring such that the semigroup generated by r and b contains a constant function?*

The problem is a natural one from the point of view of automata theory. Think of an automaton as a black box with finitely many internal states and certain allowed transitions between these states to be controlled by a red and a blue button. Suppose that, when the machine is turned on, it jumps electronically into one of its states essentially at random. You can not make use of the automaton unless you "initialize the system" by getting the machine into some known internal state. Can the system be initialized by pushing the red and blue buttons in some prescribed way? Or more precisely, can any pair of buttons be wired to the inside of the black box in such a way that the system can be initialized?

The real origin of this problem comes neither from tourist bureaus nor from automata theory but from the theory of dynamical systems. Let X be the space of all doubly infinite paths through a graph G, and let $T : X \to X$ be the "left shift map". This makes a special type of topological dynamical system called a **topological Markov shift** or **TMS**. A coloring of G associates to every element of X a doubly infinite path through the following graph, where this time r and b label the two vertices and not edges.

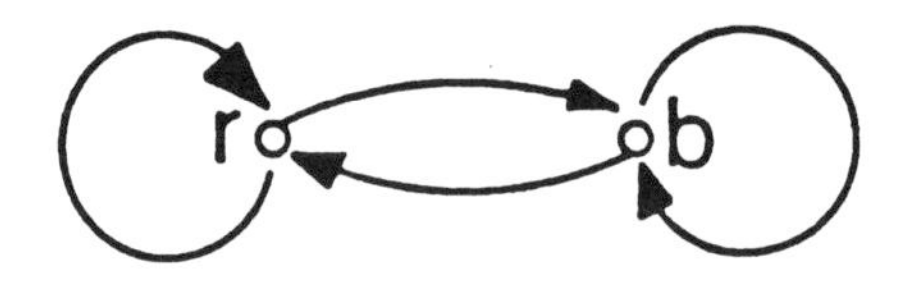

Figure 6: The TMS associated with this graph is called the full 2-shift.

If this association were one-to-one, we would say that T and the full 2-shift were **topologically conjugate**. If a resolving word w exists for this coloring, then it can be shown that this association, though not necessarily one-to-one, is almost one-to-one in the sense that, with probability one, a random path through the above graph is associated with only one point in X. In this situation we say that T and the full 2-shift are **almost topologically conjugate**. So the road coloring problem actually arose as a potential tool to prove:

Theorem. *The TMS associated with an irreducible, aperiodic graph G is almost topologically conjugate to the full 2-shift.*

The road coloring problem remains unsolved, but in 1977, R. Adler, L. Goodwyn & B. Weiss [1] circumvented it to obtain a proof of this important structural theorem in the theory of dynamical systems. This is an illustration of the closeness one often sees between a serious question of technical mathematics and a puzzle of recreational mathematics.

What is known about the road coloring problem? It is conjectured that the answer to the question is yes, such a coloring always exists. The first thing to recognize is that if G has an edge which is a self-loop, that is if there is a road circling from one intersection to itself, then a coloring and a resolving word can be found. The easiest way to explain why is to illustrate this by drawing an example:

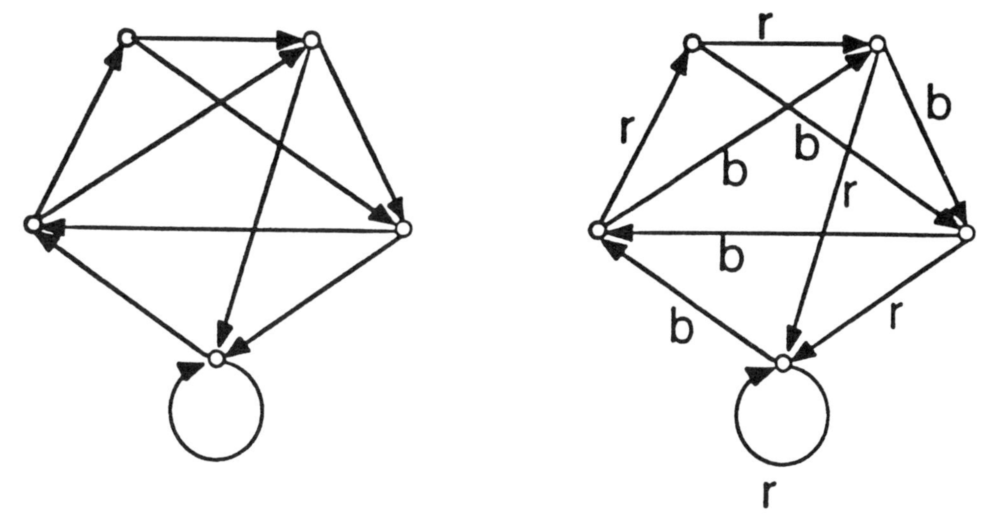

Figure 7: A graph with a loop has a coloring and a resolving word.

Here rrr is a resolving word. In general, when a self-loop is present, one can color the self-loop and other edges pointing toward the self-loop red in such a way that a word made entirely of r's is resolving.

In a beautiful and intricate paper by G.L. O'Brien **[3]**, it is proved that the problem is solvable provided that G has a cycle of prime length which does not contain all the vertices of G. The idea of O'Brien's coloring is to mimic the simple idea one uses when G has a self-loop. Color the edges in the cycle red and permit no other red cycles as you color the graph. The red edges form a subgraph of G that looks like:

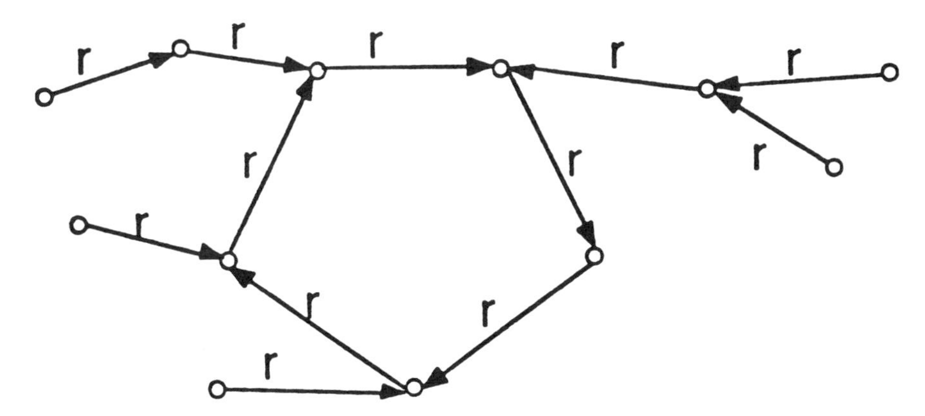

Figure 8: A unicyclic subgraph of red edges.

The remaining edges are of course colored blue. Write C for the set of vertices in the cycle. Let $\mathcal{F}$ be the semigroup of functions from $V(G)$ to $V(G)$ generated by r and b. The function $rrr\cdots r$ has range C. O'Brien shows that the aperiodicity assumption guarantees that some function $f \in \mathcal{F}$ has range which is a proper subset of C. He then completes his proof with the following lemma.

Lemma. *The semigroup G of functions from C to C generated by r (restricted to C) and f (restricted to C) contains a constant function.*

The proof is elementary, short, yet difficult, and uses the primality of $|C|$ in an essential way.

This lemma suggests the following solitaire game. Imagine a circular stand for p equally spaced marbles, where p is a prime number.

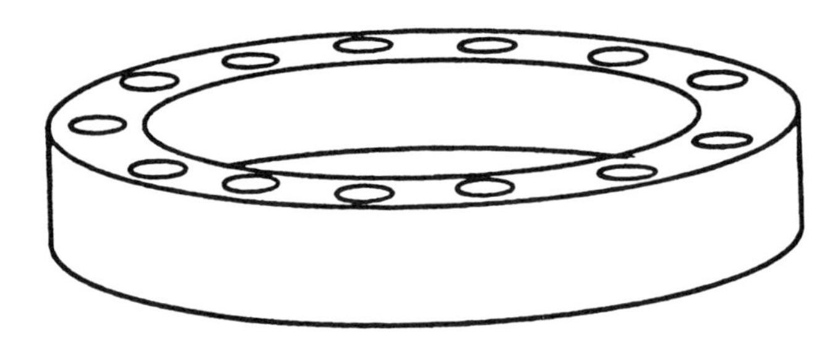

Figure 9: A circular stand for a prime number of marbles.

Inside this circular stand fits a disk on which arrows are drawn.

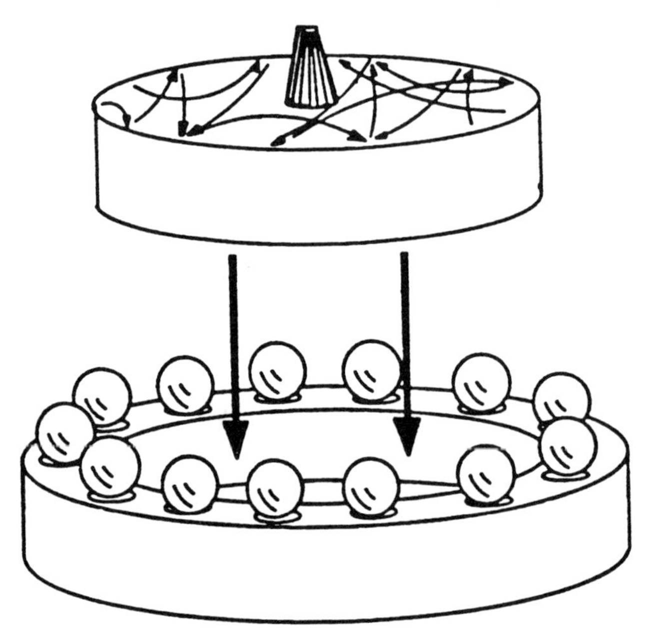

Figure 10: A disk fits inside the circular stand.

The initial position for the game puts a marble in each space. A move in the game consists of spinning the disk to some desired position and then moving all the marbles (simultaneously) to their new positions indicated by the arrows on the disk. If two marbles are asked to occupy the same position by a move, one of the marbles gets bumped (removed). Can all the marbles but one be removed by a succession of moves? O'Brien's lemma says yes, provided that the arrows on the disk do not simply permute the positions (in which case no marbles ever get bumped).

The only other paper on the subject is by J. Friedman [**2**], in which the problem is solved when the graph has a cycle whose length is relatively prime to a certain number $W(G)$, called the weight of the graph. This partial result is neither stronger nor weaker than O'Briens's result, and even in conjunction with O'Brien's result falls short of solving the general case.

Another interesting angle on the road coloring problem arises from the way Adler, Goodwyn & Weiss circumvented it to obtain their theorem. Imagine a snake, whose length is n times as long as an edge of G, slithering along the graph G. A snake's position at any time corresponds to a finite path through G of length n. Here is an example with $n = 3$.

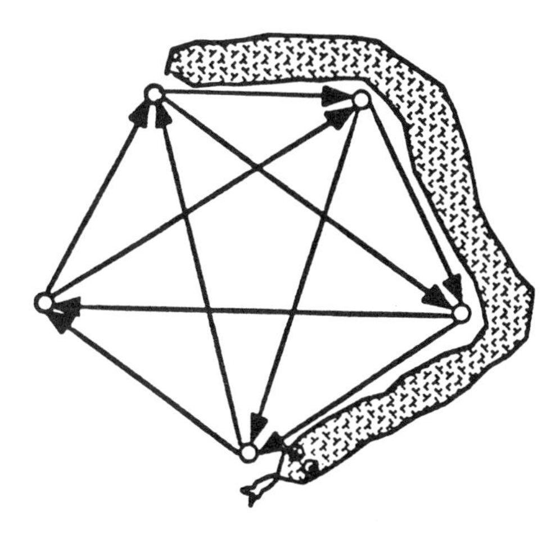

Figure 11: A snake of length 3.

The snake moves by slithering forward along the graph; from each position he has two options. Suppose, in place of the idea of coloring the edges of the graph, we color the two options for the snake, one red and one blue, in the following way. We supply the snake with a catalogue (i.e., a list) of all possible legal snake positions. For each position, the catalogue associates the color red to one of the two options and blue to the other. When the snake is given a sequence of instructions (each either r or b), he slithers along the graph, at each move opting for the fork in the road that the catalogue tells him is associated with that color. Since the catalogue lists every possible position of the snake separately, the new position of the snake depends not only on where the snake's head is but also on the entire position of the snake. The snake's body is a sort of short term memory on which the snake can rely to decide which road to choose; this is what gives the snake an advantage over our beleaguered tourist. Adler, Goodwyn & Weiss in fact showed that the snake's version of the road coloring problem can be solved for any irreducible, aperiodic graph G, provided that the snake is long enough. This trick is known to the specialist in topological dynamics as "passing to a higher block representation." To the specialist in stochastic processes, this is "relaxing the Markov condition to an n-step Markov condition." To the graph theorist, this is "passing to the nth iterated line graph of G."

Many problems of recreational mathematics are generated by the process of searching for new results in technical mathematics. Often, as in this case, the problem becomes separated from its source and develops a life of its own. One day, there may be a housewife, a high school student, or perhaps a mathematician who hears of this puzzle and solves it. The problem is within the grasp of the amateur but is (or at least at one time was) of vital interest to the professional. This illustrates that there is not such a wide gap between the puzzlist and the scientist, that doing puzzles and doing science are not disjoint activities, that recreational mathematics can be as much mathematics as it is recreation.

REFERENCES

1. R.L. Adler, L.W. Goodwyn & B. Weiss, Equivalence of topological Markov shifts, *Israel J. Math.*, **27**(1977) 49-63.

2. J. Friedman, On the road coloring problem, to appear in *Trans. Amer. Math. Soc.*

3. J.L. O'Brien, The road-colouring problem, *Israel J. Math.*, **39**(1981) 145-154.

The George Washington University,
2130 H Street, NW
Washington, DC 20052-0001

Fourteen Proofs of a Result About Tiling a Rectangle*

Stan Wagon

1 Introduction

In [2] (see also [5]) N. G. de Bruijn proved a result about packing n-dimensional bricks into an n-dimensional box that, when $n = 2$, implies that if an $a \times b$ rectangle is tiled with copies of a $c \times d$ rectangle, then each of c, d divides one of a, b. By a *tiling* we mean a covering with interior pairwise-disjoint sets. De Bruijn's proof has been generalized to yield the following more general theorem (illustrated in Figure 1), which implies his result on bricks (in the case $n = 2$, divide each side of the box by c (resp., d)).

Theorem 1 *Whenever a rectangle is tiled by rectangles each of which has at least one integer side, then the tiled rectangle has at least one integer side.*

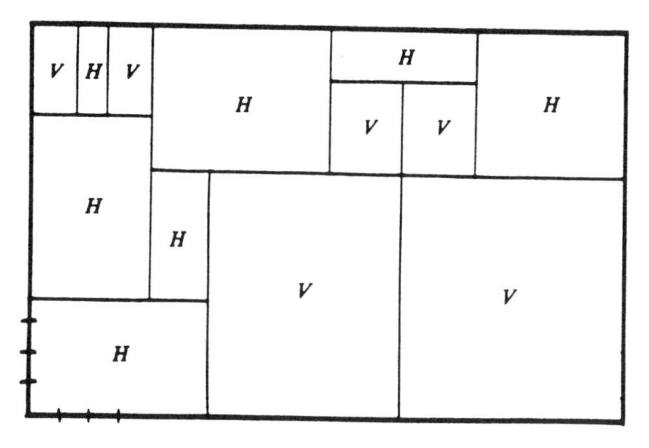

Figure 1: An example of a tiling in which each tile has at least one integer side. The tiles labelled "H" have integer width; those labelled "V" have integer height.

At the 1985 Summer Meeting of the MAA in Laramie, Wyoming, Hugh Montgomery mentioned this theorem and the proof using double integrals, in the hope of stimulating a search for more elementary proofs. That he did, as proofs have been forthcoming from various countries. Indeed, the variety of techniques that have been brought to bear is striking. Paul Erdős [1, p. 87]

*Reprinted with permission from the *American Mathematical Monthly* 94 (August-September 1987) pp. 601–617. Winner of the Lester R. Ford award.

has suggested that "[God] has a transfinite book of theorems in which the best proofs are written." It is by no means clear which of the many proofs that follow is the best (the criteria for inclusion in the book are not readily available!). Perhaps none of these proofs is in the book, and the "best" proof has yet to be discovered. Even if simplicity is taken as the criterion, it is not completely clear which proof wins—the checkerboard and bipartite graph proofs seem to be the top candidates. And if strength is taken into account, that is, the ability to yield, perhaps with modification, more general results, then the situation is complicated. Variations of the theorem are true on the cylinder and torus, in higher dimensions, and for multiple tilings, but no one of the proofs is best in terms of its ability to generalize. Before reading Section 3 the reader might enjoy trying to predict which of the proofs are most likely to generalize.

Max Zorn has pointed out that Dehn considered similar questions in 1903. Dehn [3, p. 327] proved, as a corollary to a rather different sort of investigation, that if a rectangle is tiled as in Theorem 1, then one of the sides is rational.

2 The Proofs

The *width* (resp., *height*) of a rectangle denotes its horizontal (resp., vertical) dimension. Given a tiling as in Theorem 1, let R denote the ambient rectangle. Let a tile with integer width be called an *H-tile* ("horizontal tile"); the other tiles, necessarily having integer height, are called *V-tiles* ("vertical tiles"). It is often assumed that R is in *standard position,* that is, its lower left corner is at the origin and its sides are parallel to the coordinate axes in the x-y plane.

(1) *Complex double integral* (extends original method of de Bruijn) First observe that $\int_a^b \sin 2\pi x\,dx = 0$ if and only if one of $a \pm b$ is an integer and $\int_a^b \cos 2\pi x\,dx = 0$ if and only if one of $a-b$, $a+b-1/2$ is an integer. It follows that for any rectangle T in the x-y plane with sides parallel to the axes,

$$\iint_T e^{2\pi i(x+y)} dA = 0$$

if and only if at least one side of T has integer length. Now, the hypothesis implies that the double integral over each tile vanishes and therefore, by additivity of integrals, the double integral over R is zero.

This implies that either the width or height of R is an integer. ■

(2) *Real double integral* (variation of complex double integral proof) Assume R is an $a \times b$ rectangle in standard position. As in the preceding proof, $\iint_T \sin 2\pi x \sin 2\pi y dA = 0$ for each tile T. Therefore, the double integral over R vanishes, which, because R has a corner at $(0,0)$, implies that at least one of a, b is an integer. (One could use other integrands as well, for example, $(x-[x]-1/2), (y-[y]-1/2)$.) ■

(3) *Checkerboard* (Richard Rochberg, Washington University; Sherman K. Stein) Place R in standard position. Color the square lattice generated by a $(1/2) \times (1/2)$ square with lower left corner at $(0,0)$ in black/white checkerboard fashion. Since each tile has an integer side, each tile contains an equal amount of black and white. Therefore, the same is true of R. But then R must have an integer side for otherwise it can be split into four pieces (see Figure 2), three of which

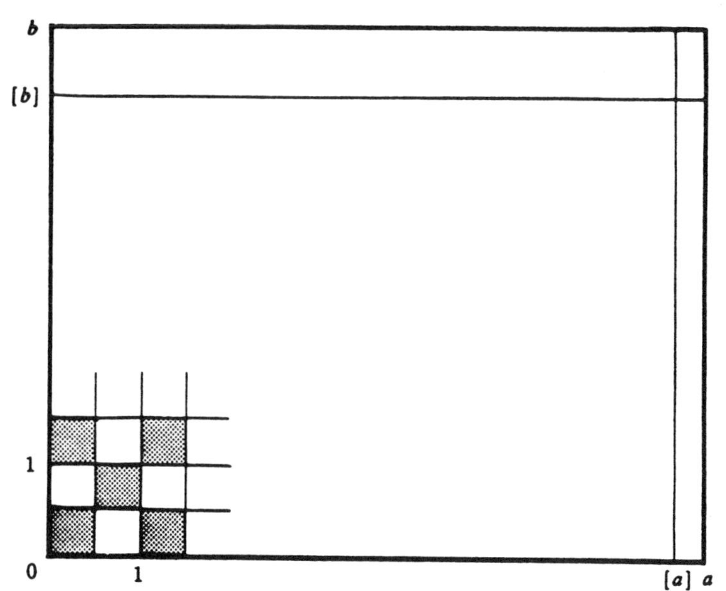

Figure 2: If neither a nor b is an integer then the upper right corner has more black than white.

have equal amounts of black and white while the fourth does not. (This proof is derived from the preceding proof by using the integrand $(-1)^{[2x]}(-1)^{[2y]}$.) ■

(4) *Counting squares* (Imre Z. Ruzsa, Mathematical Institute of the Hungarian Academy of Sciences, Budapest; Peter Gilbert, Digital Equipment Corp., Nashua, NH) Place R in standard position and let $\{x_i\}$ (resp., $\{y_j\}$) be the set of x-coordinates of vertical (resp., y-coordinates of horizontal) boundary lines of tiles. Construct an auxiliary tiling (of a possibly new rectangle R') by translating all line segments in R's tiling as follows. If a segment is on a line corresponding to an integer value of x_i or y_j, it is not moved. If it is a vertical segment lying on $x = x_i$ where x_i is not an integer, translate it rightward or leftward to the line $x = [x_i] + 1/2$. Similarly, vertical segments on $y = y_j$ are translated up or down to $y = [y_j] + 1/2$, if y_j is not an integer. This construction may reduce the number of tiles, but this is unimportant.

Now, if the conclusion is false, then R' is a rectangle in standard position having both side-lengths equal to one-half of an odd integer. Hence, R' contains an odd number of squares in the (uncolored) checkerboard described in the previous proof. But the hypothesis implies that each tile in R' has an even number of squares, contradiction. ■

(5) *Polynomials* (Adrien Douady, École Normale Supérieure, Paris) Place R in standard position and construct an auxiliary tiling in a way similar to the preceding proof. Choose a parameter t and translate only those segments having a noninteger coordinate. Translate vertical segments on $x = x_i$ rightward to $x = x_i + t$, and horizontal segments upward to $y = y_j + t$. If t comes from a sufficiently small interval $[0, \varepsilon]$, this construction yields a tiling of R', with the same number of tiles as in R.

Now, if the conclusion is false then R' is an $(a + t) \times (b + t)$ rectangle, whence its area is a quadratic polynomial in t. But the hypothesis implies that each $w \times h$ tile in R becomes, in R', a tile of one of the forms $w \times (h \pm t)$, $(w \pm t) \times h$, $w \times h$. In all cases the area of the modified tile is a linear or constant function of t, and, hence, the same is true of the area of R'. Since t can take on any value in an interval, this contradicts the quadratic representation of the area. ■

(6) *Prime numbers* (Raphael Robinson, Univ. of California, Berkeley) We claim that for each prime p, either the height or width of R is within $1/p$ of an integer. It follows that one of these is an integer. To prove the claim, scale the entire tiling up by a factor of p in each direction, and consider the tiling obtained by replacing all tile-corners (x, y) in the scaled-up tiling by $([x], [y])$. This yields an integer-sided rectangle by integer-sided rectangles, each of which has one side a multiple of p. Therefore, the area of the large integer-sided rectangle is a multiple of p, whence one of its sides must be a multiple of p. Moreover, the dimensions of this rectangle differ from the dimensions of the scaled-up rectangle by less than 1. It follows that R has a side that differs from an integer by less than $1/p$. ■

(7) *Eulerian graph* (Michael S. Paterson, Univ. of Warwick, Coventry, England) Let Γ be the graph whose vertices are the corners of all the tiles, with two vertices joined whenever they correspond to the ends of a horizontal side of an H-tile or the vertical side of a V-tile. Multiple edges may exist. To make the picture clearer (and to see that Γ is planar), curve the edges a little in the direction of the tile defining the edge (see Figure 3). All vertices (except the corners of the large rectangle) lie on either 2 or 4 rectangles, and hence on either 2 or 4 edges in Γ. The corner vertices lie on 1 edge. It follows that a walk along edges that begins at one corner and does not repeat any edges will not terminate until it hits another corner, thus proving Theorem 1. ■

(8) *Bipartite graph* (variation of Eulerian path proof) Place R in standard position, let S be the set of corners of tiles having both coordinates integers, and let T be the set of tiles. Form a bipartite graph on $S \cup T$ by connecting each point in S to all tiles of which it is a corner. There is an even number of edges because the hypothesis implies that each tile has 0, 2, or 4 corners in S. But each point in S that is not a corner of R lies on either 2 or 4 tiles. Since $(0, 0)$, which lies on only one tile, is in S, there must be another point in S lying on an odd number of tiles. This can happen only if another corner of R lies in S, which means that either the width or height of R is an integer. ■

(9) *Induction* (Raphael Robinson) The proof will be by induction on the number of H-tiles in a tiling in which each H-tile has width 1 and each V-tile has height 1. Since tiles may be split in their designated direction, this case suffices. Choose any H-tile T_0 (if there is none the result is immediate). If there are H-tiles whose lower border shares a segment with the upper border of T_0,

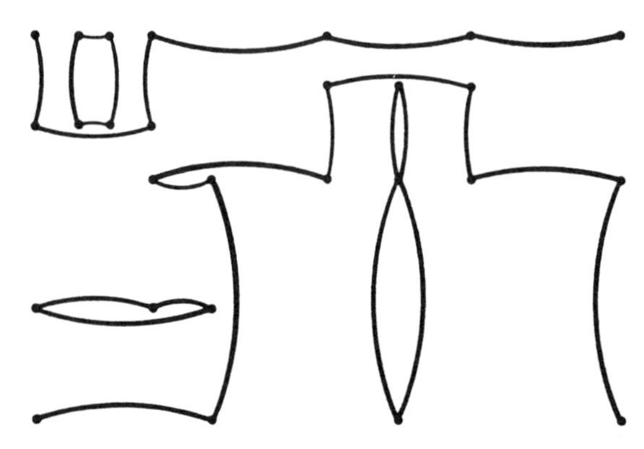

Figure 3: The near-Eulerian graph Γ arising from the tiling in Figure 1.

choose one and call it T_1. Otherwise only V-tiles share this border, and we may expand T_0 upward 1 unit. This does not increase the number of H-tiles, and the cut vertical tiles still have height 1. Continue expanding T_0 upward until either the top of the rectangle is reached, or a choice of an abutting H-tile T_1 is possible. Then continue upward similarly from T_1 to get T_2, etc. This yields a chain $T_0, T_1, \ldots, T_m$ of H-tiles from T_0 to the top of R. We can work downward from T_0 similarly, thus getting a chain

$$T_{-n}, \ldots T_{-1}, T_0, T_1, \ldots T_m$$

of H-tiles stretching from bottom to top. Remove these tiles and slide the rest together to get a rectangle with fewer H-tiles; induction applied to this smaller rectangle yields the result for the original rectangle. ■

(10) *Induction, variation* (Richard Bishop, Univ. of Illinois; Stan Wagon) Define a V*-link* to be a maximal vertical line segment in the tiling whose interior is not crossed by any horizontal line segment. Define H*-link* similarly. A link is *reducible* if it is a V-link (resp., H-link) having only H-tiles (resp., V-tiles) on one of its sides. In the tiling of Figure 1 there are lots of reducible links, for example, the V-link separating the large V-tile in the center from the two H-tiles on its left. It suffices to show that all tilings have a reducible link. For if we are given, say, a reducible V-link with only H-tiles bordering it on the right, let w be the width of the narrowest of these H-tiles. Then expand all tiles bordering the V-link on the left w units rightward (see Figure 4). Since heights are unchanged, V-tiles remain V-tiles; since widths are changed by the addition or subtraction of w, H-tiles remain H-tiles. But this expansion reduces the number of tiles by at least 1, as required for the induction.

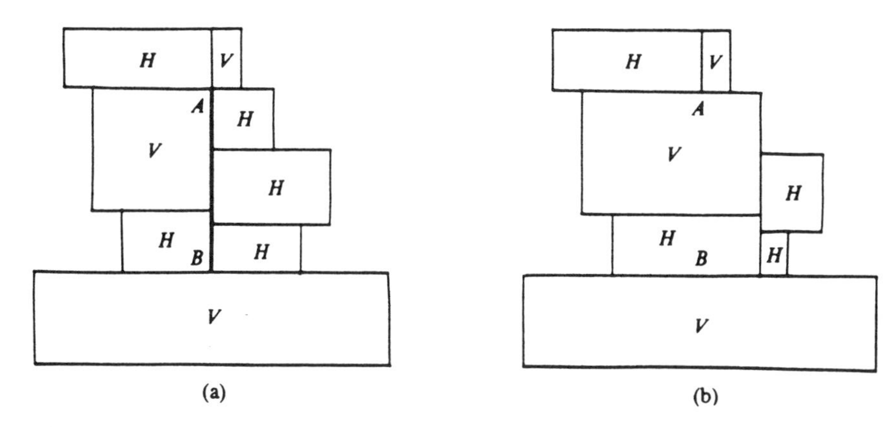

Figure 4: In (a), AB is a reducible V-link. Figure (b) shows the tiling after the tiles adjacent to AB on the left have been expanded rightward.

(11) *Minimal cut-set* (Paul Seymour, Bell Communications Research, Morristown, NJ) Define a graph Γ as follows. The vertices are all horizontal line segments in the tiling, and two vertices are connected by m edges if there are m tiles (*either* H-tiles or V-tiles) connecting the corresponding segments. The exterior of R is considered as a tile, thus adding an additional edge connecting the top and bottom vertices. The tiling yields an embedding of Γ in the plane, since the vertical bisectors of the tiles can be used to form the edges (Figure 5). The edge corresponding to the additional tile can be drawn as in Figure 5, though it is more natural to preserve symmetry by embedding on the surface of a sphere instead.

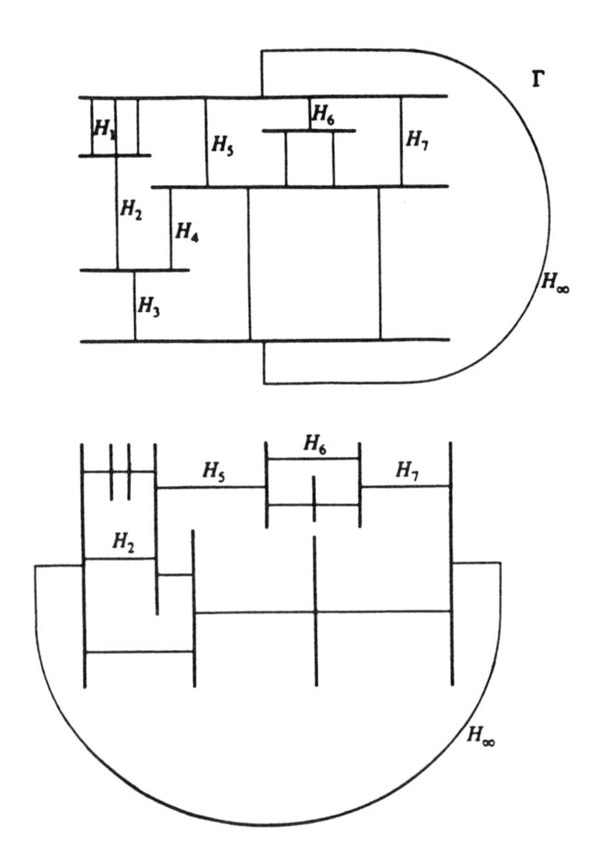

Figure 5: The upper diagram represents the graph Γ of the minimal cut-set proof corresponding to the tiling in Figure 1. Horizontal segments are vertices, and vertical segments are edges, marked with an H if they arise from an H-tile in the tiling. The lower diagram shows the representation of the dual graph, Γ^*, using V-links in the tiling and horizontal segments as edges. The edges in S' and S^*, corresponding to a minimal cut-set in Γ, are H_2, H_5, H_6, H_7, and H_∞.

Let Γ^* be the dual graph of Γ; the vertices of Γ^* are the faces of Γ and two vertices in Γ^* are connected by an edge if the corresponding faces in the planar embedding of Γ are incident. The faces of Γ have a simple structure: each face arises from part of a vertical segment in the tiling—a V-link, in the terminology of the preceding proof (see Figure 5)—and all tiles adjacent to the V-link. And if two faces in Γ are incident along an edge, then there is a tile whose vertical boundaries lie on the V-links corresponding to the faces.

Now let S be the set of edges in Γ corresponding to H-tiles, together with the exterior edge from top to bottom. If the removal of S does not disconnect the top vertex from the bottom vertex, then there is a top-to-bottom path with all vertical steps integers, as desired. Otherwise, let S' be a minimal subset of S whose removal disconnects the top from the bottom in Γ, and let S^* be the set of edges in Γ^* corresponding to edges in S'. By a well-known theorem for planar graphs [7, Thm. 15C], S^* is a cycle in Γ^*. Moreover, since S' must contain the exterior edge, S^* induces a path from the left boundary of R to the right boundary, which has every horizontal step of integer length. Therefore, the width of the rectangle is an integer. ■

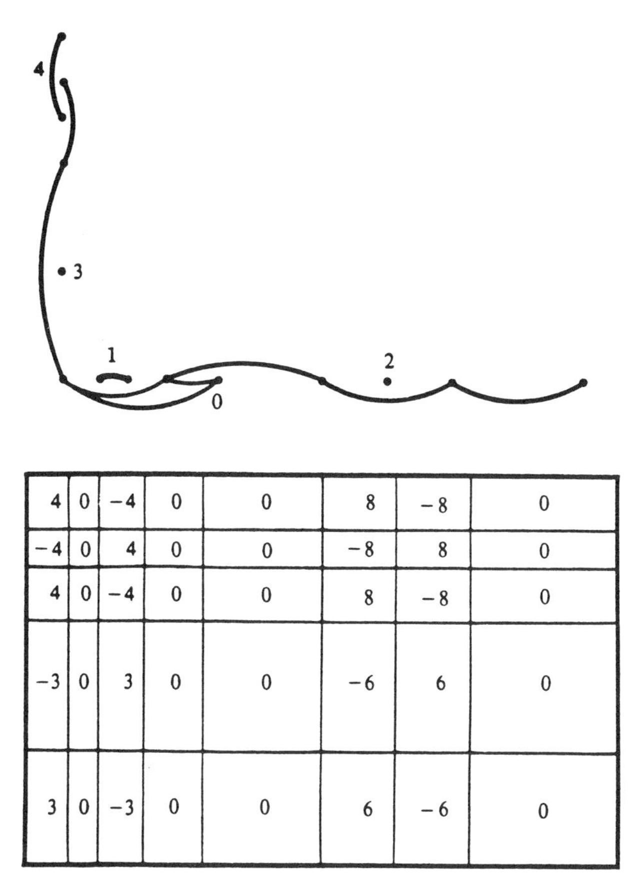

4	0	−4	0	0	8	−8	0
−4	0	4	0	0	−8	8	0
4	0	−4	0	0	8	−8	0
−3	0	3	0	0	−6	6	0
3	0	−3	0	0	6	−6	0

Figure 6: The graph (with components numbered 0–4) and grid of the step function proof, using the tiling of Figure 1.

(12) *Sweep-line* (Gennady Bachman, Univ. of Illinois; Mihalis Yannakakis, Bell Labs, Murray Hill, NJ) Assume R is an $a \times b$ rectangle in standard position, and that b is not an integer. Let $\{R_i\}$ be the set of tiles, but assume that the closed segment forming the bottom border of each has been removed. Let a_i, b_i be the width and height, respectively, of R_i. Define $f : [0, b] \to [0, a]$ by setting $f(t)$ equal to the sum of all a_i such that R_i intersects the line $y = t$ and the y-coordinate of the top of R_i is not an integer. Then $f(0) = 0$ and it is easy to check that whenever f changes its value then it does so in a way that it remains an integer; as the "sweep-line" crosses a horizontal line in the tiling the difference between f's gains and losses is an integer. Therefore, $f(b)$ is an integer. But since b is not an integer $f(b)$ is simply the sum of the widths of all tiles touching the top, that is, $f(b) = a$. ■

(13) *Step functions* (Melvin Hochster, Univ. of Michigan; Attila Máté, Brooklyn College) Place the rectangle in standard position. Then define a graph Γ whose vertices are all points on the x-axis such that some tile has a vertical boundary at that value (call these x_i, in increasing order), and all points on the y-axis that occur as top or bottom coordinates of some tile (y_i). Connect two vertices on the x-axis if some H-tile spans the interval and connect two vertices on the y-axis if some V-tile spans the interval (see Figure 6). The goal is to show that the origin lies in the same connected component of Γ as either $(a, 0)$ or $(0, b)$.

Assign, in an arbitrary way, distinct numbers to the connected components in Γ. Then define a step function on $[0, a]$ by defining f on the interval (x_i, x_{i+1}) to be the number of the component that contains x_{i+1} less the number of the component that contains x_i. Note that the sum of the f-values on the intervals between two vertices connected by an edge is 0. Define g similarly on $[0, b]$. Now, refine the tiling into a grid by drawing all lines $x = x_i$ and $y = y_i$, and observe that $f(x)g(y)$ is constant in the interior of each rectangle in the grid. Moreover, the sum of these products over all grid rectangles is 0. But this sum is just the product of $\sum\{f(I) : I$ an interval between consecutive vertices on the x-axis$\}$ with $\sum\{f(J) : J$ an interval between consecutive vertices on the y-axis$\}$. Therefore one of these sums vanishes, which implies that the origin and one of $(a, 0)$, $(0, b)$ lie in the same component. ■

(14) *Sperner's lemma* (James Schmerl, Univ. of Connecticut) Assume the conclusion is false and R is placed in standard position. Triangulate R by drawing a diagonal in each tile. Then label all vertices in the tiling as follows: (x, y) is labelled A if $x \in \mathbf{N}$, B if $x \notin \mathbf{N}$, and C if neither x nor y is an integer. Then by a variation to Sperner's lemma (see [6, Lemma 2]), the number of triangles labelled ABC is odd. But the hypothesis implies that no triangle is so labelled, contradiction. ■

3 Generalizations

A first reaction to these proofs might be that they are not all different, since many of them have similar ingredients. In some cases this view is valid; the real double-integral proof is a specialization of the complex double-integral proof, and the checkerboard proof is a discretization of the real double-integral proof using a $\{\pm 1\}$-valued function instead of a product of sines. Also, the two induction proofs are closely related, as are the Eulerian path and bipartite graph proofs. But an examination of various generalizations brings out differences in all the other proofs (see Appendix).

A natural generalization of Theorem 1 is to the case where the integers are replaced by other groups of reals. Consider a tiling of R where each tile has one *designated* side, not necessarily of integer length (a tile with designated width (resp., height) is called an H-tile (resp., V- tile)). The goal here is to show that R has either its width in the (additive) subgroup of R generated by widths of H-tiles or its height in the group generated by heights of V-tiles. For example, if each tile has either integer width or algebraic length, then R has either integer width or algebraic length. The Eulerian path, minimal cut-set, sweep-line, step function, and polynomial proofs, as well as the variation to the induction proof, all yield this generalization with essentially no modifications. The bipartite graph proof works if S is the set of tile-corners having both coordinates in the corresponding groups. The induction proof can be made to work in this case, if one excises only part of the chosen horizontal tiles, corresponding to the width of the narrowest member, thus reducing the number of horizontal tiles.

Note that although the Eulerian path, minimal cut-set, and step function proofs all work by finding a path in a certain graph, there are essential differences. The first two use graphs that are planar, while the step function proof might construct a nonplanar graph. The Eulerian proof is the only one of the three that shows that there is a path along integer-length sides of tiles from one side of the rectangle to the opposite side. However, the step function proof seems to have the capability of discovering "paths" that the others miss (see Figure 7).

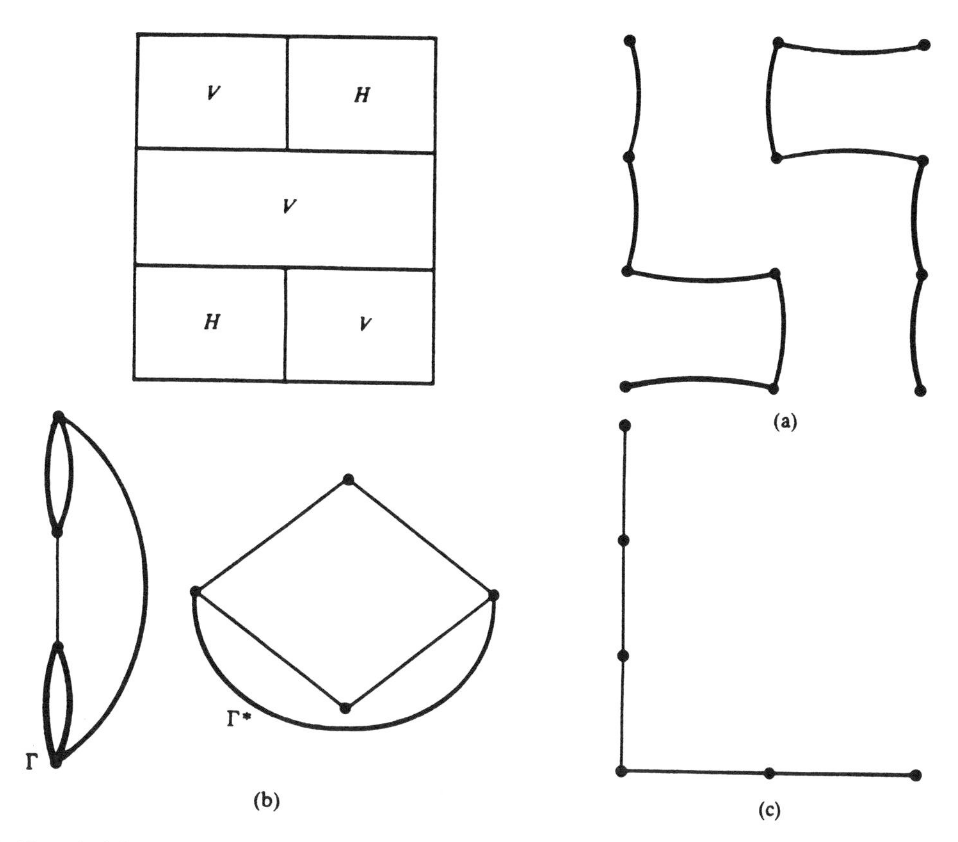

Figure 7: Graph (a) is from the Eulerian path proof, the graphs labelled (b) are Γ and Γ* from the minimal cut-set proof, and graph (c) is from the step function proof.

Ruzsa has pointed out that Theorem 1 remains true only if it is assumed that tiles having at least one corner in $\mathbf{Z}^2$ have an integer side (here we assume R is in standard position). Another way of stating this is: Each tile has either 0, 2, or 4 corners in $\mathbf{Z}^2$. Ruzsa's square-counting proof, the bipartite graph proof, and the polynomial proof yield this result with no modification. The step function and Eulerian path proofs work as well (for the latter, consider only vertices lying in $\mathbf{Z}^2$), as does the Sperner lemma proof.

As observed by several of the authors of proofs of Theorem 1, that result generalizes to higher dimensions. All the proofs, except (apparently) the minimal cut-set, sweep-line, and induction proofs, yield this generalization. Moreover, the higher-dimensional version allows k (rather than just 1) of the sides of each tile to be "designated."

Theorem 2 *Suppose a box R in $\mathbf{r}^n$ is tiled with n-dimensional boxes and each tile has at least k integer sides. Then R has at least k integer sides.*

Proof. The polynomial proof requires almost no modification. The tiling can be perturbed by moving hyperplanes t units, for small t, as in the proof of Theorem 1. If the conclusion is false then the volume of the modified box is a polynomial in t having degree greater than k. But the hypothesis implies that each tile in the auxiliary tiling is a polynomial of degree at most k, contradiction. ■

Several of the other proofs work as well. For the real integral proof, replace the integrand by the product of $t + \sin 2\pi x_r$, $r = 1, \ldots, n$. Then the integral over a box is a polynomial in t that is divisible by t^k if and only if the box has at least k integer sides. For the prime number proof, consider only primes p larger than any of the side-lengths. This guarantees that p^2 does not divide any of the side-lengths in the scaled-up box. The step function proof works if $f(x)$ and $g(y)$ are replaced by $t+f(x_1), t+f(x_2), \ldots$, as in the extension of the integral proof, and a similar approach generalizes the checkerboard proof, which works easily if $k = 1$. The square counting proof works too, though if $k > 1$ one must use an odd integer when moving the boundary hyperplanes to an integer value; then the power of two dividing the number of squares in the auxiliary rectangle corresponds to the number of integer sides.

The Eulerian path and bipartite graph proofs yield Theorem 2 if $k = 1$, since each corner (except the corners of the ambient box) still lies on an even number of tiles. For larger k one can use induction, as pointed out by Andreu Mas-Colell: the $k = 1$ case yields one integer-length side; then project to the hyperplane perpendicular to this direction and use induction on the dimension. The advantage of this inductive approach is that it yields Ruzsa's extension for $k > 1$, where it is assumed only that tiles having a corner with all coordinates integers have k integer sides.

The polynomial, Eulerian path, and step function proofs of Theorem 2 show that the group-theoretical generalization to arbitrary n and k is valid. More precisely: If an n-dimensional box is tiled by boxes, each of which has at least k designated sides, then there are at least k directions in which the side-length of the ambient box lies in the subgroup of R generated by the designated side-lengths in the direction.

Another generalization comes from considering multiple tilings of the rectangle, that is, finitely many tiles that are not necessarily pairwise disjoint, but such that each point of the ambient rectangle (except for the boundaries of the tiles) is contained in the same finite number (the *multiplicity*) of tiles. The integration proofs work in the integer case, as do the checkerboard, polynomial, square counting (replace $1/2$ by $1/p$, where p is a prime larger than the multiplicity), and prime number proofs (use primes larger than the multiplicity). The Eulerian path proof will work if, as pointed out by Paterson, one makes a directed graph, with edges directed out of the lower left and upper right corners of each tile, and into the other two corners. Then the lower left corner has out-degree equal to the multiplicity, while the vertices not equal to a corner have equal in-degree and out-degree. Hence a directed walk starting from the lower left corner will end at one of the adjacent corners of R. The Eulerian path, step function, and polynomial proofs work in the case of groups as well.

Next, we can try to generalize to the case where sides of the rectangle are identified, that is, to the cylinder or torus. Consider the cylinder first, where we assume that opposite vertical sides are identified. The direct generalization of Theorem 1 is valid, as shown by either the sweep-line proof, the induction proof (which was invented for the cylinder and torus), the variation to the induction proof, or the Eulerian path proof (modified as in the proof of Theorem 3 below).

The torus is more interesting since Theorem 1 is false. Consider an $a \times b$ flat torus, that is, an $a \times b$ rectangle in the plane with opposite sides identified. The example in Figure 8, discovered independently by Solomon Golomb and Raphael Robinson, shows that the naive generalization of Theorem 1 is false.

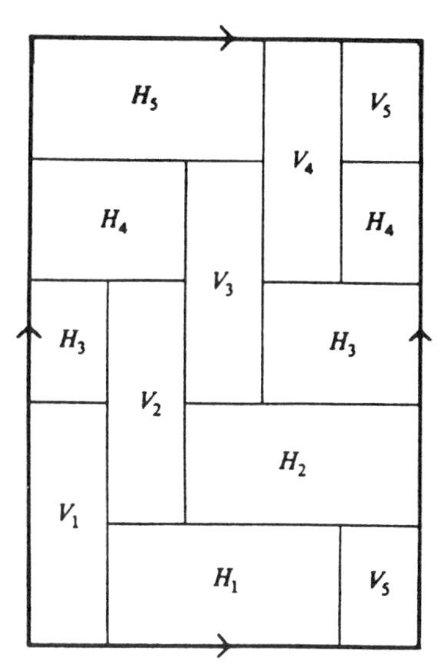

Figure 8: H-tiles are 6×3; V-tiles are 2×6; torus is 10×15. Dividing by 6 yields a tiling of a $\frac{5}{3} \times \frac{5}{2}$ torus using $1 \times \frac{1}{2}$ and $\frac{1}{3} \times 1$ tiles.

Theorem 1 does nevertheless generalize to the torus, although the statement is more complicated. The following theorem was first proved in the integer case by Robinson, whose proof used the method of the induction proof of Theorem 1 and could be extended to the case of arbitrary subgroups. The proof of Theorem 3 given below combines ideas of the Eulerian path and induction proofs, and is due to Joan Hutchinson and the author. Note the curious situation that the original result in rectangles extends to arbitrary subgroups of **R**, while the toroidal result generalizes to arbitrary subfields of **R**.

Theorem 3 (R. M. Robinson) *Suppose an a-by-b flat torus is tiled with rectangles parallel to the sides of the torus. Suppose each tile, regardless of its length or width, is designated to be either an H-tile or a V-tile and let G_H (resp., G_V) be the group generated by the widths of the H-tiles (resp., heights of the V-tiles). Then at least one of the following is true:*

(1) a is in G_H;

(2) b is in G_V;

(3) For some relatively prime integers m and n, ma is in G_H and nb is in G_V.

Proof. Let Γ be the graph associated with the tiling, as described in the Eulerian path proof. On the torus, loops can occur. To ensure that Γ embeds on the torus, curve the edges a little in the direction of the tile defining the edge; see Figure 9. As in the planar case, each vertex has degree 2 or 4. (In the degenerate cases, such as a tiling with one tile, the corners have 2 or 4 loops.) Thus each component of Γ is Eulerian. In particular, any edge lies on a simple cycle.

The proof will be by induction on N, the total number of tiles. If $N = 1$ either (1) or (2) holds. For $N > 1$ observe that if Γ has a noncontractible cycle, then one of (1), (2), or (3) follows, for we may assume that there is a simple noncontractible cycle C. If C winds exactly once in one of the directions, then (1) or (2) holds. Otherwise we may use the well-known result that

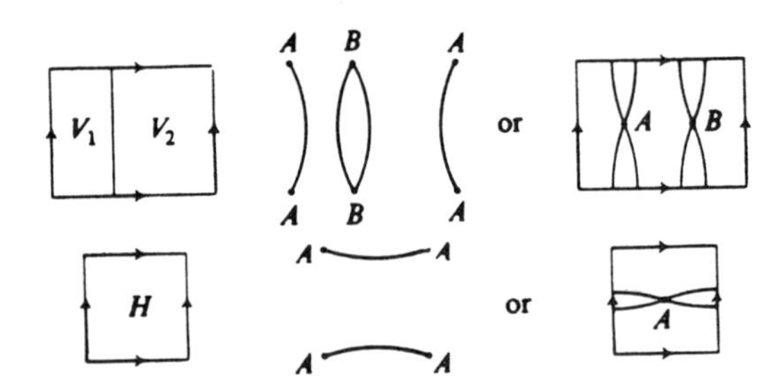

Figure 9: Two examples—one with two tiles, one with just a single tile—of graphs associated with tilings of a torus.

if C winds more than once in one direction, then its winding numbers in the two directions are relatively prime (this is a consequence of P. Lévy's "Universal Chord Theorem" which implies that if $\gcd(m, n) = d$ then a simple curve from the origin to (m, n) has a chord that is a translate of the segment from the origin to $(m/d, n/d)$ (see [4, p. 23])). This yields (3). For example, the graph of the Robinson–Golomb tiling has two cycles, each winding thrice around the horizontal direction and twice vertically, so $3a$ is in G_H and $2b$ is in G_V. However, there are tilings for which Γ has no noncontractible cycles (example in Figure 10); in such cases we shall show that the tiling can be modified so that there are fewer than N tiles.

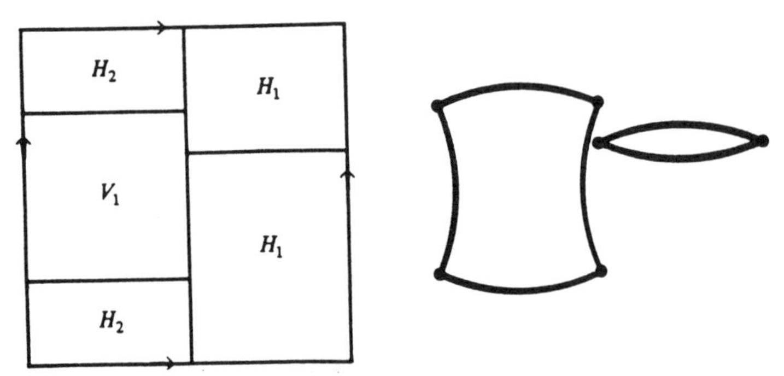

Figure 10: An example of a toroidal tiling for which the graph has no noncontractible cycle.

Suppose then that Γ has only contractible cycles. Then Γ must have a simple contractible cycle with no edges in its interior (called an *empty* cycle). For if C is a simple contractible cycle with the fewest number of edges in its interior then this number must be 0; any edge inside C would lie on a cycle having fewer edges in its interior than C does.

First suppose Γ has an empty contractible cycle which, when viewed in the tiling, has no tile in its interior. Such a cycle, viewed in the tiling, must traverse each part of its boundary twice, once in each direction. Since the cycle is simple, this means it can have no right angles, and so must look like one of the cycles in Figure 11. In either case we can modify the tiles as in Figure 4 (expanding one side to absorb the narrowest—or shortest—tile on the other side), which reduces N by at least 1, as desired for the induction.

If there is no cycle as in the preceding paragraph, then Γ has an empty contractible cycle C that does have a tile, say, an H-tile, in its interior. Because C is empty, both the top and bottom of the tile correspond to edges on C. Label the tile's corners a, b, c, d starting from the upper

Figure 11: A contractible cycle that, when viewed in the tiling, has no tile in its interior, must look like one of these two cycles.

right and going clockwise. Because C is a simple cycle, C must have the form $a \ldots bc \ldots da$. Now, adding the vertical steps in C between a and b yields that the distance from a to b lies in G_V. But then we may switch the tile from an H-tile to a V-tile, shortening the length of the cycle. We can continue shortening the cycle in this way until it no longer has a tile in its interior. Then we are in the preceding case, where it was easy to reduce the number of tiles. ■

Not too much is known for higher-dimensional tori. Robinson has generalized his prime number proof to show that if such a torus is tiled with boxes having an integer side, then the torus has at least one rational side. This can also be proved by the step function proof, which has the advantage of working for arbitrary groups.

Theorem 4 (Robinson, Máté) *Suppose an n-dimensional flat torus with side-lengths a_i, $i = 1, \ldots, n$, is tiled by n-dimensional boxes parallel to the sides of the torus, with each tile having one designated side. Let G_i be the group generated by the designated side-lengths in the ith direction. Then for at least one a_i there is a positive integer m such that ma_i is in G_i.*

Proof (Máté). We give the details assuming the case of an $a \times b$ flat torus in $\mathbf{R}^2$; the extension to higher dimensions will be clear. Assume the $a \times b$ rectangle is in standard position and the origin is the corner of a tile. Extend the tiling periodically to the whole plane and define a graph Γ as in the step function proof: If (x, y) is a corner of a tile in any copy of the torus then the points $(x, 0)$ and $(0, y)$ are vertices. Connect two vertices on the x-axis if the interval they define is spanned by an H-tile in the tiling of the plane; this includes the case of tiles straddling a vertical boundary of a torus. Define edges on the y-axis likewise using V-tiles.

It is sufficient to prove the following claim, for if Γ has infinitely many vertices on, say, the x-axis that are in the same connected component, then that component must contain two vertices of the form $(x, 0)$, $(x + ma, 0)$. This implies that ma is in G_1.

Claim. The graph Γ has an infinite connected component.

Proof of claim. To prove the claim, assume it is false. Then all components are finite and we may define a function C_1 on the points on the x-axis corresponding to vertices in Γ by letting $C_1(x)$ be the least t such that $(t, 0)$ is in Γ and in the same component as $(x, 0)$. Define C_2 for vertices on the y-axis similarly. Now define a step function f on the x-axis by letting f equal $C_1(x'') - C_1(x')$ on the interval between any two consecutive vertices x', x'' in Γ.

As before, the sum of f-values over intervals subdividing a single H-tile in the tiling of $\mathbf{R}^2$ is zero. Moreover, because of the periodicity of the tiling of the plane, $C_1(x+a) = C_1(x)+a$ and f is periodic with period a. It follows that the sum of f-values over intervals subdividing a single H-tile in the original torus in standard position is 0. These properties also hold for the step function g, defined using C_2 analogously to f. To conclude, proceed as in Theorem 1 to refine the tiling into a grid and observe that the sum of the fg values over a tile vanishes, whence the sum over the entire $a \times b$ torus vanishes. But this is a contradiction since this sum equals ab: the sum of f (resp., g) over the intervals in $[0, a]$ (resp., $[0, b]$) is simply $C_1(a) - C_1(0) = a$ (resp., $C_2(b) - C_2(0) = b$). ■

The preceding argument also works on a box in $\mathbf{R}^n$ where some, but not all, sides are identified. The result then states that either there is an "unidentified" direction of the box whose side-length is in the subgroup generated by designated lengths in that direction, or there is an "identified" direction for which an integer multiple of the side-length is in the group corresponding to that direction. In the case of the standard torus or cylinder this is not best possible, but the proofs in those cases do not generalize to higher dimensions. Unlike the proof of Theorem 3, the preceding argument works for multiple tilings, and so yields something for multiple tilings of the standard two-dimensional torus: If each tile has one integer side then at least one side of the torus is rational (and similarly in higher-dimensional multiple tilings).

4 Summary and open questions

The various generalizations considered here do a fairly complete job of distinguishing the proofs. If one calls two proofs equivalent provided they work on the same set of generalizations then, unless new modifications are found, the only equivalences are (1) $\sim$ (2) $\sim$ (3) and (9) $\sim$ (10). The two most powerful proofs seem to be the Eulerian path and step function proofs. The former fails only on high-dimensional tori, multiple tilings of the standard torus, and the $k > 1$ case of Theorem 2; the latter works in all cases, except the cylinder and torus, where it does not yield the best possible result. A definitive comparison will have to wait until the true situation in higher dimensions is resolved; see Problem (a) below.

Problem (a). Can the seemingly weak statement about tilings of higher-dimensional tori or cylinders be improved, or is it best possible? The simplest unsolved case is that of a box with left and right faces identified. Does there exist a box having dimensions $\alpha \times \beta \times \gamma$, with α rational and β and γ irrational, such that it can be tiled with boxes each of which has a side of unit length?

Problem (b) (S. Golomb). For which triples (a, b, k) can the $a \times b$ torus be tiled using copies (vertical or horizontal) of a $1 \times k$ tile?

Remarks. De Bruijn's original result characterized the rectangles that could be tiled using copies of a $1 \times k$ tile: either the height or the width of the rectangle is a multiple of k. This is true for an $a \times b$ torus if k is a prime power. This was first proved by Robinson and Golomb using coloring techniques; it also follows from Theorem 3 above if one divides everything by k and observes that the relatively prime coefficients m and n cannot both absorb a power of the same prime. It follows that $k = 6$ is the smallest number for which there is a triple (a, b, k) as in Question (a) with neither a nor b divisible by k. An unsolved special case of Question (a) is the problem of determining which $a \times b$ tori can be tiled with copies of a 1×6 tile. Golomb has shown that the 10×15 torus is the smallest example. Further results on tiling a torus with integer-sided rectangles have been obtained by A. Clivio, "Tilings of a torus with rectangular boxes," *Discrete Math.* 91 (1991) 121–139.

Problem (c). What is the situation regarding double tilings of the standard torus where each tile has at least one integer side? Is it true that either one side of the torus is an integer or both sides are rational?

Acknowledgements. The author is grateful to David Gale, Solomon Golomb, Joan Hutchinson, Hugh Montgomery, and Raphael Robinson for their collaboration and enthusiasm and to the authors of the proofs for permission to include them.

References

1. Donald J. Albers & G. L. Alexanderson, *Mathematical People,* Birkhäuser, Boston, 1985.

2. N. G. de Bruijn, Filling boxes with bricks, *Amer. Math. Monthly* 76 (1969) 37–40.

3. M. Dehn, Über Zerlegung von Rechtecken in Rechtecke, *Math. Ann.* 57 (1903) 314–332.

4. Hugo Hadwiger, Hans Debrunner, and Victor Klee, *Combinatorial Geometry in the Plane,* Holt, Rinehart and Winston, New York, 1964.

5. Richard Johnsonbaugh, De Bruijn's packing problem, in *Two-Year College Mathematics Readings,* ed. W. Page, Mathematical Association of America, 1981, pp. 258–262.

6. D. G. Mead, Dissection of the hypercube into simplices, *Proc. Amer. Math. Soc.* 76 (1979) 302–304.

7. Robin J. Wilson, *Introduction to Graph Theory,* Academic Press, New York, 1972.

Department of Mathematics
Macalester College
St. Paul, MN 55105-1899

Appendix to justify claim that proofs are different:

Proofs:

(1) Complex double integral
(2) Real double integral
(3) Checkerboard
(4) Counting squares
(5) Polynomials
(6) Prime numbers
(7) Eulerian path
(8) Bipartite graph
(9) Induction
(10) Induction, variation
(11) Minimal cut-set
(12) Sweep-line
(13) Step functions
(14) Sperner's Lemma

Generalizations:

1. Plane
2. Plane, Ruzsa hypothesis
3. Plane arbitrary groups
4. n-dimensions, $k = 1$
5. n-dimensions, $k > 1$
6. n-dimensions, $k > 1$, Ruzsa hypothesis
7. Cylinder
8. Torus
9. Plane, multiple tiling
10. Plane, multiple tiling, arbitrary groups
11. High-dimensional torus
12. Torus, multiple tiling

Proof number	Works in cases
1, 2, 3	1, 4, 5, 9
4	1, 2, 4, 5, 9
5	1, 2, 3, 4, 5, 9, 10
6	1, 4, 5, 9, 11
7	1, 2, 3, 4, 7, 8, 9, 10
8	1, 2, 3, 4, 9, 10
9, 10	1, 3, 7, 8
11	1, 3
12	1, 3, 7
13	1, 2, 3, 4, 5, 9, 10, 11, 12
14	1, 2, 3, 4

Tiling R^3 with Circles and Disks*

J. B. Wilker

Abstract

While it is not possible to tile R^2 with circles, several constructions are known for tiling R^3 with circles or with homeomorphs of circles subject to various constraints. We extend this work on circles and obtain as well some analogous results on tiling R^2 and R^3 with disks.

1 Introduction

A collection of circles or of disks gives a tiling of R^3 if each point of R^3 belongs to one and only one of the sets in question. Recent interest in tiling R^3 with circles ([2], [4], [5]) has prompted the question of what similar things can be done with disks. We address this question after setting the stage by reviewing a number of known constructions involving circles and adding some new remarks about them. We gratefully acknowledge discussions of parts of this material with R. Grinnell and B. Monson.

2 Circle tilings

The most famous tiling of this sort is surely Hopf's fibration of the 3-sphere. If R^4 is coordinatized by pairs of complex numbers then the 3-sphere

$$S^3 = \left\{(z_1, z_2) : |z_1|^2 + |z_2|^2 = 1\right\}$$

is invariant under the Hopf flow

$$(z_1, z_2) \to (e^{it}z_1, e^{it}z_2), \qquad t \in [0, 2\pi)$$

and the orbits of this flow lying on S^3 are great circles such as $(e^{it}, 0)$, $t \in [0, 2\pi)$. Since each point of S^3 belongs to one and only one of these great circles they can be taken together to form a remarkable tiling of S^3 by congruent circles of radius 1.

Stereographic projection of this tiling into R^3 gives an interesting partition of R^3 into circles and one line. (See [7] for a description of the Hopf fibration that begins with this partition.) While it is often allowed to consider a line or a point as a degenerate circle we shall not do this here. Under our strict convention the stereographic image of the Hopf fibration cannot be considered

*Reprinted with permission from *Geometriae Dedicata* 32(1989) 203–209.

a tiling of R^3 by circles. Nevertheless there does exist a quite surprising R^3-analogue of the Hopf fibration. Conway and Croft [1] have used the axiom of choice to give a nonconstructive proof that R^3 admits a tiling by congruent circles of radius 1. One may object that a tiling produced with the help of the axiom of choice is impossible to visualize and therefore somewhat unsatisfactory. But in point of fact, the argument only requires consideration of circles in c planes through each point and if these planes are taken from a one-parameter family which is almost horizontal, the circles of the resulting tiling will be almost horizontal as well. Thus there is at least one sense in which the completed tiling is very easy to visualize. On the other hand, it seems impossible to control the way in which the circles of the Conway–Croft tiling are linked and this does stand in marked contrast to the circles of the Hopf fibration where every pair of circles is linked.

In any event it seems worthwhile to produce explicit tilings of R^3 by circles. One approach taken independently by J. Schaer (private communication) and A Szulkin [5] depends on the fact that a 2-sphere less 2 points can be tiled by circles. Both Schaer and Szulkin begin by partitioning R^3 into a single point 0 and a nest of 2-spheres centered at this point. Then to cover 0 and to put 2 puncture marks into each 2-sphere of the nest they select their own suitable sequence of circles beginning with one through 0. In each case they complete the tiling of R^3 by tiling each twice punctured 2-sphere of the nest. These tilings include circles of every radius $r > 0$. To come closer to the objective of tiling with congruent circles set by the Conway–Croft example we allow ourselves the freedom of tiling with homeomorphs of the circle.

3 Tilings with topological circles

In [3] Bankston and McGovern produced a tiling of R^3 by simple closed curves whose arc lengths are bounded above and we shall modify their construction to ensure that the arc lengths are bounded away from 0 as well.

The construction begins with the observation that R^3 trivially admits a tiling by half-open cubes $[0,1)^3$ and each of these cubes is homeomorphic to an open ball $x^2 + y^2 + z^2 < 1$ together with part of its boundary $x^2 + y^2 + z^2 = 1, x < \frac{1}{2}$. The open ball less the z-axis is a union of cylinders, each of which can be tiled with circles parallel to the (x, y)-plane. After the z-axis is completed to a simple closed curve by adding the meridian semicircle $(-\sin\psi, 0, \cos\psi)$, $0 \le \psi \le \pi$, all that remains of our set is a slit disk in the boundary portion and this can be tiled with simple closed curves that wind once around the slit.

The deficiency in this tiling is that the circles used on the cylinders $x^2 + y^2 = r^2$ have arc length tending to 0 with r. To remedy this we retile these segments of cylinders by beginning with the infinite cylinder$x^2 + y^2 = r^2$ tiled with ellipses lying in planes parallel to $z = r^{-1}x$. These ellipses each cover a range of z-values $\Delta z = 2$ and so can be wound a uniform of times around the cylinder by applying the twist given in cylindrical coordinates by $(r, \theta, z) \to (r, \theta + kz, z)$. With $k = n\pi$ we get n revolutions of the double strand for a lower bound on arc length of $\approx 4\pi rn$ which is not diminished when the infinite cylinder is compressed by $z \to hz/(|z| + 1)$, $h = \sqrt{1 - r^2}$, to fit its place in the open ball $x^2 + y^2 + z^2 < 1$.

Of course to complete this example we must apply a suitable homeomorphism to deform the half-open ball back into the half-open cube $[0,1)^3$. While this clearly can be done with sufficiently small local distortion to maintain good bounds on arc length the procedure begs the general question of determining precise results on homeomorphisms with such arc length constraints. Also, we still require the construction of a tiling of R^3 by simple closed curves with constant arc length, or even better, by congruent simple closed curves. T. Zamfirescu has suggested the interesting additional constraint of requiring the tiles to be smooth, that is continuously differentiable with nonvanishing derivative.

On another front, there is the question of linking. With circles of unbounded arc length one could imitate the Hopf fibration and ask for every pair to be linked. With circles of bounded arc length this is impossible and one might go to the opposite extreme and seek tilings without any links.

An example without links but using unbounded simple closed curves was given by Kakutani and reported by Bankston and Fox [2]. We include an explicit version of this tiling.

The basic idea is to write R^3 as a disjoint union of two sets b and c such that (i) B admits a link-free tiling by simple closed curves; (ii) no curve in B can link one in C; and (iii) there is a homeomorphism $f : R^3 \to C$ whose iterates give a partition $R^3 = \bigcup_{n=0}^{\infty} f^n(B)$. In these circumstances the iterates of f extend the link-free tiling of B to a link-free tiling of all of R^3. For an explicit example we can take B to be the region $\max\{|x|, |y|\} \geq 1$ tiled with squares surrounding the complementary square cylinder C and $f : R^3 \to C$ to be the mapping

$$f(x, y, z) = \left(\frac{x}{|x| + 1}, \frac{z}{|z| + 1}, y + |z| + 1 \right).$$

4 The impossible

Having given some examples of what might be called improbable tilings of R^3 we make the transition from circles to disks by discussing a number of constructions that are actually impossible.

First we mention that it is impossible to tile R^2 with circles. Following an elegant proof kindly supplied by Vahe Minassian we let C_1 be any circle of an alleged tiling C. Then we let C_2 be the circle of C that covers the center of C_1 and therefore has radius at most half that of C_1. Continuing in this way we determine a sequence of circles C_n in C bounding nested closed disks D_n which must converge to a point because their radii tend to 0. The circle of C through this point obviously intersects a sufficiently small circle C_n and this contradicts the fact that C is a tiling.

It is apparently harder to prove that R^2 does not admit a tiling by arbitrary simple closed curves. For if we drop the requirement that our curves be rectifiable then the class includes Osgood curves and an individual curve of this sort can pass through a set of points of positive area. (See [6] for a simple construction of examples of these exotic curves and their n-dimensional analogues.) In spite of the apparent threat posed by Osgood curves the preceding argument can in fact be generalized to show that R^2 does not admit a tiling by any family of simple closed curves. The key point is that the Jordan curve theorem applies to say that an arbitrary simple closed curve bounds a closed disk consisting of the curve itself together with its interior points. Thus any family of mutually disjoint simple closed curves can be partially ordered by inclusion of these disks and the Hausdorff maximal principle guarantees the existence of a maximal linearly ordered subset of them. The intersection of the disks in such a nest must include points in the interior of every curve belonging to the nest. if the original family is alleged to be a tiling we get a contradiction because the tiling curve through the point we have just located defeats the maximality of the nest.

The first problem in connection with tilings by disks is to decide exactly what a disk is to be. If we allow homeomorphs of $(0,1)^2$ or $[0,1)^2$ then it is uninterestingly easy to tile R^2 and hence R^3. We therefore define a disk to be an arbitrary homeomorph of $[0,1]^2$. As an indication that this definition is appropriate we have

Theorem 1 *R^2 does not admit a tiling by disks.*

Proof. A disk in R^2 is a compact set with non-void interior. Because of the presence of interior points any collection of mutually disjoint disks has at most countably many members. Because compact sets are bounded a tiling of R^2 by disks must induce on any line in R^2 a partition of that line into a countable infinity of closed sets. The proof of the theorem is completed by the following

Lemma. *Every countably infinite union of mutually disjoint closed subsets of a line must omit c points of that line.*

Proof. Consider such a union of closed sets $\bigcup_{n=1}^{\infty} F_n$ and let J be a closed interval on the line that meets F_1 and F_2. let L and R be the points of $J \cap F_1$ and $J \cap F_2$ that are closest together and define I to be the interval $I = [L, R]$.

Let F_m, $m \geq 3$, be the first of the remaining sets to meet I; let P_L and P_R be the points of $I \cap F_m$ closest to L and R respectively; let $I_L = [L, P_L]$ and $I_R = [P_P, R]$.

By continuing this process we derive nests of closed intervals indexed by sequences of L's and R's. Only the nests corresponding to sequences that are eventually constant can intersect in points belonging to one of the sets F_n. Each of the c remaining nests gives at least one point not in any F_n and it follows that $\bigcup_{n=1}^{\infty} F_n$ omits c points as required.

5 Tiling R^3 with disks

The fact that R^3 admits tilings by disks can be proved by a single remark [3]: $[0,1)^3$ is homeomorphic to $[0,1]^2 \times [0,1)$. Because our present interest is geometrical as well as topological we present a more explicit theorem based on this idea.

Theorem 2 *R^3 can be tiled by non-planar hexagonal disks, each arbitrarily close in shape to the planar region bounded by a regular hexagon.*

Proof. We begin with the observation that R^3 can be tiled with integer lattice translates of $[0,1)^3$. Then we concentrate on $[0,1)^3$ and show that it can be tiled by non-planar but piecewise linear hexagonal disks H_t, $t \in [0,1)$, centered on the body diagonal of $[0,1)^3$ at (t,t,t) and admitting the symmetry of 120° rotations about this body diagonal. There many ways of effecting this tiling but for convenience we fix a parameter $u \in [0,1)$ and let H_t have vertices

$$v_1(t) = (u + t(1-u), 0, 0), v_2(t) = (u + t(1-u), u + t(1-u), 0)$$

and the images $v_3(t)$, $v_4(t)$, and $v_5(t)$, $v_6(t)$ of these two points under the 120° rotations mentioned above. Then H_t is the crinkled hexagonal disk made up of six successively abutting plane triangles

$\Delta_i(t)$ with vertices $v_i(t)$, $v_{i+1}(t)$ and (t,t,t). By fixing the parameter u arbitrarily close to 1 we can make the bounding hexagons arbitrarily close to being mutual congruent. Finally, to make the non-planar hexagonal disks H_t and their translates in the tiling of R^3 each arbitrarily close in shape to a planar regular hexagonal disk H, we place a copy of H in the plane $x+y+z=0$ with its center at the origin and apply to all of R^3 a nonsingular affine distortion which maps the basis vectors e_1, e_2, e_3 as close as we please to the alternate vertices v_1, v_3, v_5 of H.

To help visualize the completed packing of Theorem 2 we note that the orthogonal projection of a single layer of cells of the cubic packing joined edge to edge and running perpendicular to the vector $(1,1,1)$ gives hexagonal close packing in the plane $x+y+z=0$. If the diameter of H is $\frac{2}{3}\sqrt{6}$ the affine distortion in the proof of Theorem 2 can be achieved without altering this projection and the distorted cells, each carrying its load of disks, resemble wafer thin hexagonal bipyramids lying in hexagonally close packed sheets which repeat $A, B, C, A, B, C, \ldots$ as one moves in the $(1,1,1)$ direction.

The result of Theorem 2 and a diagram in [1] suggested the possibility of tiling R^3 with planar disks.

Theorem 3 *R^3 admits a tiling by rhombi with edge length* 1.

Proof. (i) In this part we elaborate on the Conway–Croft tiling of the (x,y)-plane with closed unit line segments.

The column $0 \le x \le 1$, $y > 0$, can be tiled with unit segments parallel to the x-axis; we shall speak of the bounding ray $x = 0$, $y > 0$, as the leading edge of this column and the missing segment $0 \le x \le 1$, $y = 0$ as its foot. The transformation $(x,y) \to (x + y\cos\theta, y\sin\theta)$ maps out tiles $(\pi/2)$-column into a tiled θ-column whose leading edge and foot meet at the angle θ.

The open first quadrant is an open $(\pi/2)$-wedge with vertex at the origin; the lines $\theta = (1 - (1/2^n))(\pi/2)$, $n = 1,2,3,\ldots$, comprise a fan in this wedge. By inserting a tiled $(1/2^n)(\pi/2)$-column with its leading edge on the nth line of this fan and its open foot resting on the previous column we completely tile the quarter disk $x^2 + y^2 \le 1$, $x > 0$, $y > 0$ and partition the untiled portion of the first quadrant into open $(1/2^n)(\pi/2)$-wedges, $n = 1,2,3,\ldots$. Continuing in this manner into each of these wedges we complete a tiling of the open first quadrant. If we reflect this tiled quadrant in the line $x = -\frac{1}{2}$ and fill the gap with a tiled $(\pi/2)$-column we obtain a tiled open halfplane. Finally if we reflect this tiled halfplane in the line $y = -\frac{1}{2}$ and fill the gap with a doubly infinite column of unit segments we obtain a tiling of the whole plane.

Now if we cross each segment in the tiled (x,y)-plane with the segment $-1 \le z \le 0$ we obtain a slab $-1 \le z \le 0$ in R^3 tiled with unit squares. This tiled slab will comprise part of the desired tiling of R^3.

(ii) The portion $-1 \le z \le 0$, $x > 0$, of the tiled slab produced in part (i) can be broken off from the rest of the slab. We place a congruent copy of this fragment on top of the slab in the position $0 \le y \le 1$, $z > 0$, and refer to it as a tiled $(\pi/2)$-wall. The transformation $(x,y,z) \to (x, y + z\cos\theta, z\sin\theta)$ maps our tiled $(\pi/2)$-wall into a tiled θ-wall; however, the tiles in the θ-wall are not all squares but rhombi with small angle between $(\pi/2)$ and θ depending on how close to horizontal or vertical its preimage lies in the $(\pi/2)$-wall.

The outline of these walls appears in the (y,z)-plane in just the same way that the outline of the columns of part (i) appeared in the (x,y)-plane. Accordingly our work in part (i) can be carried over as instructions for fitting appropriate θ-walls into appropriate dihedral wedges in order to achieve a tiling of the halfspace $z > 0$ by rhombi. This tiled halfspace fits together with its mirror image in the plane $z = -\frac{1}{2}$ and the tiled slab of part (i) to complete the required tiling of R^3.

A number of problems have been mentioned at appropriate points in the paper.

We close with the question: Can the tiling in Theorem 3 be improved to a tiling with congruent squares?

References

1. J. H. Conway & H. T. Croft, Covering a sphere with congruent great-circle arcs, *Proc. Camb. Phil. Soc.* 60 (1964) 787–800.

2. Paul Bankston & Ralph Fox, Topological partitions of Euclidean space by spheres, *Amer. Math. Monthly* 92 (1985) 423–424.

3 Paul Bankston & Richard J. McGovern, Topological partitions, *General Topology and App.* 10 (1979) 215–229.

4. A. B. Kharazishvili, Partition of a three-dimensional space into congruent circles, *Soobshch. Akad. Nauk. Gruzin SSR* 119 (1985) 57– 60.

5. Andrzej Szulkin, R^3 is the union of disjoint circles, *Amer. Math. Monthly* 90 (1983) 640–641.

6. J. B. Wilker, Space curves that point almost everywhere, *Trans Amer. Math. Soc.* 250 (1979) 263–274.

7. J. B. Wilker, Inversive geometry and the Hopf fibration, *Stud. Sci. Math. Hungar.* 21 (1986) 91–101.

Scarborough College
University of Toronto,
West Hill, Ontario M1C 1A4

Part 2

Games & Puzzles

Introduction to Blockbusting and Domineering

Elwyn Berlekamp

An expanded version of the paper I presented at the Strens Conference was published in the *Journal of Combinatorial Theory Series A* in September 1988 under the title "Blockbusting and Domineering." The present paper serves as an introductory advertisement to that one. Parts of the present paper are identical to "Two-person, perfect information games" which was published by the American Mathematical Society as the proceedings of a conference that was held in June 1988 to commemorate the work of the late John von Neumann.

Domineering is a two-person, perfect-information game which is played on some subset of a (possible enlarged) checkerboard. Each player has a bountiful supply of 2×1 dominoes. At each turn, Left places one of his dominoes onto any available space on the board in a verticaL position; Right then places one of his dominoes onto the board in a hoRizontal position. The game continues until one player is unable to move and that player is then the loser.

Figure 1 shows a domineering game in progress, with Left's turn to play. The continuation of the game is shown in Figure 2. Left decides to move at location 1; Right, at location 2; Left, at location 3; then Right, at location 4; then Left, at location 5, after which the position is shown in Figure 3.

Figure 1: A Domineering game . . .

Figure 2: . . . continued.

The position shown in Figure 3 is easily decomposed into several regions. Each of these regions can be assigned a value according to the number of moves advantage that it provides to Left. For

Figure 3: Domineering values.

example, a 2×1 region provides a place where Left can place one domino, but Right cannot place any. Hence we say that the value of a 2×1 region is $+1$; it constitutes a one-move advantage for Left. Similarly, a 4×1 region provides a place where Left can place two dominoes while Right cannot move at all. Evidently, a 4×1 region has value $+2$. Similarly, a 1×2 region has a value of -1 because it offers a one-move advantage to Right. A 1×1 region has value 0 because it offers no moves to either player.

Regions of more complicated shapes have more complicated values. For example, each of the two regions of Figure 3 that is shaped like an L lying on its side has a value of $-1/2$. We shall justify this assertion later. At the moment, however, we merely wish to show that we can compute the value of the entire board simply by summing the values of the regions therein. Thus, the value of the game position shown in Figure 3 is $+2 +2 +1 -1 -1 -1 -1 -1/2 -1/2 -1 = -1$. Since this is negative, the game can be won by Right.

As in *Winning Ways,* we say that a game is

- Positive iff Left can win (no matter who plays first)
- Negative iff Right can win (no matter who plays first)
- Null iff second player can win (no matter whether he is Left or Right)
- Fuzzy iff first player can win (no matter whether he is Left or Right)

Every Domineering region has a value, and the value of a sum of regions is equal to the sum of the corresponding values. Sometimes, as in Figure 3, the values are familiar numbers. However, other positions, such as those which appear in Figure 1, have the more complicated values shown in Figure 4.

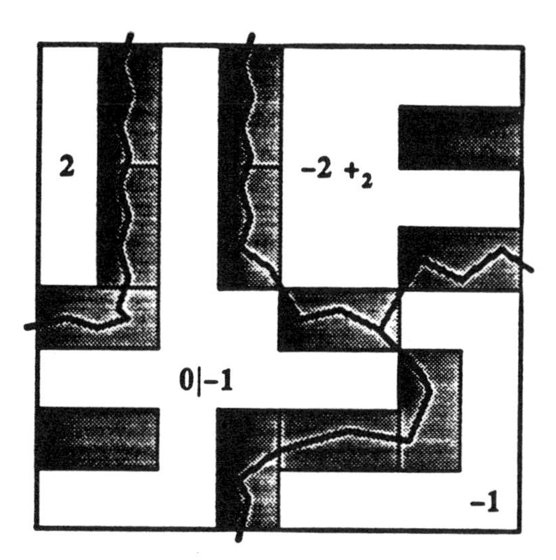

Figure 4: Some more complex values.

In general, we define a game, recursively, as a pair of sets of games. Symbolically, these two sets are separated by a vertical slash:

$$G = \{G^L \mid G^R\}$$

In words, G is defined as the game whose Left followers comprise the set G^L and whose Right followers comprise the set G^R. We usually simply denote these sets by listing their elements, separated by commas.

The simplest game is the one whose sets of Left and Right followers are both empty. This game is denoted by "0":

$$0 = \{ \mid \}$$

In general, we may define the **birthday** of a game as one more than the birthday of any of its followers, with the starting proviso that the game 0 has birthday 0. There are then three different games born on Day 1:

$$\begin{aligned} 1 &= \{0 \mid \} \\ -1 &= \{ \mid 0\} \\ * &= \{0 \mid 0\} \end{aligned}$$

The game "$*$" is called "star." It appears in Domineering as

$$\begin{aligned} &= \left\{ \quad \middle| \quad \right\} \\ &= \{0 \mid 0\} \\ &= * \end{aligned}$$

Altogether, there are 4 games born on Day 0 or Day 1. A game born on or before Day 2 can have any subset of these 4 games as its Left followers and any subset of these four games as its Right followers. This gives $2^4 = 16$ possible sets of Left followers and 16 possible sets of Right followers, for 256 choices altogether. Among these 256 different choices we find such games as these:

$$\{-1 \mid 1\}$$
$$\{-1 \mid *\}$$
$$\{-1, * \mid\}$$

Each of these three games has the property that the second player can win. It is not hard to show that adding such a game into a sum has no effect on the outcome. More specifically, if I can win a game H, and if G is a game in which the second player can win, then I can also win on $G + H$. My strategy is simple: I never play on G except when my opponent does, and then I always respond immediately on G according to the winning strategy for the second player there.

This argument justifies calling *any* game which the second player can win a **Null** game, and denoting its value by zero:

$$\begin{aligned} \{-1 \mid 1\} &= 0 \\ \{-1 \mid *\} &= 0 \\ \{-1, * \mid \} &= 0 \end{aligned}$$

Similarly, we may define the **negative** of a game G as

$$-G = \{-G^R \mid -G^L\}.$$

It is not hard to show that $G + (-G)$ is a game which the second player can win, i.e., its value is zero. It is then reasonable to say that two games, G and H are **equal** if and only if $G - H$ is a game which the second player can win. This definition of equality partitions the 256 games born on or before Day 2 into 22 equivalence classes, such as the following:

$$\begin{aligned} 2 &= \{0, 1 \mid \} = \{1 \mid \} \\ 1/2 &= \{0 \mid 1\} = \{-1, 0 \mid 1\} \\ \uparrow = \text{“up”} &= \{0 \mid *\} = \{0, * \mid *\} \\ *2 &= \{0, * \mid 0, *\} \\ \pm 1 &= \{1 \mid -1\} \\ -1/2 &= \{-1 \mid 0\} = \{-1 \mid 0, 1\} \end{aligned}$$

The game $-1/2$ appeared in Figure 3:

$$\begin{aligned} &= \{-1 \mid 0 + 0, 1\} \\ &= \{-1 \mid 0\} \end{aligned}$$

It is now also possible to justify the reasonableness of our choices of names. To justify calling $\{0 \mid 1\}$ by the name "1/2", we should show that

$$\tfrac{1}{2} + \tfrac{1}{2} - 1 = 0$$

in the sense that second player can win from

$$\tfrac{1}{2} + \tfrac{1}{2} - 1.$$

We may also define inequality in the obvious manner.

$$G > H$$

means that Left can win on $G - H$ no matter who goes first, while

$$G \geq H$$

means that Left, going second, can win on $G - H$.

These basics, which I have just described intuitively, can also be stated more formally as the following axioms of combinatorial game theory. This formulation of the axioms is due to John Conway.

Conway's Axioms

1. Game
   $G = \{G^L \mid G^R\}$
2. Sum
   $G + H = \{G + H^L, G^L + H \mid G + H^R, G^R + H\}$
3. Minus
   $-G = \{-G^R \mid -G^L\}$
4. Partial Order
   $G \geq H$
   unlesss (unless and only unless)
   $H \geq$ some G^R
   or
   some $H^L \geq G$

To see the correspondence between Axiom 4 and the intuitive view that we discussed above, we might restate Axiom 4 as

$$G - H \geq 0 \quad \text{unlesss}$$
$$G^R - H \leq 0 \quad \text{for some } G^R$$
$$\text{or}$$
$$G - H^L \leq 0 \quad \text{for some } H^L.$$

This assertion can now be directly interpreted as saying that Left, going second on $G - H$, can win unless and only unless Right, going first, can win on $G - H$. For, if Right has a winning opening move on $G - H$, then it must be to a position of the form $G^R - H$ or $G - H^L$ and, in either case, such a position must then be one from which Right can win going second.

The axioms define an Abelian group of objects called games. One interesting subgroup of games, called **numbers**, can be defined as those games which satisfy two additional restrictions:

- all G^L and all G^R are numbers
- no $G^R \geq$ any G^L

If G^R is empty, then the value of the game is the least integer greater than all G^L; if G^L is empty the value of the game is the greatest integer less than all G^R. If both G^L and G^R are nonempty, then it turns out that the value of such a game is the **simplest** number in the interval between G^L and G^R.

Simplicity can be defined in several different ways which yield equivalent results. One approach is to say that among integers, the smaller the magnitude, the simpler the integer; among nonintegers, the smaller the power of two in the denominator, the simpler the number. Another approach is to define simplicity explicitly in terms of the game's birthday: the integers n and $-n$ are born on Day n; any fraction between 0 and 1 which is represented by an odd number divided by 2^{-n} is also born on Day n. The earlier the birthday, the simpler the game.

Abstractly, examples of numbers include the following:

$$\tfrac{1}{8} = \{0 \mid \tfrac{1}{4}\}$$

$$\cdots$$

$$2^{-(n+1)} = \{0 \mid 2^{-n}\}$$

$$\tfrac{3}{4} = \{\tfrac{1}{2} \mid 1\}$$

$$\tfrac{7}{8} = \{\tfrac{3}{4} \mid 1\}$$

$$\cdots$$

$$3 = \{2 \mid \}$$

$$4 = \{3 \mid \}$$

$$\cdots$$

$$1\tfrac{1}{2} = \{1 \mid 2\}$$

$$2\tfrac{3}{4} = \{2\tfrac{1}{2} \mid 3\}$$

$$\cdots$$

And here are some numbers that are not in canonical form:

$$\tfrac{1}{2} = \{0 \mid \tfrac{3}{4}\}$$

$$1 = \left\{0 \mid \{1 \mid 1\}\right\}$$

$$\tfrac{3}{4} = \left\{\tfrac{5}{8} \mid 1\right\}$$

Many domineering positions are numbers:

$\frac{3}{4}$ =

$\frac{7}{8}$ =

$1\frac{1}{2}$ =

0 =

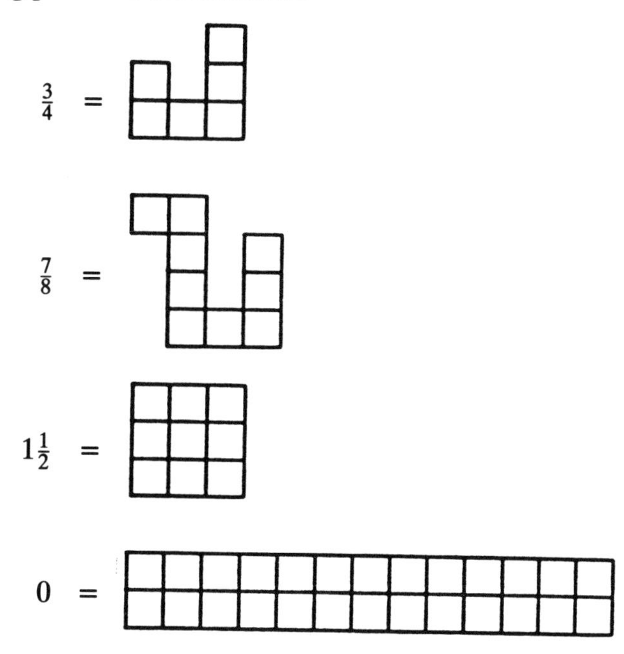

These assertions are by no means obvious. The proof that the 2×13 board has value 0 is well beyond the scope of the present introduction; it appears in "Blockbusting and Domineering."

It is instructive to examine the ordering of sums of some of the games born on or before Day 2. For example, we find that

$$* + * = 0$$

This means that, in some sense, $*$ has characteristic 2.

The game, $\{0 \mid *\} = \uparrow$ (pronounced "up"), and its negative, $\{* \mid 0\} = \downarrow$ (pronounced "down"), have several interesting properties. These games occur in Domineering:

$$
\begin{aligned}
&= \{0 + *, * + 0 \mid 0 + *, 0 + 0 + 0 + 0, * + 0\} \\
&= \{* \mid 0, *\} \\
&= \downarrow
\end{aligned}
$$

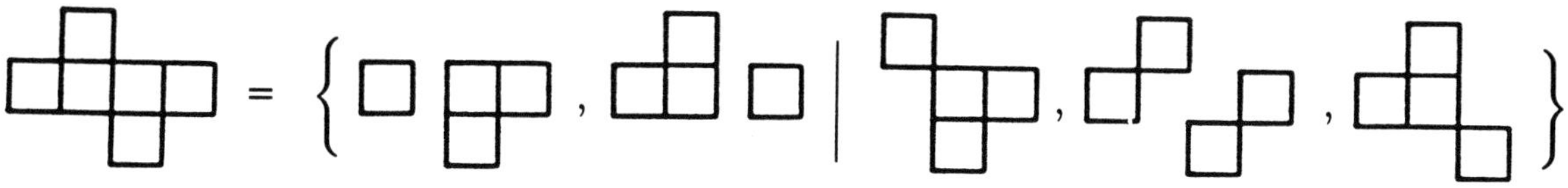

It is easily seen that Left can win the game, $\uparrow$, no matter who goes first. Hence,

$$\uparrow > 0.$$

Therefore,

$$0 < \uparrow < \uparrow + \uparrow < \uparrow + \uparrow + \uparrow < \cdots$$

It is convenient to denote these sums by

$$2 \cdot \uparrow = \uparrow + \uparrow$$
$$3 \cdot \uparrow = \uparrow + \uparrow + \uparrow$$
$$\cdots$$

If n is any positive integer, then we still find that

$$n \cdot \uparrow < 1.$$

In other words, $\uparrow$ is a positive infinitesimal and we write

$$0 < \uparrow \ll 1$$

$*$ is also an infinitesimal. In fact,

$$2 \cdot \downarrow < * < 2 \cdot \uparrow$$

although each of the games $*$, $* + \uparrow$, and $* + \downarrow$ is fuzzy.

The 4×2 rectangular Domineering game has the value

$$\{0 \mid \{0 \mid -2\}\}$$

We christen this game "tiny-2". More generally, if x is any nonnegative number we define the game

$$\text{"tiny-}x\text{"} = \{0 \mid \{0 \mid -x\}\}.$$

In particular, $\uparrow =$ "tiny-0".

It can be shown that, if $0 < x < y$, then

$$0 < \text{"tiny-}y\text{"} \ll \text{"tiny-}x\text{"}.$$

In particular, "tiny-2" is a higher order infinitesimal than $\uparrow$.

It turns out that there are many types of infinitesimals other than the tinys. However, $\uparrow$ has the lowest order of them all, in the sense that given any such infinitesimal, one can find an integer multiple of $\uparrow$ that is larger.

The position in the upper right corner of Figure 4 has value equal to

$$-2 + \text{"tiny-2"}$$

which we write as -2_{+2}. Just as elementary school teachers tell us to omit the implied "plus" sign in numbers like $2\frac{1}{4}$, game theorists find it convenient to omit plus signs between numbers and infinitesimals. This differs from the common convention in physics or engineering, where omitted arithmetic operators are assumed to be "times". In game theory, we assume omitted arithmetic operators are "plus".

The finite games which are numbers form a dense set in the real line and any finite game can be compared with the set of numbers. A typical game will be less than all sufficiently large positive numbers, greater than all sufficiently small negative numbers and "confused" with all numbers in some intermediate range (in the sense that the difference between the game under consideration and any number in that range is a "fuzzy" game which can be won by the first player). This range is called the "confusion interval". For some games, like $\uparrow$, the confusion interval can be empty: $\uparrow$ is less than all positive numbers but greater than zero and all negative numbers. For other games, the confusion interval may consist of a single number: $*$ is confused only with 0; it is greater than all negative numbers and less than all positive numbers.

However, some other games have bigger confusion intervals. Several such games appear on Day 2, including the following:

$$\{1 \mid -1\} \qquad \{0 \mid -1\} \qquad \{1 \mid *\}$$

The confusion intervals of these games are $[-1, 1]$, $[-1, 0]$, and $(0, 1]$, respectively. Games whose confusion intervals consist of the single number 0 are infinitesimals. Games whose confusion intervals contain more than one number are said to be "hot".

Many interesting hot games appear in Domineering, including those shown in Figure 5.

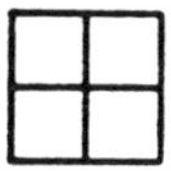

$$= \{1 \mid -1\} = \pm 1$$

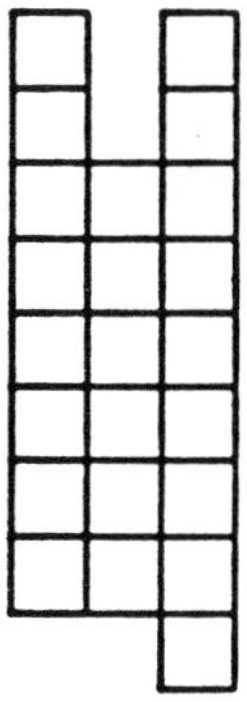

$$= \left\{5\tfrac{3}{4} \mid \left\{\{5\tfrac{1}{2} \mid 3\tfrac{1}{4}\} \mid \{3 \mid \tfrac{1}{2}\}\right\}\right\}$$

$$= 5\tfrac{3}{4} \,|||\, 5\tfrac{1}{2} \mid 3\tfrac{1}{4} \,||\, 3 \mid \tfrac{1}{2}$$

Figure 5:

(As seen in the second example in Figure 5, instead of writing multiple levels of brackets, we often find it convenient to omit all brackets entirely and to indicate the precedence of slashes by using multiple slashes. Double slash takes precedence over single slash; triple slash takes precedence over double slash, etc.)

We now have this complicated-looking expression:

$$5\tfrac{3}{4} \,|||\, 5\tfrac{1}{2} \mid 3\tfrac{1}{4} \,||\, 3 \mid \tfrac{1}{2}$$

This expression becomes tractable when it is viewed as an **overheated number**. Abstractly, such games are defined by applying "heating" and "overheating" operators to numbers. It turns out that, under appropriate restrictions too complicated to detail here, such operators are linear and order-preserving. To emphasize the linearity of these operators, we denote them by integral signs. Since combinatorial game theory is totally discrete, there is no risk of confusion with the conventional meanings of integrals.

Heating and overheating are discussed in Chapter 6 of *Winning Ways.* In "Blockbusting and Domineering," those concepts are refined and extended to the following:

Heating and Overheating

Let G be a game whose canonical form is

$$G = \{G^L \mid G^R\}$$

then we define "G heated by t" as

$$\int^t G = \begin{cases} G, & \text{if } G \text{ is a number} \\ \{t + \int^t G^L \mid -t + \int^t G^R\}, & \text{otherwise} \end{cases}$$

and for $s > 0$, we define "G overheated from s to t" as

$$\int_s^t G = \begin{cases} Gs = \overbrace{s + s + \cdots + s}^{G \text{ times}}, & \text{if } G \text{ is an integer} \\ \left\{t + \int_s^t G^L \mid -t + \int_s^t G^R\right\}, & \text{otherwise.} \end{cases}$$

For example,

$$\begin{aligned} \int_1^{1*} \tfrac{3}{4} &= \int_1^{1*} \{\tfrac{1}{2} \mid 1\} \\ &= \left\{1* + \int_1^{1*} \tfrac{1}{2} \,\middle\|\, -1* + \int_1^{1*} 1\right\} \\ &= \left\{1* + \int_1^{1*} \{0 \mid 1\} \,\middle\|\, -1* + 1\right\} \\ &= \left\{1* + \left\{1* + \int 0 \,\middle|\, -1* + \int 1\right\} \,\middle\|\, *\right\} \\ &= \left\{1* + \{1* \mid *\} *\right\} \\ &= 2 \mid 1 \,\|\, *. \end{aligned}$$

The heating and overheating allow us to express many Domineering positions in another form. For example,

$$5\tfrac{3}{4} \,|||\, 5\tfrac{1}{2} \,|\, 3\tfrac{1}{4} \,||\, 3 \,|\, \tfrac{1}{2} = 4\tfrac{5}{8} + \int^{9/8} * - \int^{9/8} \int_{1/4}^{1/4*} \tfrac{7}{8}.$$

In "Blockbusting and Domineering," it is shown that the values of many positions which occur in $n \times 3$ Domineering can be expressed explicitly in the form

$$\underline{a} + \int^{9/8} * + \int^{9/8} \int_{1/4}^{1/4*} \underline{b}$$

or as

$$\underline{a} + \int^{9/8} \int_{1/4}^{1/4*} \underline{b}$$

where $\underline{a}$ and $\underline{b}$ are numbers.

Similarly, many positions which occur in $n \times 2$ Domineering can be expressed explicitly in the form

$$\underline{a} + \int^{3/4} * + \int^{3/4} \int_{1/2}^{1/2*} \underline{b}$$

or as

$$\underline{a} + \int^{3/4} \int_{1/2}^{1/2*} \underline{b}$$

In both $n \times 3$ and $n \times 2$ Domineering, the $\underline{b}$'s, the numbers which appear under the heating and overheating operators, are essentially the same as the numbers which occur in a simpler, but closely related, game called **Blockbusting**.

Blockbusting is a partizan game which Left and Right play on an $n \times 1$ strip of squares called **parcels.** Each player, in turn, claims one previously unclaimed parcel and colors it with his color: bLue for Left; Red for Right. The game ends when all parcels have been colored and Left's score is then equal to the number of parcel boundaries which have been colored blue on both sides. No points are awarded for blue-red or red-red adjacencies. Evidently, Left seeks to maximize the number of blue-blue adjacencies while Right seeks to minimize this number.

Three types of positions occur in Blockbusting: LnL, LnR, RnR, which denote a strip of n squares bordered on both ends by blue, a strip of n squares bordered on opposite ends by different colors and a strip of n squares bordered on both ends by red. The respective values of these positions are

$$n \cdot * + \int_1^{1*} x_n, \qquad n \cdot * + \int_1^{1*} y_n, \qquad n \cdot * + \int_1^{1*} z_n$$

where $n \cdot * = 0$ or $*$ according as n is even or odd and x_n, y_n, z_n are as in the following table:

n	x_n	y_n	z_n
0	1	0	0
1	$1*$	$\frac{1}{2}$	0
2	1	$\frac{3}{4}$	0
3	$1\frac{1}{2}$	$\frac{7}{8}$	$\frac{1}{2}$
4	$1\frac{3}{4}$	1	$\frac{1}{2}$
5	$1\frac{7}{8}$	$1\frac{1}{4}$	$\frac{3}{4}$
6	2	$1\frac{1}{2}$	1
7	$2\frac{1}{4}$	$1\frac{3}{4}$	1
8	$2\frac{1}{2}$	$1\frac{7}{8}$	$1\frac{1}{2}$

Except for the initial values shown above the dark line, these numbers are arithmetic-periodic with period 5 and saltus 1.

By evaluating the overheated numbers, the solution of Blockbusting can also be expressed as follows:

n	LnL	LnR	RnR
0	1	0	0
1	$2 \mid 0$	$1 \mid 0$	$*$
2	1	$2 \mid 1 \parallel *$	0
3	$2 \mid 1$	$3 \mid 2 \parallel 1* \mid\mid\mid 0$	$1 \mid 0$
4	$3 \mid 2 \parallel 1*$	1	$1* \mid *$
5	$4 \mid 3 \parallel 2* \mid\mid\mid 1$	$2 \parallel 1* \mid *$	$2* \mid 1* \parallel 0$
6	2	$2* \mid 1*$	1

References

1. E. R. Berlekamp, J. H. Conway, & R. K. Guy, *Winning Ways for Your Mathematical Plays,* Academic Press, London, 1982.

2. E. R. Berlekamp, Two-person, perfect-information games, *The legacy of John von Neumann (Hempstead, NY, 1988)*, 275–287, *Proc. Sympos. Pure Math.,* **50**(1990), American Mathematical Society, Providence, RI; MR **91g**:90180.

3. E. R. Berlekamp, Blockbusting and domineering, *J. of Combin. Theory Ser. A,* **49**(1988) 67–116.

University of California
Berkeley, CA 94720

A Generating Function for the Distribution of the Scores of all Possible Bowling Games*

Curtis N. Cooper & Robert E. Kennedy

1 Introduction

Assuming that you know the terminology, rules of play, and the method of scoring, we will determine the number of ways that any particular score can occur in an ordinary game of ten-pin bowling. For example, it is clear that there is exactly one way that a perfect score of 300 can be obtained (all strikes) and exactly one way that a score of 0 can be obtained (all gutter balls). It may not be easy to see, however, that there are exactly 50613244155051856 ways of obtaining a score of 100.

Consider the following "line" of a bowling game:

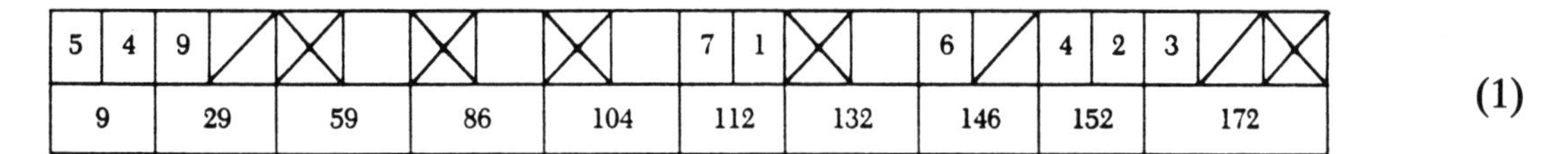

(1)

The lower number in each frame is the cumulative score. The upper numbers in each frame are the number of pins knocked down by the first and second ball respectively. The "/" indicates that the remaining pins were knocked down by the second ball and is called a "spare." The "X" indicates that all ten pins were knocked down by the first ball and is called a "strike."

Note that the game given in (1) can be represented by the sequence of nine ordered pairs and an ordered triple,

$$(5,4),(9,1),(10,0),(10,0),(10,0)(7,1),(10,0)(6,4),(4,2),(3,7,10). \tag{2}$$

With (2) as a model, we define a "bowling game" as a sequence of the form

$$(x_1,y_1),(x_2,y_2),\ldots,(x_9,y_9),(x_{10},y_{10},z_{10}), \tag{3}$$

where the terms of (3) represent the ten frames of a game and each component of a term denotes the number of pins knocked down by that ball. Here, x_i, y_i, and z_{10} are nonnegative integers where

$$x_i + y_i \leq 10 \qquad \text{for } i = 1,\ldots,9, \tag{4}$$

*Reprinted with permission from *Mathematics Magazine,* vol. 63, 4(October 1990) pp. 239–243. ©1990 The Mathematical Association of America.

and with somewhat more involved conditions given in (5) on x_{10}, y_{10}, and z_{10}.

It was shown in [1] that the number of all possible bowling games is $(66^9)(241)$ which is approximately 5.7×10^{18} and that the mean of all possible bowling games is approximately 80.

Thus, to determine the exact distribution of all possible bowling games by a computer generation of all possibilities would require in excess of 180 years even if every computer operation would take less than one-billionth of a second to perform. We will avoid this problem by constructing a generating function which determines the distribution of the scores of all possible bowling games. To do this, we will use the following sets where all components are nonnegative integers:

$$\begin{aligned} A &= \{(x,y) : x + y \leq 9\} \\ B &= \{(x,y,0) : (x,y) \in A\} \\ &\quad \cup \{(x, 10-x, z) : x \leq 9; z \leq 10\} \\ &\quad \cup \{(10, y, z) : y \leq 9; y + z \leq 10\} \\ &\quad \cup \{(10, 10, z) : z \leq 10\}. \end{aligned} \tag{5}$$

Therefore, A is the set of frames in which a "mark" (spare or strike) is not made, and the set B is the set of all possibilities for the tenth frame. We wish to determine a polynomial function of the form

$$P(t) = \sum_{i=0}^{300} s_i t^i, \tag{6}$$

where s_i if the number of ways that a score of i is made. For example, $s_0 = 1$, $s_{300} = 1$, and $s_{100} = 506132441550518 56$. Thus, $P(t)$ is a generating function for the set of all possible bowling scores.

2 States

To aid in finding the function given in (6), we define four "states" in the process of calculating the score of a game. They are called the "OPEN state," the "SPARE state," the "STRIKE state," and the "DOUBLE state." These are defined as follows:

1. The OPEN state describes that the current frame is open. (7)
2. The SPARE state describes that a spare has been made in the current frame.
3. The STRIKE state describes that a strike has been made in the current frame and that either an open or spare was made in the previous frame.
4. The DOUBLE state describes that a strike was made in both the current and previous frames.

We observe here that all capital letters are used in reference to a state so as to emphasize that the state of a frame is not the same as what was rolled in that frame. For example, a frame in which a strike is rolled is not necessarily in the STRIKE state since the previous frame may not have been an open or a spare.

As we bowl a game and keep score, we pass from one state to another state. By convention, we will assume that each game starts with a 0th frame which is in the OPEN state with an accumulated score of 0. To clarify the above terminology, consider the bowling line given in (1). The 0th, 1st, 6th, and 9th frames are in the OPEN state, the 2nd and 8th frames are in the SPARE state, the 3rd and 7th frames are in the STRIKE state, while the 4th and 5th frames are in the DOUBLE state. Since the current state of a frame will determine the contribution of the next frame to the accumulated score, we do not need to define a state for the 10th frame.

3 Generating functions of transitions

Here, we will determine the generating functions for the 16 possible transitions from one state to another state. First, it is clear that the transitions

OPEN	to	DOUBLE,	
SPARE	to	DOUBLE,	(8)
STRIKE	to	STRIKE,	

and

DOUBLE to STRIKE

cannot occur and hence may be considered as having a generating function of 0. To determine the generating function of a transition where (x, y) is the second state, we define the "value" of this transition as the contribution of the second state to the accumulated score. The following list gives the value of each of the other 12 transitions in terms of x and y.

OPEN	to OPEN	has a value of	$x + y$	
OPEN	to SPARE	has a value of	10	
OPEN	to STRIKE	has a value of	10	
SPARE	to OPEN	has a value of	$2x + y$	
SPARE	to SPARE	has a value of	$x + 10$	
SPARE	to STRIKE	has a value of	20	
STRIKE	to OPEN	has a value of	$2x + 2y$	(9)
STRIKE	to SPARE	has a value of	20	
STRIKE	to DOUBLE	has a value of	20	
DOUBLE	to OPEN	has a value of	$3x + 2y$	
DOUBLE	to SPARE	has a value of	$x + 20$	
DOUBLE	to DOUBLE	has a value of	30	

Thus considering the information in (8) and (9), we see that the generating functions for the 16 transitions are given by the following transition matrix, T.

$$\begin{pmatrix} \sum_{(x,y)\in A} t^{x+y} & 10t^{10} & t^{10} & 0 \\ \sum_{(x,y)\in A} t^{2x+y} & \sum_{x=0}^{9} t^{x+10} & t^{20} & 0 \\ \sum_{(x,y)\in A} t^{2x+2y} & 10t^{20} & 0 & t^{20} \\ \sum_{(x,y)\in A} t^{3x+2y} & \sum_{x=0}^{9} t^{x+20} & 0 & t^{30} \end{pmatrix} \tag{10}$$

The rows of the matrix T represent the first state while the columns of T represent the second state in the order OPEN, SPARE, STRIKE, and DOUBLE.

For example, the entry in the third row and second column of T is the generating function for the transition STRIKE to SPARE.

In addition, the column matrix

$$C = \begin{pmatrix} \sum_{(x,y,z)\in B} t^{x+y+z} \\ \sum_{(x,y,z)\in B} t^{2x+y+z} \\ \sum_{(x,y,z)\in B} t^{2x+2y+z} \\ \sum_{(x,y,z)\in B} t^{3x+2y+z} \end{pmatrix} \tag{11}$$

represents the contribution made by the 10th frame depending on whether the 9th frame is in the OPEN, SPARE, STRIKE, or DOUBLE state.

Since T^9 will be the matrix that gives the generating functions for all possible scores commencing with a given state and terminating with another state through nine transitions, it follows that the generating function $P(t)$ in (6) will be the entry in the one-by-one matrix

$$RT^9C \tag{12}$$

where $R = (1,0,0,0)$.

Fortunately, we do not have to actually calculate and simplify the matrix expression in (12). An Apple IIe Pascal program was written which uses (12) and determines the coefficient of each term of (6) and hence finds the exact number of ways that each bowling score can occur. This program is available upon request. The distribution of all possible bowling scores generated by this program is listed in appendix A.

APPENDIX A

Distribution of Bowling Scores

Score	Count	Score	Count	Score	Count
0	1	50	11193770355829009	100	50613244155051856
1	20	51	13810930667765157	101	45887089510794122
2	210	52	16878453276117746	102	41483436078768079
3	1540	53	20435326129713654	103	37397371704961189
4	8855	54	24515635362932954	104	33621048067136846
5	42504	55	29146610869639549	105	30144388614623696
6	177100	56	34346628376654913	106	26955619314626157
7	657800	57	40123251227815383	107	24041709119775647
8	2220075	58	46471404549689351	108	21388640692533960
9	6906900	59	53371780703441318	109	18981680119465910
10	20030010	60	60789577452586487	110	16805547548715206
11	54627084	61	68673668434334934	111	14844654231857239
12	141116637	62	76956298564663402	112	13083276623221517
13	347336412	63	85553384395717227	113	11505812292077067
14	818558424	64	94365480254213528	114	10096971927616045
15	1854631380	65	103279445170253902	115	8842020009154293
16	4053948342	66	112170812747354087	116	7726929590817265
17	8574134256	67	120906827121834566	117	6738528470417086
18	17590903116	68	129350064451661348	118	5864552560171552
19	35084425512	69	137362512979745598	119	5093653838062639
20	68153183370	70	144809940796620325	120	4415377510495980
21	129156542039	71	151566341291631624	121	3820097597373727
22	239128282128	72	157518221668013078	122	3298981687014508
23	433093980298	73	162568486673578693	123	2843905747206868
24	768175029950	74	166639683923175378	124	2447444695948898
25	1335679056261	75	169676402232105648	125	2102793053565659
26	2278764308864	76	171646676234883305	126	1803790254604935
27	3817721269708	77	172542309343731946	127	1544848145184291
28	6285424931278	78	172378125687965848	128	1320992367181792
29	10176048813473	79	171190226627438257	129	1127775864826813
30	16210652213304	80	169033430825208027	130	961294388171457
31	25423690787719	81	165978103316094584	131	818085023387881
32	39274771758064	82	162106654714921075	132	695128788327698
33	59789973730461	83	157509948809043576	133	589753122859383
34	89736657900900	84	152283892386077931	134	499630252931260
35	132834787033075	85	146526364181517039	135	422696870992462
36	194006223597572	86	140334651650668803	136	357151976811922
37	279661205716974	87	133803399444707801	137	301400973036441
38	398018151390200	88	127023103852577896	138	254052574077937
39	559449136091831	89	120079021507938035	139	213889601295347
40	776838931567572	90	113050455155943519	140	179862464456172
41	1065940588576732	91	106010240661754449	141	151065169242834
42	1445705502357343	92	99024411737621323	142	126722015973414
43	1938561121705315	93	92151904402003308	143	106169469752641
44	2570605432880903	94	85444345654857875	144	88840622360686
45	3371684590465908	95	78945863453573001	145	74252067274687
46	4375319099346208	96	72693023944120045	146	61990415093876
47	5618445228564793	97	66714881583314335	147	51701385089887
48	7140942201229333	98	61033240145235763	148	43082666091665
49	8984922304030443	99	55663091133973346	149	35870481552300

150	29843343433392	200	1526313637	250	37965
151	24808172866872	201	1239515641	251	31193
152	20607116162379	202	1007719386	252	26131
153	17101443169235	203	818568928	253	21406
154	14181008701762	204	666193896	254	17422
155	11747089496422	205	542061609	255	13613
156	9723545122578	206	442072320	256	10696
157	8040378083433	207	360234562	257	7975
158	6644452641044	208	293886739	258	6005
159	5486702080236	209	239045260	259	4374
160	4529003381568	210	194337731	260	3534
161	3736165201688	211	157306293	261	3016
162	3081105018158	212	127325163	262	2635
163	2539255963377	213	102799565	263	2264
164	2091793858275	214	83194097	264	1933
165	1721930513702	215	67300605	265	1603
166	1416734360140	216	54691522	266	1323
167	1164733232308	217	44477808	267	1045
168	957190045595	218	36317458	268	810
169	785911852914	219	29606794	269	585
170	645295369580	220	24117404	270	406
171	529489941608	221	19554213	271	277
172	434606120455	222	15820964	272	258
173	356481490646	223	12736481	273	227
174	292487050484	224	10258846	274	206
175	239755303889	225	8244157	275	173
176	196550315542	226	6659561	276	150
177	160954253448	227	5381526	277	115
178	131791387388	228	4385243	278	90
179	107847709116	229	3576841	279	53
180	88241591630	230	2930385	280	26
181	72162948863	231	2376760	281	15
182	59038079745	232	1924226	282	15
183	48284335855	233	1541327	283	14
184	39509743432	234	1231527	284	14
185	32308399043	235	975760	285	13
186	26423428886	236	777090	286	13
187	21582203262	237	617547	287	12
188	17624621529	238	498228	288	12
189	14368737009	239	404981	289	11
190	11720626558	240	335065	290	11
191	9552812749	241	275998	291	1
192	7790240907	242	226966	292	1
193	6351933169	243	183727	293	1
194	5185250585	244	148442	294	1
195	4232118751	245	117291	295	1
196	3457204258	246	93525	296	1
197	2821392492	247	73010	297	1
198	2302090127	248	57960	298	1
199	1874802017	249	45826	299	1
				300	1

References

1. C. N. Cooper & R. E. Kennedy, Is the mean bowling score awful?, *J. Recreational Math.* 18(3) (1985–86). Reprinted in this volume on pp. 151–154.

Department of Mathematics
Central Missouri State University,
Warrensburg, MO 64093-5045

Is the Mean Bowling Score Awful?*

Curtis N. Cooper & Robert E. Kennedy

As the title of this article suggests, in the following discussion we are concerned with the game of bowling. We will be primarily interested in the process of keeping score. We will not be able to help you improve your bowling average, but we might make you feel better about your average by calculating the mean bowling score of all possible games.

To clarify this, we will demonstrate how to calculate the score for a particular bowling game. Consider the sequence of frames:

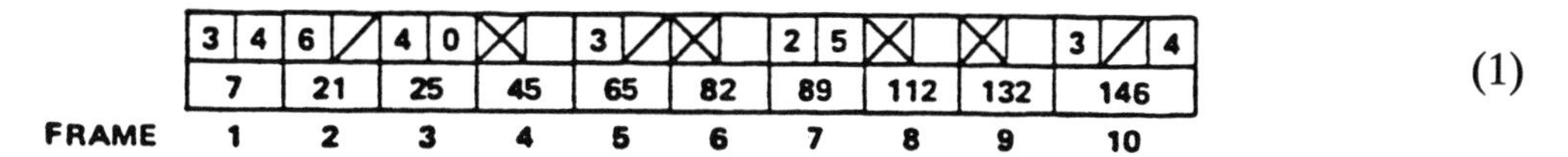

(1)

The lower number in each frame is the cumulative score. The upper numbers in each frame are the number of pins knocked down by the first and second ball, respectively. The "/" indicates that the remaining pins were knocked down by the second ball and is called a "spare." The "X" indicates that all ten pens were knocked down by the first ball and is called a "strike." The rules for calculating the cumulative score for a given frame are:

(2)

- For frames one through nine:

  (i) If neither a spare nor a strike is thrown, then the total number of pins knocked down by the two balls is added to the previous cumulative score.

  (ii) If a spare is thrown, then the number of pins knocked down by the next ball thrown, plus 10, is added to the previous cumulative score.

  (iii) If a strike is thrown, then the number of pins knocked down by the next two balls thrown, plus 10, is added to the previous cumulative score.

- For the tenth frame:

  (iv) The total number of pins knocked down is added to the previous cumulative score. This is enhanced by the convention that if a spare is thrown in the tenth frame, an extra ball is allowed, while if a strike is thrown, two extra balls are given as a bonus.

*Reprinted, with permission, from the *Journal of Recreational Mathematics,* vol. 18, no. 3, 1985–1986. ©1986 Baywood Publishing Co., Inc.

Most of these possibilities are exemplified in (1). Frame seven demonstrates (i), frame two demonstrates (ii), frame six demonstrates (iii), while frame ten demonstrates (iv) and the bonus given by making a spare in the tenth frame.

The final score of 146 is probably not a "bad" score for a person who bowls only occasionally, but is far from the "perfect" score of 300. Note that the game given by (1) can be represented by the sequence of nine ordered pairs and the ordered triple,

$$(3,4),(6,4),(4,0),(10,0),(3,7),(10,0),(10,0),(3,7,4). \tag{3}$$

With the above example as a model, we define a "bowling game," g, as a sequence of the form

$$(x_1,y_1),(x_2,y_2),\ldots,(x_9,y_9),(x_{10},y_{10},z_{10}), \tag{4}$$

where the terms of (4) represent the ten frames of a game and each component of a term of (4) denotes the number of pins knocked down by that ball. In what follows we use the following sets to facilitate the determination of the arithmetic mean of the scores of all possible bowling games. Here, all variables are nonnegative integers.

$$\begin{aligned} A &= \{(x,y) : x+y \le 10\}, \\ B &= \quad \{(x,y,0) : x+y \le 9\} \\ &\quad \cup \{(x,10-x,z) : x \le 9 \text{ and } z \le 10\} \\ &\quad \cup \{(10,y,z) : y \le 9 \text{ and } y+z \le 10\} \\ &\quad \cup \{(10,10,z) : z \le 10\}, \end{aligned} \tag{5}$$

and

$$G = \text{the set of all possible bowling games.}$$

Note that the first nine frames of any game are elements of A while the tenth frame is an element of B. Since the number of elements in A is 66 and the number of elements in B is 241, it follows that the number of elements in G is

$$\#(G) = (66^9)(241), \tag{6}$$

which is approximately 5.7×10^{18}.

For $i = 1,2,3,4,5,6,7$ and 8, consider the expression

$$x_i + y_i + \left[\frac{x_i+y_i}{10}\right] x_{i+1} + \left[\frac{x_i}{10}\right] y_{i+1} + \left[\frac{x_i+x_{i+1}}{20}\right] x_{i+2} \tag{7}$$

As usual, the square brackets indicate the integer part operator. We see that (7) is the number of pins added to the previous cumulative score to arrive at the cumulative score in frame i since (7) takes into consideration the definitions of sets A and B and the scoring rules (i), (ii), and (iii) given in (2). Likewise, the expression

$$x_9 + y_9 + \left[\frac{x_9+y_9}{10}\right] x_{10} + \left[\frac{x_9}{10}\right] y_{10} + x_{10} + y_{10} + z_{10} \tag{8}$$

is the number of pins added to the cumulative score of frame eight to give the final score, since (8) takes into consideration the definitions of A and B and the scoring rules (i), (ii), (iii), and (iv) given in (2).

Thus, the score, $s(g)$, of the game given by (4) is determined by the formula

$$\sum_{i=1}^{8} \left(x_i + y_i + \left[\frac{x_i + y_i}{10}\right] x_{i+1} + \left[\frac{x_i}{10}\right] y_{i+1} + \left[\frac{x_i + x_{i+1}}{20}\right] x_{i+2}\right)$$
$$+ x_9 + y_9 + \left[\frac{x_9 + y_9}{10}\right] x_{10} + \left[\frac{x_9}{10}\right] y_{10} + x_{10} + y_{10} + z_{10}. \tag{9}$$

We now proceed to calculate the arithmetic mean of the scores of all possible games. That is, we will determine the value of the quotient,

$$\frac{\sum_{g\in G} s(g)}{\#(G).} \tag{10}$$

To compute the numerator of (10), we will need the following equalities which follow immediately from the definitions of sets A and B. Their derivation will not be given here.

$$\begin{aligned} \textstyle\sum_{(x,y)\in A} x &= \textstyle\sum_{(x,y)\in A} y = 220 & \textstyle\sum_{(x,y,z)\in B} x &= 1420, \\ \textstyle\sum_{(x,y,z)\in B} y &= 1090, \qquad \text{and} & \textstyle\sum_{(x,y,z)\in B} z &= 825. \end{aligned} \tag{11}$$

Then by formula (9), after regrouping and changing the order of summation, we have

$$\begin{aligned} \sum_{g\in G} s(g) = {} & \sum_{i=1}^{9}\sum_{g\in g}(x_i + y_i) + \sum_{i=1}^{8}\sum_{g\in G}\left[\frac{x_i + y_i}{10}\right] x_{i+1} + \sum_{i=1}^{8}\sum_{g\in G}\left[\frac{x_i}{10}\right] y_{i+1} \\ & + \sum_{i=1}^{8}\sum_{g\in G}\left[\frac{x_i + x_{i+1}}{20}\right] x_{i+2} + \sum_{g\in G}\left[\frac{x_9 + y_9}{10}\right] x_{10} \\ & + \sum_{g\in G}\left[\frac{x_9}{10}\right] y_{10} + \sum_{g\in G}(x_{10} + y_{10} + z_{10}). \end{aligned} \tag{12}$$

From (11), and the definitions of sets A, B, and G, it follows that for $i = 1, 2, 3, \ldots, 9$,

$$\sum_{g\in G}(x_i + y_i) = 66^8(241)(440), \tag{13}$$

and for $i = 1, 2, 3, \ldots, 8$,

$$\sum_{g\in G}\left[\frac{x_i + y_i}{10}\right] x_{i+1} = 66^7(241)(11)(220), \quad \text{and} \quad \sum_{g\in G}\left[\frac{x_i}{10}\right] y_{i+1} = 66^7(241)(220), \tag{14}$$

and for $i = 1, 2, 3, \ldots, 7$,

$$\sum_{g\in G}\left[\frac{x_i + x_{i+1}}{20}\right] x_{i+1} = 66^6(241)(220). \tag{15}$$

The remaining terms of (12), which represent the tenth frame, are given by

$$\sum_{g\in G}\left[\frac{x_8+x_9}{20}\right]x_{10}=66^7(1420),\qquad \sum_{g\in G}\left[\frac{x_9+y_9}{10}\right]x_{10}=66^8(11)(1420),\tag{16}$$

$$\sum_{g\in G}\left[\frac{x_9}{10}\right]y_{10}=66^8(1090),\quad\text{and}\quad \sum_{g\in G}(x_{10}+y_{10}+z_{10})=66^9(3335).$$

Thus, from (13), (14), (15), and (16), we have that

$$\frac{\sum_{g\in G}s(g)}{\#(G)}=\frac{\begin{array}{c}9(66^8)(241)(440)+(8)(66^7)(241)(11)(220)+8(66^7)(241)(220)+\\(7)(66^6)(241)(220)+66^7(1420)+66^8(11)(1420)+66^8(1090)+66^9(3335)\end{array}}{66^9(241)},\tag{17}$$

which is approximately 80 (for precision bowlers, it's 79.7439...).

Thus, the mean bowling score is indeed awful even if you are just an occasional bowler. Even though this information is interesting, there are more difficult questions about the game of bowling that could be asked. For example, you might wish to determine the standard deviation of the set of bowling scores and hence know more about the distribution of the set of all bowling scores. But the exact determination of the distribution of the set of scores is, in our opinion, a difficult problem. For example, given an integer k between 0 and 300, how many different bowling games have the score k? This, we leave as an open problem.

Department of Mathematics,
Central Missouri State University,
Warrensburg, MO 64093

Recreation and Depth in Combinatorial Games*

Aviezri S. Fraenkel

Abstract

Combinatorial games terminating in a finite number of moves by one of the two players winning and the other losing, behave as expected: precisely one of the players has a winning strategy. Despite this tame front, combinatorial games present many challenging problems. In fact, they generally lie in higher complexity classes than existential problems such as the Traveling Salesman problem. A small selection of problems is discussed, which is designed, nevertheless, to give the typical flavor of the high complexities of combinatorial games.

1 Setting the Stage

For our purpose here, the class of *combinatorial games* consists of the collection of two-player games with perfect information (no hidden information such as in certain card games), no chance moves (no dice) and outcomes restricted to Lose, Win and, sometimes, Draw. Examples are Nim, Hex, Chess, Go, ...

The most basic result on combinatorial games (simply *games* in the sequel) is the following

FACT. In every game precisely one of the two players has a winning move, or else there is a winning move for neither player but both can maintain a draw.

Indeed, suppose that a certain position u_0 is neither a Previous-player-win (P*-position*), nor a Next-player-win (N*-position*). Then there is a *follower* u_1 (a position reachable from u_0 in one move) which is also neither a P nor an N-position. Because if any follower u_1 would be a P-position, then u_0 would be an N-position; and if all followers of u_0 would be N-positions, then u_0 would be a P-position. Moreover, the only non-losing move for a player moving from u_0 is to go to such u_1. Thus there is an infinite sequence $u_0, u_1, \ldots$ with u_{i+1} a follower of u_i, such that u_i is neither a P nor an N-position, and a best move from u_i is to go to a position of the form u_{i+1} ($i \geq 0$). In particular, the outcome is a draw. ∎

This Fact and the easy strategy for Nim suggest that combinatorial games are well-understood, if not dull. It seems to take perverse Kafka or Thomas Mann-like "Bedenken" to suspect any unconventional behavior from this respectable family, no member of which crosses the thresholds of Las Vegas and Atlantic City. It may therefore come as somewhat of a surprise to discover that there is an abundance of difficult problems and counterintuitive results in this area.

*Part of the work was done while visiting at the Universities of Calgary and Montréal. The author wishes to thank Professors Richard K. Guy and Anton Kotzig for partial support from their Canadian Research Grants NSERC 69-0695 and A-9232.

On second thought, the complexity of games is not really surprising: Many hundreds of existential decision problems, such as the Traveling Salesman problem, Hamiltonian circuit and various scheduling problems, are NP-complete. The question whether player I has a winning strategy in a given game (without even asking what the strategy *is*), takes the form: "Does there exist a move of player I such that for every move of player II there exists a move of player I such that for every move of player II, ..., there exists a winning move of player I?" This question involves a sequence of quantifiers, rather than a single existential quantifier as in the Traveling Salesman Decision problem ("Does there exist a tour of cost $\leq C$?"), which is NP-complete. This explains why games, if not polynomial, tend to be complete in complexity classes higher than NP; typically they are complete in Pspace, Extime or even Exspace. Thus we should perhaps, rather, be surprised at the above Fact, which seems to suggest that games are easy.

For the sake of the uninitiated we present informally some complexity issues. A problem π is in PO (POlynomial) if there is a fixed positive integer c so that the problem can be solved in time not exceeding $|\pi|^c$ ("polynomial time"), where $|\pi|$ is the input size of π. It is in NP (Nondeterministic Polynomial) if, *given* a solution to π, its validity can be verified in time polynomial in $|\pi|$. (We do not require, however, that *finding* a solution be polynomial.) The problem π is in Pspace (Polynomial Space) if, given a solution to π, its validity can be verified in (memory) space polynomial in $|\pi|$.

A problem π is *complete* in some class $\mathcal{C}$ of problems if it belongs to the hardest problems of $\mathcal{C}$. The NP-complete problems are *equivalent* in the sense that if one of them is polynomial, then all are; and if one has a lower nonpolynomial bound, then all are nonpolynomial. A major unsolved problem is which of these two possibilities holds, but in practice all NP-complete problems are nonpolynomial at present, since nobody has a polynomial solution for any of them—any such solution would imply NP = P.

Since P $\subseteq$ NP $\subseteq$ Pspace, the problems complete in Pspace are at least as hard as the NP-complete problems. The equivalence and doubt stated for NP-complete problems hold also for Pspace-complete problems. The problems complete in Extime (Exponential Time), have a *provably* exponential lower bound on their solution time. The same holds with respect to space for the Exspace-complete problems. Very few problems are known to be complete in Extime or Exspace. Examples of the former are Fischer & Rabin [1974], and for games: Fraenkel & Lichtenstein [1981]; and for the latter, Robson [1984]. For the general theory of computational complexity see Garey & Johnson [1979], and the NP-completeness columns of Johnson, e.g., Johnson [1983].

Below we mainly discuss problems of classifying games into their natural complexity classes. Though the selection of games is of necessity rather small, it should nevertheless give the flavor of the problems and results in this area. For further results and problems see e.g., Conway [1976], Berlekamp, Conway & Guy [1982], Garey & Johnson [1979; mainly Appendix A8], Johnson [1983] and Fraenkel [1980].

Figure 1 schematically depicts part of a classification of games into complexity classes, where "Wonderland" contains games we are still wondering about their true niche in the classification system. A line such as "Annihilation Games" connecting PO and Pspace-complete, indicates that some annihilation games are polynomial, others Pspace-complete.

The play of a game is called *normal* if the player making the last move wins. It is called *misère* if the player making the last move loses.

Probably the best-known and simplest combinatorial game is Nim: A finite collection of piles

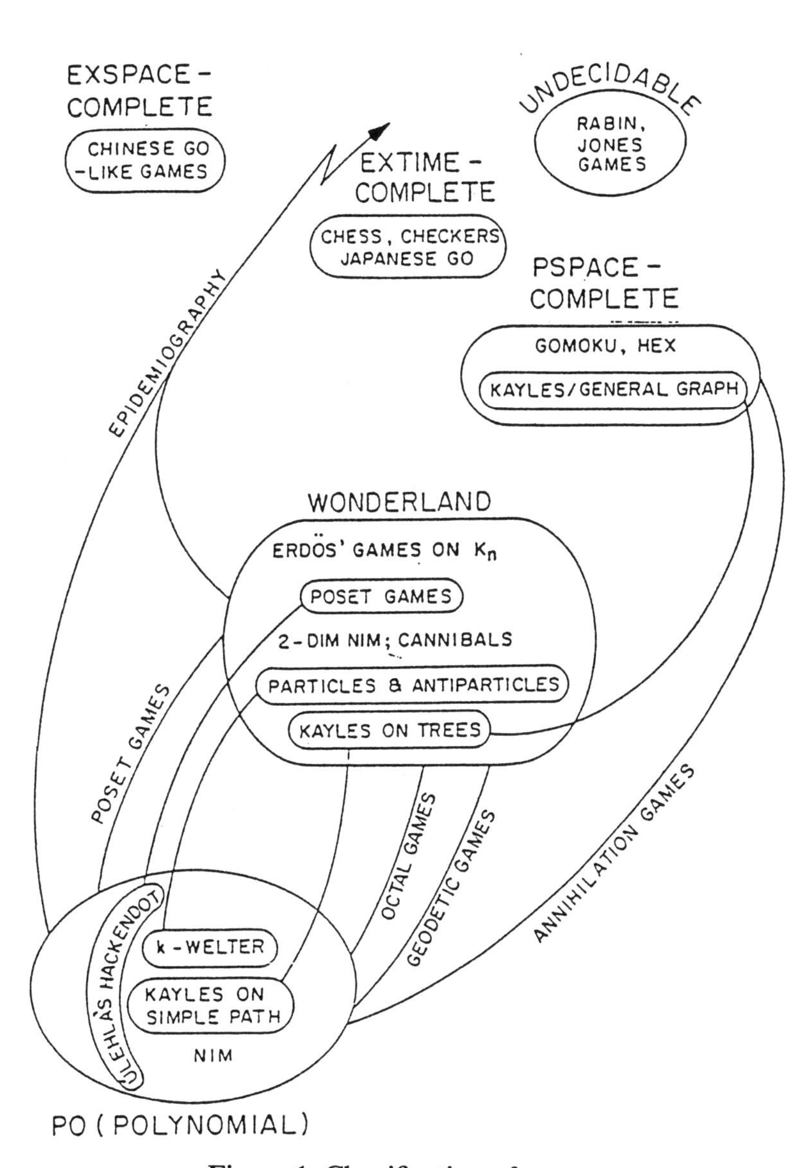

Figure 1: Classification of games.

$$
\begin{pmatrix}
1 & 1 & 1 & \cdots & 1 & 1 & 1 \\
1 & 1 & 1 & \cdots & 1 & 1 & 1 \\
 & \vdots & & & & \vdots & \\
1 & 1 & 1 & \cdots & 1 & 1 & 1 \\
1 & 1 & 1 & \cdots & 1 & 1 & 1 \\
0 & 1 & 1 & \cdots & 1 & 1 & 1
\end{pmatrix}
$$

Figure 2: The initial position of Gale's Nim.

of tokens is given. Each player at his turn removes tokens from a single pile, possibly the entire pile, but at least one token. For normal play, if the Nim-sum of the pile-sizes is 0, then the position is a P-position; otherwise it is an N-position. The *Nim-sum* of a collection of binary numbers is their XOR, i.e., addition mod 2 in each column.

2 Poset Games

The general setting of this subfamily of games is of two players alternately pointing to an as yet unremoved element of a given partially ordered set, removing the element together with all elements "larger" than it.

(i) **Gale's Nim.** This game is played on an $m \times n$ matrix all of whose elements are 1, except for the element at $(0,0)$, which is 0. A player pointing to element (i,j) removes all the "larger" elements, that is, he removes the entire north-east sector with south-west corner (i,j) (indicated by a dotted line in Figure 2). The main content of Gale [1974] is that player I can win in normal play unless $m = n = 0$. (Actually in Gale's version the element at $(0,0)$ is 1 and misère play is considered; but the argument below holds for both normal and misère play. We put 0 at $(0,0)$ only in order to have nontrivial normal play.)

Here is the argument: If player I can win by removing the largest element, namely (m,n), we are done. Otherwise taking (m,n) is a losing move. That is, player II has a winning response, e.g., removing (i,j). Then player I can take (i,j) in his first move and win. ∎

This proof has the following properties: (a) It is nonconstructive. In particular, it appears to be unknown at present whether player I has a polynomial winning strategy. (b) It is not, on the other hand, a proof by contradiction, as many such "strategy stealing" arguments go. It is perfectly possible that player I's winning move is to take an element other than the largest. In fact, computer experiments suggest that for normal play, player I's winning move is indeed usually other than taking the largest element. See also Gardner [1973, Jan., Feb., May], who calls the game "chomp".

The above argument that player I can win in both normal and misère play, applies to a number of generalizations. Here are some of them.

(a) The given $m \times n$ matrix may consist of 0's and 1's, with 1 at (m,n) and 0 at $(0,0)$. (In each move, at least one 1 is taken.)

(b) Instead of a 2-dimensional matrix we may like to play on an n-dimensional lattice parallelotope of size $d_1 \times d_2 \times \cdots \times d_n$ in the nonnegative octant, with 0 at the origin and 1 at $(d_1 - 1, \ldots, d_n - 1)$.

(c) Combining (a) and (b), we can construct sparse n-dimensional parallelotopes representing trees of maximal out-degree n. For trees there is actually a polynomial winning strategy (see (iii) below). Other special parallelotopes may be of interest.

(d) Each entry at a lattice point of the parallelotope may be an arbitrary nonnegative integer, with 0 at the origin and some positive integer at $(d_1 - 1, \ldots, d_n - 1)$. Each player points at a lattice point and decreases the integers of any nonempty subset of the "north-east" sector defined by the lattice point pointed at.

(ii) **The Power Game**. For fixed positive integers n and k, with $k \le n$, let

$$A_n^k = \{B \subseteq \{1, \ldots, n\} : 0 < |B| \le k\}.$$

A player at his turn points to an as yet unremoved set in A_n^k, removing it together with all the sets containing it. The argument given in (i) shows that player I can win A_n^n. However, for normal play, to which we now restrict attention, player II can clearly win A_3^2: If player I removes $\{3\}$ (and so also $\{13\}$ and $\{23\}$), player II removes $\{12\}$ and conversely (see Figure 3). More generally, the conjecture is that A_n^k is a P-position if and only if $n \equiv 0 \pmod{k+1}$.

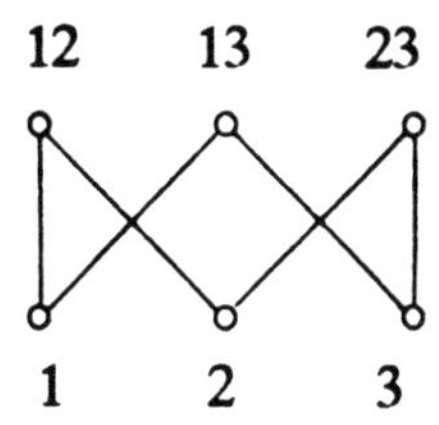

Figure 3: The game A_3^2.

Gale and Neyman [1982], who made this conjecture, proved it for the case $k = 2$. Apart from a few small additional cases settled by this proof, the general problem is still open. For example, the nature of the positions A_7^3 and A_6^5 is already unknown.

Note that if the conjecture is true, then the unique winning move in A_n^n is to take the largest element, much unlike the observed behavior in Gale's Nim. Further note that straightforward computer experiments seem hopeless even for small n and k, since the game graph has close to 2^{2^n} vertices. (The *game graph* of a game is a directed graph (*digraph*) whose vertices are the game positions, and there is a directed edge (u, v) if and only if there is a move from position u to position v.)

The Sprague–Grundy function (g-function for short) assigns to every vertex of a digraph a nonnegative value, which is the smallest nonnegative integer not appearing among its followers. Thus the g-function of the game graph of a Nim-pile of size n is n. The 0's of g are the P-positions of the game. Moreover, the g-function of a game which is a finite disjoint sum of games, such as several Nim piles, is the Nim-sum of the g-function of the component games. For more on the g-function, see e.g., Berge [1985], Berlekamp, Conway & Guy [1982] and Conway [1976].

The g-function of the first few values of the power-game has been computed at the end of Fraenkel & Scheinerman [1989]. Based on it, one is tempted to make the stronger conjecture that $g(A_n^k)$ is the smallest nonnegative remainder obtained when dividing n by $k+1$. This is in fact true for $k \le 2$, and perhaps even for $k = 3$. But it is wrong for $k = 4$, since $g(A_4^4) = 1$, as can be verified easily by hand-computation. This and another indicator—non-boundedness of the g-function on hypergraphs—perhaps cast some doubt on the Gale-Neyman conjecture.

(iii) **Hackendot**. Given a forest of rooted trees, each tree directed away from the root. Player I selects a vertex u in one of the trees T with root r, removing the unique path of vertices $(r, \ldots, u)$, leaving a forest T^u. The case where the initial forest is a tree is depicted in Figure 4 (with directions suppressed). Von Neumann pointed out that the first player can win (follows in the same way as for Gale's Nim, where now the largest element is the tree's root). Unlike the lack of constructive knowledge on the two preceding games, however, there exists a rather interesting polynomial strategy for this game, due to Úlehla [1980]: For a forest F, define the "Root Imparity" function

$$\mathrm{Rip}(F) = \begin{cases} 0 & \text{if the number of roots which are } P\text{-positions of the graph } F \text{ is even} \\ 1 & \text{otherwise.} \end{cases}$$

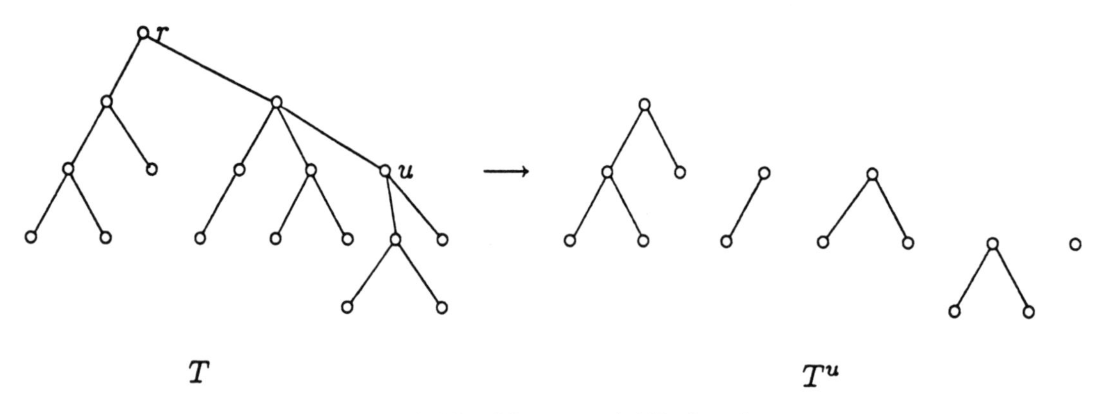

Figure 4: Von Neumann's Hackendot.

Also define an operator L: the vertices of LF are the vertices which are N-positions of the graph F, and (u, v) is an edge of LF if it was an edge in F or if there was an intervening P-vertex (which has been omitted in the transformation $F \to LF$). Then the Sprague-Grundy function g of the game is given by

$$g(F) = \sum_{n \geq 0} 2^n \mathrm{Rip}(L^n F).$$

Example. Let F be a simple path p of length n. Hackendot on p is clearly the same as Nim on a Nim-pile of size n. Úlehla's algorithm applied to the case $n = 6$ is illustrated in Figure 5, which yields $g(p) = 6$ as desired.

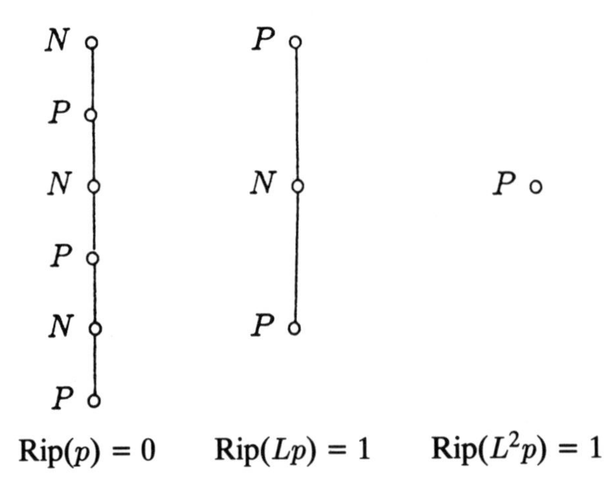

Figure 5: Úlehla's algorithm applied to a simple path.

Unfortunately, Úlehla's algorithm, the validity proof of which is not very intuitive, does not seem to generalize easily even to a DAG (Directed Acyclic Graph).

(iv) **Complexity of Poset Games.** Each of the games described in (i)–(iii) has a natural representation as a DAG, called *poset graph*. Its vertices are the poset elements, and there is a directed edge (u, v) if and only if $u > v$ and there is no w such that $u > w > v$. Here $>$ denotes the order relation of the partial order. The poset graphs of Gale's Nim for $m = 2, n = 3$ and of the power game A_3^3 are shown in Figure 6. The poset graph of Hackendot is the given forest of the game.

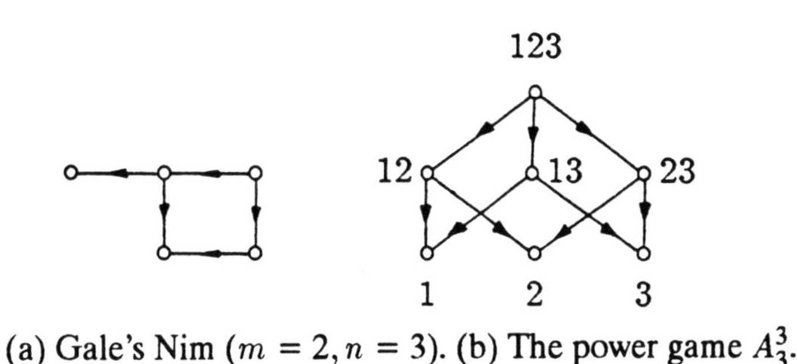

(a) Gale's Nim ($m = 2, n = 3$). (b) The power game A_3^3.

Figure 6: Poset graphs.

With any poset we may associate the *transitive closure graph* $G = (V, E)$ of its poset graph. This is a digraph whose vertices V comprise the poset elements, and $(u, v) \in E$ if and only if $u > v$. The transitive closure graphs of the poset graphs of Figure 6 are shown in Figure 7.

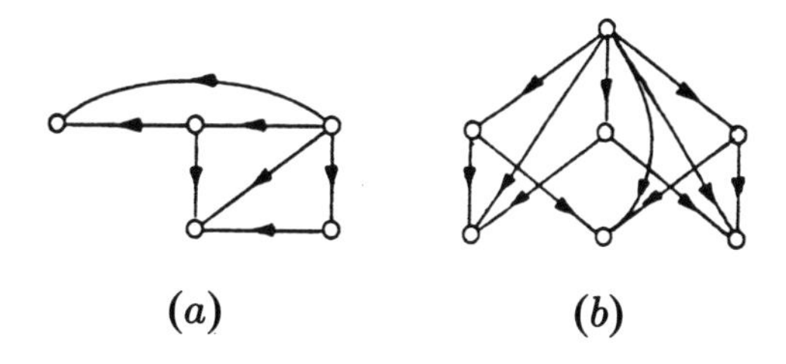

(a) (b)

Figure 7: The transitive closure graphs of the above poset graphs.

The graph *Directed Kayles* is played on a finite DAG, $G = (V, E)$. A move consists of labeling an as yet unlabeled vertex u, provided none of its followers has yet been labeled; that is, an unlabeled vertex u can be labeled if and only if v is unlabeled for every $(u, v) \in E$. The game *Kayles* is played similarly on a finite undirected graph, where a move consists of labeling an as yet unlabeled vertex, none of whose neighbors has yet been labeled. Schaefer [1978] has shown that the decision problem of whether player I can win in Kayles in normal play is Pspace-complete. A simple (identity) reduction from Kayles shows that Directed Kayles on a cyclic digraph is also Pspace-complete in normal play. The thing to note is that any poset game is equivalent to Directed Kayles played on the transitive closure graph of the poset graph.

Recasting the known facts and the many problems of poset games in the language of the complexity of certain digraph games, we can say that Úlehla's result shows that Directed Kayles is polynomial when played on the transitive closure of forests. It is Pspace-complete on general cyclic digraphs. Poset games, whose transitive closure graphs constitute more general DAG's than closure of forests, lie somewhere in the middle. Where precisely?

3 Geodetic Games

Let $G = (V, E)$ be a finite graph. The set L of labeled vertices is initially empty. Two players I and II move alternately, by choosing an unlabeled vertex $u \in V - L$; then u itself and all vertices on shortest paths between u and any vertex of L are adjoined to L. When $L = V$, the game is over.

The following results are known (Fraenkel & Harary [1989]).

(i) If G is a tree, then normal play can be reduced to Úlehla's Hackendot. But we cannot extend this polynomial strategy even to forests.

(ii) If $G = C_n$ is a simple cycle with n vertices, then player I can win if and only if $n = 2^k - 1$ in normal play; $n = 2^k$ in misère play.

(iii) If $G = K_m \circ R_n$, where R_n consists of rays of length n emanating from each of the m vertices of the complete graph K_m, then player II can win polynomially for all $m, n \geq 2$! (The graph $K_3 \circ R_2$ is depicted in Figure 8.)

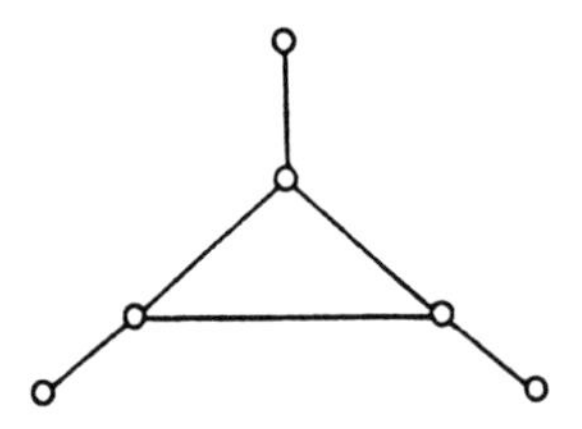

Figure 8: $K_3 \circ R_2$.

The case of different length rays emanating from K_m can probably also be settled polynomially but for general graphs the problem is open.

4 Epidemiography

Let $f\colon \mathcal{Z}^+ \to \mathcal{Z}^0$ be a function from the natural numbers into the nonnegative integers. Let $G = G_1$ be a digraph with a uniquely labeled vertex u_1. *Dancing Mania* on G is played as follows. Player I begins by removing the label from u_1, labeling in G_1 a follower u_2 of u_1, and adjoining $f(1)$ isomorphic copies of G_1, in each of which u_2 is the unique labeled vertex. The players alternate turns. At step k, a player selects a labeled vertex u_i in any one of the existing copies of G, removes the label from u_i, labels some follower u_{i+1} of u_i, and adjoins $f(k)$ isomorphic copies of G in each of which u_{i+1} is the unique labeled vertex. The player first unable to move loses and his opponent wins, since he manages to kill the epidemic.

A particularly simple case is when G is a simple directed path, giving rise to the game called *Nimania*. Consider Nimania for the case $f(k) = k$. It is equivalent to the following game: Given a positive integer n. In his first move, player I subtracts 1 from n. If $n = 1$, the result is the empty set and the game ends with player I winning. If $n > 1$, one additional copy of the resulting number $n - 1$ is adjoined, so at the end of the first move there are two (indistinguishable) copies of $n - 1$ (denoted by $(n-1)^2$). At the kth stage, a player selects a copy of a positive integer m of the present position, and subtracts 1 from it. If $m = 1$, the copy is deleted. If $m > 1$, k copies of $m - 1$ are adjoined to the resulting $m - 1$ ($k \geq 1$). We consider normal play throughout.

Examples. (i) $n = 1$. As we saw above, player I wins. (ii) $n = 2$. Player I moves to 1^2, player II to 1, hence player I again wins. (iii) $n = 3$. A self-explanatory diagram (Figure 9) shows that again player I can win. Unlike the above two cases, however, not all moves of player I are winning: Player I has to select his moves carefully to win, following the lower branches of the diagram.

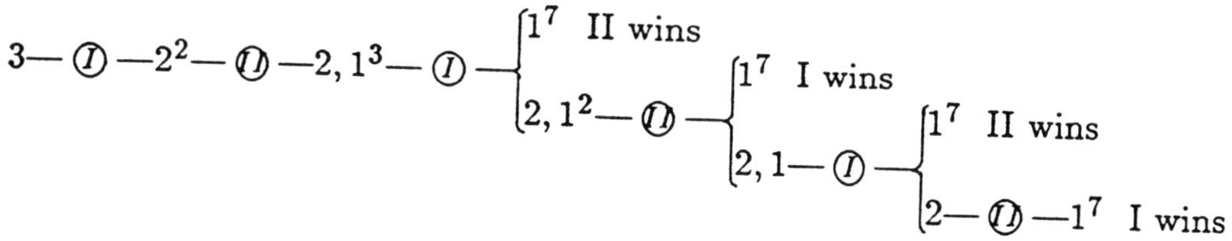

$$3 - \textcircled{I} - 2^2 - \textcircled{II} - 2,1^3 - \textcircled{I} - \begin{cases} 1^7 \ \text{II wins} \\ 2,1^2 - \textcircled{II} - \begin{cases} 1^7 \ \text{I wins} \\ 2,1 - \textcircled{I} - \begin{cases} 1^7 \ \text{II wins} \\ 2 - \textcircled{II} - 1^7 \ \text{I wins} \end{cases} \end{cases} \end{cases}$$

Figure 9: A proof that player I wins $n = 3$ of Nimania. The numbers in circles indicate the player making the move.

An attempt to resolve the problem for $n = 4$ by a similar diagram construction is rather frustrating, because of the size and the many branches of such a diagram. In fact, the longest branch has now length $41 \times 2^{39} - 2$. In general, for $f(k) = k$, the maximum length of Dancing Mania on a finite connected DAG is an Ackerman function. In Fraenkel & Nešetřil [1985] it is proved, nevertheless, that for $f(k) = k$, Nimania is a first player win for every $n \geq 1$!

Nimania$_p$ is played as Nimania, except that in the former we may label any vertex lying on a simple path up to p edges away from a presently labeled vertex. Consider Nimania$_2$. For $n = 1$, player I clearly wins. For $n = 2$, player I can move to 0 winning swiftly, or he can move to 1^2, in which case he also wins, though only after having prolonged the malady. For $n = 3$ player I can move to 1^2 and again win.

What is the prognosis for Nimania$_2$? Can player I win again for all n? Surprisingly, the above cases are the only ones for which player I can win. For every $n > 3$, player II can in fact win!

Dancing Mania on an epidemiograph is called *Common Mania*, or *Coma* for short. The term *epidemiograph*, coined by Joel Spencer, designates a connected finite DAG. In Fraenkel, Loebl & Nešetřil [1988] it is shown that Coma terminates for every function $f\colon \mathcal{Z}^+ \to \mathcal{Z}^0$, even if f grows arbitrarily fast, and the following is shown.

A function f will be called *parity preserving* [*reversing*] if $f(k) \equiv k \pmod 2$ [$k+1 \pmod 2$] for all $k \in \mathcal{Z}^+$. Roughly speaking, if the epidemiograph G on which Coma is played contains no triangle on its leaves (no a, b, c with (a, b), (b, c), (a, c) edges, $d_{\text{out}}(c) = 0$), and if f is parity preserving [reversing], then player I [II] can win. A special case is Nimania. If, on the other hand, every leaf has a triangle, and if f is parity preserving [reversing], then player II [I] can win. A special case is Nimania$_p$ for every $p \geq 2$.

These results have the interesting property that the P, N-pattern depends only on the local structure near the leaves of the epidemiograph. Moreover, the computation of a winning move is *robust*, in that it depends only on the structure of the game graph near its leaves. This has in fact been shown for the above case where there is no triangle near the leaves. In this case the winner can be in a state of semi-coma throughout most of the play, becoming alert only whenever play nears its end.

If $f(k)$ is odd valued for all k, then the first player can in fact win while sleeping throughout the entire play: *every* move wins for him. If, however, $f(k)$ is even valued for all k, the winner has to stay alert at all times; even a single slip may cause him to lose. Coma on an epidemiograph with $f(k)$ even valued is in fact governed by the Sprague-Grundy function. The special case $f(k) = 0$ for all k gives the classical games. Thus epidemiography is a generalization of classical games.

A function f is called *phase preserving* [*reversing*] if any parity change occurs only after a move of player II [I]. It is *t-alternating* if after at most $2t$ moves there is a parity change. In Fraenkel & Lorberbom [1989] it is proved that player II [I] can win if f is a phase preserving [reversing] t-alternating function — for *every* sufficiently large digraph. Moreover, the winning strategy is "semi-robust" in that during most of the play the winner need only build up a sufficient supply of a certain configuration; and only in the end stages does he have to follow a definite winning strategy.

The epidemiography investigations originate from the study of long games, especially the Hercules-Hydra game, and the improvability of combinatorial statements; see Nešetřil & Thomas [1987] and Loebl & Nešetřil [1988].

5 k-Welter

The game of Welter is Nim with distinct-size piles. It can be thought of as being played on a semi-infinite strip with squares numbered $0, 1, 2, \ldots$ and a finite number of coins distributed initially on distinct squares. A move consists of selecting a coin and moving it to any *unoccupied* lower square. Thus in Figure 10, the coin on square 7 can be moved only to any of the squares 6, 5, 2 or 0. The game ends when the coins are jammed in the lowest squares ($0, 1, 2, 3$ in Figure 10). A full analysis of the P-positions for normal play is given in Conway [1976, Ch. 13]. Berlekamp [1972] wrote about Welter: "Its solution is easy to state, difficult to prove, even harder to generalize to almost any other nonisomorphic game (no matter how small the difference between the other game and Welter's Nim), ..."

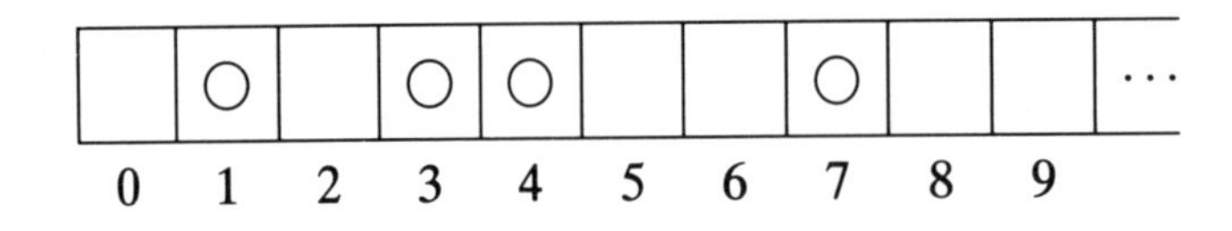

Figure 10: The game of Welter.

Consider the generalization where each move is restricted to traversing at most k squares down the line, where k is a fixed positive integer. This is a generalization, since given any position of Welter with the furthest coin at square n, k-Welter with $k \geq n$ reduces to this game of Welter.

In Kahane & Fraenkel [1987] and Duvdevani & Fraenkel [1989], the following partial results were obtained.

(i) The Sprague–Grundy function for the case of two coins was determined, enabling a polynomial computation.

(ii) A polynomial algorithm was given which permits computation of the P, N-pattern for all positions of 3-coin k-Welter for every k.

(iii) The Sprague–Grundy function for any number of coins with $k = 2^m - 1$ $(m \geq 1)$ was determined. The computation is polynomial.

(iv) For every k, the Sprague-Grundy function of k-Welter was proved to be invariant under translation of coins mod$(k + 1)$.

Is the general case polynomial?

6 Wythoff's Game and Moore's Nim

In Wythoff's game, two piles of tokens and a positive integer a are given. The moves are of two types: a player may remove any positive number of tokens from a single pile, or he may take from both piles, say k (> 0) from one and ℓ (> 0) from the other, provided that $|k - \ell| < a$. We consider normal play. The case $a = 1$ is the classical Wythoff game [1907], reportedly played in China under the name *tsianshidsi*. See Yaglom and Yaglom [1967], Connell [1959], and Silber [1976]. For the case of general a, see Fraenkel [1982].

The known results are polynomial characterizations of the P-positions. Two old unresolved problems are: (i) Decide whether a finite disjoint sum of Wythoff games has a polynomial strategy. (ii) Generalize the game to more than two piles. The crux of these problems stems from the fact that the move of taking k tokens from both piles preserves Nim-sum (since $k \oplus k = 0$), therefore no polynomial characterization of the non-zero values of the g-function is known.

The class of *Nimhoff games* lie in-between the easy game of Nim and the difficult Wythoff game. It designates games in which the move of taking the same number of tokens from both piles is restricted in various ways. We mention here just a few of the results. Given n piles. Any positive number of tokens may be removed from a *single* pile. In addition the following moves may be made.

(i) **Cyclic Nimhoff.** Removal of $(a_1, \ldots, a_n)$ from the n piles is permitted provided $\sum_{i=1}^{n} a_i \not\equiv 0 (\text{mod } h)$, where h is a fixed positive integer. Then the g-function of any position $(b_1, \ldots, b_n)$ is given by

$$g(b_1, \ldots, b_n) = h \sum_{i=1}^{n}{}' \lfloor b_i/h \rfloor + \sum_{i=1}^{n} b_i (\text{mod } h),$$

where Σ' denotes Nim-sum. The combination of Nim-sum with ordinary summation also occurs in the analysis of Welter's game. The special case $n = 2$, $h = 3$ is the "king-rook" game, mentioned e.g. in Gardner [1989].

(ii) **Balanced Nimhoff.** For a fixed positive integer k, removal of $a_s = a_t = 2^k$ for some $s \neq t$ is permitted ($a_i = 0$ for all $i \notin \{s, t\}$). Define the *k-Nim-sum* of a and b by $a \circledk b = a \oplus b \oplus a^k b^k$, where $\oplus$ denotes Nim-sum, and x^k is the k-th binary bit from the right of the binary representation of x. In other words, $a \circledk b$ is $a \oplus b$ unless $a^k = b^k = 1$, in which case it is $a \oplus b$ with the least significant bit complemented. Then the g-function of balanced Nimhoff is

$$g(b_1, \ldots, b_n) = b_1 \circledk \cdots \circledk b_n.$$

Further results, given in Fraenkel & Lorberbom [1988], are connected with the function macs(a, b), which is the number whose binary encoding consists of the common right bits of a and b, and with

$$\text{macs}_k(a, b) = \max\{d : a \oplus b = (a - ik) \oplus (b - ik) \text{ for every } i \leq d\}.$$

(These functions are similar to the "mating function" used by Conway [1976, Ch. 13] in his analysis of Welter's game.)

We remark that also in Moore's Nim (take any positive number of tokens from up to k piles for fixed k) a polynomial algorithm for computing the P-positions is known, but no polynomial characterization of the non-zero g-values is known. Moore's Nim also admits the Nim-sum preserving move as in Wythoff's game. Some results for Moore's Nim are given in Jenkyns & Mayberry [1980].

For misère play of Wythoff's game, the P-positions can again be characterized in an interesting way (see Fraenkel [1984]), but the Sprague-Grundy function computation seems again difficult. What about Moore's Nim?

7 Partizan Subtraction Games

A game is *impartial* if the admissible moves of both players are identical for all game positions. Otherwise it is *partizan*. The above mentioned games are all impartial; the present game is partizan.

Given a positive integer n. The two players subtract integers from n, and play ends when the number has been reduced to 0. Left subtracts any number belonging to a set S, and Right subtracts any number belonging to a set T. We consider normal play in the game $G(S,T)$, and say that S *dominates* T ($S \succ T$) if there is a nonnegative integer n_0 such that Left can win both as first and as second player for every $n \geq n_0$.

In Fraenkel & Kotzig [1987], the following example is given. Let $S_1 = \{1,2,6\}$, $S_2 = \{1,3,5\}$, $S_3 = \{2,3,4\}$ and consider the games $G_1 = (S_1, S_2)$, $G_2 = (S_2, S_3)$, $G_3 = (S_3, S_1)$. From Table 1 we see that the dominance relation is intransitive: $S_1 \succ S_2 \succ S_3 \succ S_1$. Thus Left can be a "winning gentleman": He can let Right select one of S_1, S_2, S_3, then selecting himself one of the two remaining sets; and he can then let Right choose who will make the first move. Left can still always win if $n \geq 12$.

It is further proved that partizan subtraction games are periodic, and pure periodicity is investigated. We can consider a subtraction game as being played on a row of tokens, where tokens are removed from one of the ends of the row. What happens if tokens can be removed from the middle of a row, that is, the row may be split after the move? The impartial version, known as *octal games*, has been investigated by Guy & Smith [1956]; see also Austin [1976], Conway [1976] and Berlekamp, Conway & Guy [1982]. But not much seems to be known about the partizan version of general octal games.

8 Games in which the Loser Wins

The Fact in §1 implies that for any game terminating in a finite number of moves in one of the players winning and the other losing, precisely one of the players has a winning strategy.

Who has this winning strategy? What is it like? These types of questions are not at all covered by the Fact, giving rise to the possibility of strange games. In fact, the winning strategy may be undecidable (Rabin [1957]), or our present knowledge may be insufficient to determine who is the winner. In Jones [1982], a game is given in which player II can win if and only if there are

Table 1: A curious dominance relation.

S_1	G_1	S_2	S_2	G_2	S_3	S_3	G_3	S_1
P	0	P	P	0	P	P	0	P
N	1	N	N	1	P	P	1	N
N	2	P	N	2	N	N	2	N
N	3	N	N	3	N	N	3	N
N	4	P	N	4	N	N	4	P
N	5	N	N	5	P	N	5	P
N	6	P	N	6	P	N	6	N
N	7	P	N	7	P	N	7	N
N	8	P	N	8	P	N	8	P
:	9	:	:	9	:	N	9	P
:	10	:	:	10	:	N	10	P
						N	11	N
						N	12	P
						N	13	P
						N	14	P
						:	15	:
						:	16	:

infinitely many primes of the form $n^2 + 1$, an old unsolved problem of number theory. Examples of a different flavor are given in Fraenkel & Jones [1989].

Here is an example of a game in which the winner loses, loosely speaking. Given the polynomial

$$p = x_1x_3 - x_2x_4 + x_1 + x_3 + 1.$$

Two players alternate in selecting positive integers > 1 for the variables as follows: Player I selects x_1 such that $x_1 + 1$ is not prime; player II selects x_2 such that $x_2 \neq x_1 + 1$; player I selects x_3; player II selects x_4. Player II wins if and only if $p = 0$.

The game has length 4. We claim that player II has a winning strategy, but, roughly speaking, in any actual play—which should not take undue time!—if player II can win, then governments will topple and generals' heads roll.

The polynomial can be written in the form

$$p = (x_1 + 1)(x_3 + 1) - x_2x_4.$$

Both players can verify probabilistically in polynomial time that $x_1 + 1$ is indeed composite, with probability of error less than, say, the probability that a dropped spoon will rise to the ceiling rather than fall to the floor. See Rabin [1976]. Now player II can clearly win by selecting x_2 as any factor of $x_1 + 1$, and then letting $x_4 = (x_1 + 1)(x_3 + 1)/x_2$. However, if player I selects $x_1 + 1$ as the product of two large primes of the same order of magnitude, then if player II can compute a factor in reasonable time, then he can break the RSA cryptosystem (see Rivest, Shamir & Adleman [1978]).

References

1. R. Austin [1976], Impartial and partisan games. M.Sc. Thesis, Univ. of Calgary.

2. C. Berge [1985], *Graphs*, North-Holland, Amsterdam.

3. E. R. Berlekamp [1972], Some recent results on the combinatorial game called Welter's Nim, Proc. 6th Annual Princeton Conference on Information Science and Systems, 203–204.

4. E. R. Berlekamp, J. H. Conway and R. K. Guy [1982], *Winning Ways for Your Mathematical Plays* (two volumes), Academic Press, London.

5. I. G. Connell [1959], A generalization of Wythoff's game, *Canad. Math. Bull.* **2**, 181–190.

6. J. H. Conway [1976], *On Numbers and Games*, Academic Press, London.

7. N. Duvdevani & A. S. Fraenkel [1989], Properties of k-Welter's game, *Discrete Math.*, in press.

8. M. J. Fischer & M. O. Rabin [1974], Super-exponential complexity of Presburger arithmetic, in: *Complexity of Computation* (R. M. Karp, ed.), Amer. Math. Soc., Providence, RI.

9. A. S. Fraenkel [1980], From Nim to Go, *Annals of Discrete Math.* **6**, Proc. Symp. on Combinatorial Designs and Their Applications, Colorado State Univ., Fort Collins, Colorado, June 1978 (J. Srivastava, ed.), pp. 137–156.

10. A. S. Fraenkel [1982], How to beat your Wythoff games' opponents on three fronts, *Amer. Math. Monthly* **89**, 353–361.

11. A. S. Fraenkel [1984], Wythoff games, continued fractions, cedar trees and Fibonacci searches, *Theoretical Computer Science* **29**, 49–73.

12. A. S. Fraenkel & F. Harary [1989], Geodetic contraction games on graphs, *Internat. J. Game Theory*, in press.

13. A. S. Fraenkel & J. P. Jones [1989], Polynomial time computable winning strategies in polynomial games, preprint.

14. A. S. Fraenkel & A. Kotzig [1987], Partizan octal games: partizan subtraction games, *Internat. J. Game Theory* **16**, 145–154.

15. A. S. Fraenkel & D. Lichtenstein [1981], Computing a perfect strategy for $n \times n$ chess requires time exponential in n, *J. Combin. Theory Ser. A* **31**, 199–214.

16. A. S. Fraenkel, M. Loebl & J. Nešetřil [1988], Epidemiography-II. Games with a dozing yet winning player, *J. Combin. Theory Ser. A* **49**, 129–144.

17. A. S. Fraenkel & M. Lorberbom [1988], Nimhoff games, Technical Report CS88-19, Department of Applied Mathematics and Computer Science, The Weizmann Institute of Science, Rehovot 76100, Israel, Nov. 1988.

18. A. S. Fraenkel & M. Lorberbom [1989], Epidemiography with various growth functions, *Discrete Math.*, in press.

19. A. S. Fraenkel & J. Nešetřil [1985], Epidemiography, *Pacific J. Math.* **118**, 369–381.

20. A. S. Fraenkel & E. R. Scheinerman [1989], A deletion game on hypergraphs, *Discrete Appl. Math.*, in press.

21. D. Gale [1974], A curious Nim-type game, *Amer. Math. Monthly* **81**, 876–879.

22. D. Gale & A. Neyman [1982], Nim-type games, *Internat. J. Game Theory* **11**, 17–20.

23. M. Gardner [1973], *Mathematical Games*, Scientific American.

24. M. Gardner [1989], *Penrose Tiles to Trapdoor Ciphers*, W. H. Freeman and Co., New York.

25. M. R. Garey & D. S. Johnson [1979], *Computers and Intractability: A Guide to the Theory of NP-Completeness*, W. H. Freeman and Co., San Francisco.

26. R. K. Guy & C. A. B. Smith [1956], The G-values of various games, *Proc. Cambridge Phil. Soc.* **52**, 514–526.

27. T. A. Jenkyns & J. P. Mayberry [1980], The skeleton of an impartial game and the Nim-function of Moore's Nim$_k$, *Internat. J. Game Theory* **9**, 51–63.

28. D. S. Johnson [1983], The NP-Completeness Column: An Ongoing Guide, *J. of Algorithms* **4**, 397–411 (9th quarterly column; on games).

29. J. P. Jones [1982], Some undecidable determined games, *Internat. J. Game Theory* **11**, 63–70.

30. J. Kahane & A.S. Fraenkel [1987], k-Welter — A generalization of Welter's game, *J. Combin. Theory Ser. A* **46**, 1–20.

31. M. Loebl & J. Nešetřil [1988], Linearity and unprovability of set union strategies, Proc. 20th ACM Symp. Theory of Computing (Chicago, IL, May 1988), 360–366, Assoc. Comp. Math., New York, NY.

32. J. Nešetřil & R. Thomas [1987], Well quasi ordering, long games and combinatorial study of undecidability, *Contemporary Math. Proc. Symposia AMS* **65**, 281–293.

33. M. O. Rabin [1957], Effective computability of winning strategies, Contributions to the Theory of Games, Vol. 3, Annals of Math. Studies No. 39, Princeton, pp. 147–157.

34. M. O. Rabin [1976], Probabilistic algorithms, Proc. Symp. on New Directions and Recent Results in Algorithms and Complexity, Carnegie-Mellon, 1976 (J.F. Traub, ed.), 21–39, Academic Press, New York.

35. R. L. Rivest, A. Shamir & L. M. Adleman [1978], On digital signature and public key cryptosystems, *Communications ACM* **21**, 120–126.

36. J. M. Robson [1984], Combinatorial games with exponential space complete decision problems, Proc. 11th Symp. Math. Foundations of Computer Science (M.P. Chytie and V. Koubek, eds., Praha, Czechoslovakia, 1984), Lecture Notes in Computer Science, Vol. 176, pp. 498–506, Springer, Berlin.

37. T. J. Schaefer [1978], On the complexity of some two-person perfect-information games, *J. Computer and System Sciences* **16**, 185–225.

38. R. Silber [1976], A Fibonacci property of Wythoff pairs, *Fibonacci Quarterly* **14**, 380–384.

39. J. Úlehla [1980], A complete analysis of von Neumann's Hackendot, *Internat. J. Game Theory* **9**, 107–113.

40. W. A. Wythoff [1907], A modification of the game of Nim, *Nieuw Arch. Wisk.* **7**, 199-202.

41. A. M. Yaglom & I.M. Yaglom [1967], *Challenging Mathematical Problems with Elementary Solutions*, translated by J. McCawley, Jr., revised and edited by B. Gordon, Vol. II, Holden-Day, San Francisco.

Weizmann Institute of Science
Rehovot, Israel

Recreational Games Displays

During the conference two displays of recreational games and materials were maintained for the participants to examine. The first, organized by A. Fraenkel, was of combinatorial games. The other display was of puzzles created by K. Jones. Descriptions of both displays follow.

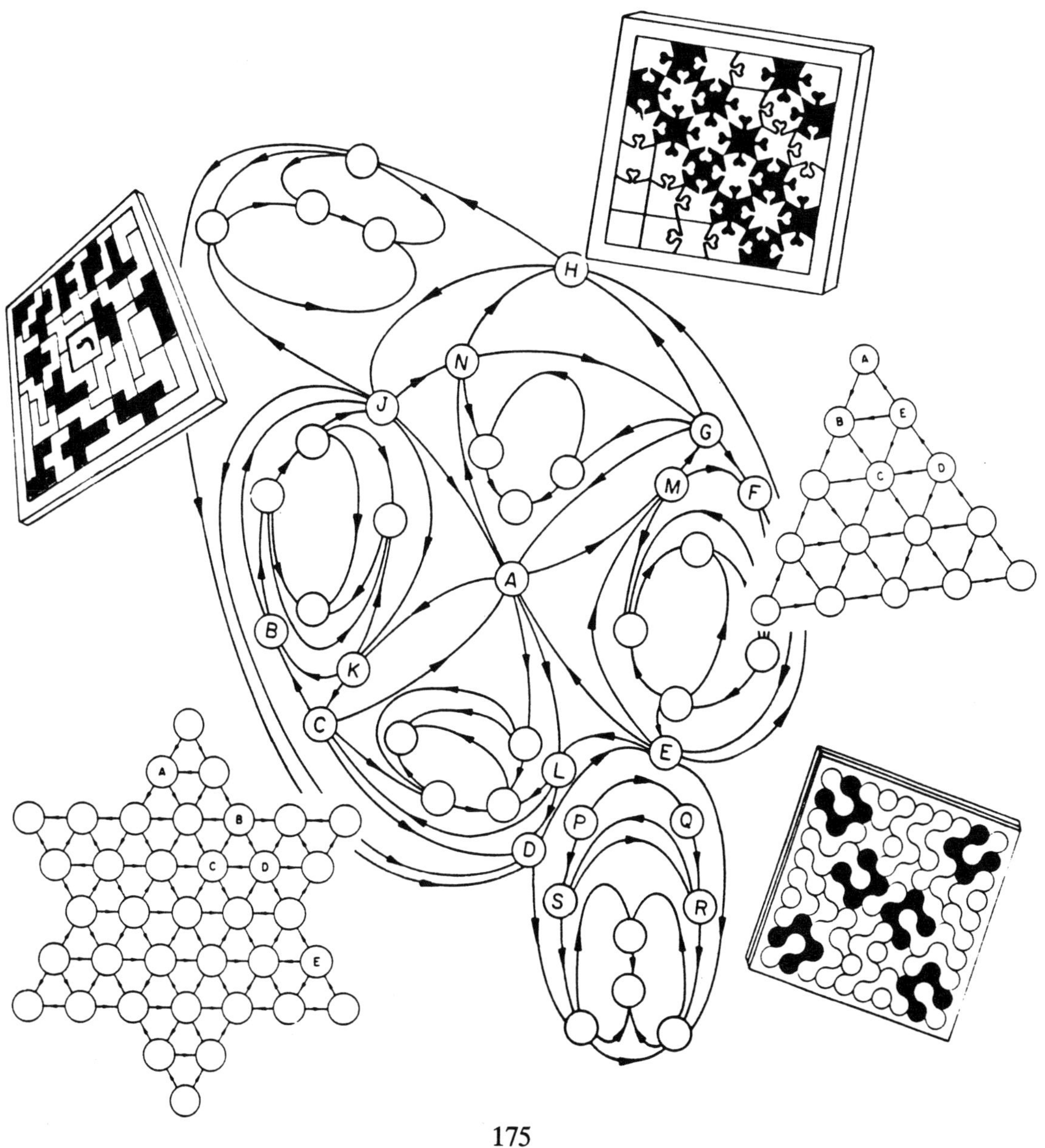

Combinatorial Games

Aviezri S. Fraenkel

Two-Dimensional Nim*

A two-person game.

Aim of Game

To make the last move while removing tokens from any straight line parallel to the edges of a given array.

The Moves

Removal of an arbitrary positive number of tokens from:

- An arbitrary row or column of the rectangular array.
- Any straight line parallel to one of the three edges of the triangular board.

The Play

(a) 15 tokens are placed on the 15 squares of the rectangular board or on the 15 triangles of the triangular board.

(b) The players play alternately. Each player at his turn makes one move.

The Winner

The winner is the player who removes the last token(s) from the array.

References

1. D. Fremlin, Well-founded games, *Eureka* **36** (1973) 33–37.

2. A. S. Fraenkel & H. Herda, Never rush to be first in playing Nimbi, *Math. Mag.* **53** (1980) 21–26.

Tsianshidsi*

A two-person game.

Aim of Game

To make the last move while removing tokens from two given piles.

The Moves

- Removal of an arbitrary positive number of tokens from a single pile.
- Removal of tokens from both piles. If n tokens are removed from one pile, the number to be removed from the other pile is one of $n, n \pm 1, \ldots, n \pm (a-1)$, where a is a given positive integer.

The Play

(a) Two piles of arbitrary size are placed on the table.

(b) The players play alternately. Every player at his turn makes one of the two possible moves.

The Winner

The winner is the player who removes the last token(s).

Remark

The case $a = 1$ is a game which, according to the Russians, is played in China under the name "tsianshidsi", which it is said, means "throwing stones" in Chinese. In the West the game is called "Wythoff's Game", after the Dutch mathematician Wythoff.

References

1. W. A. Wythoff, A modification of the game of Nim, *Nieuw Arch. Wisk.* **7** (1907) 199–202.

2. A. S. Fraenkel & I. Borosh, A generalization of Wythoff's game, *J. Combin. Theory Ser. A* **15** (1973) 175–191.

3. A. S. Fraenkel, How to beat your Wythoff games' opponent on three fronts, *Amer. Math. Monthly* **89** (1982) 353–361.

Welter

A two-person game.

Aim of Game

To make the last move while moving a token on a linear board to any lower-numbered *unoccupied* square.

The Moves

Moving any token from a square to any unoccupied lower-numbered square.

The Play

(a) Several tokens are placed on arbitrary squares, not more than one token per square.

(b) The players play alternately. Each player at his turn makes one move.

The Winner

The winner is the player who makes the last move so that all tokens are jammed on the lowest squares $0, 1, 2 \ldots$.

Remark

Try to play *k-Welter*, which is Welter with the restriction of moving a coin at most k squares down from its current position.

References

1. C. P. Welter, The theory of a class of games on a sequence of squares, in terms of the advancing operation in a special group, *Proc. Nederl. Akad. Wetensch. Ser. A* **16** (1954) 194–200.

2. J. H. Conway, *On Numbers and Games*, Academic Press, 1976 (Ch. 13).

3. E. R. Berlekamp, Some recent results on the combinatorial game called Welter's Nim, Proc. 6th Annual Princeton Conf. on Information Science and Systems, pp. 203–204, 1972.

4. J. Kahane & A. S. Fraenkel, k-Welter—a generalization of Welter's game, *J. Combin. Theory Ser. A* **46** (1987) 1–20.

Nimbi*

A two-person game invented by Piet Hein.

Aim of Game

To avoid making the last move while removing a connected portion of tokens from any straight line parallel to the edges of the given hexagonal board.

The Moves

Removal of any contiguous portion of tokens from a single line parallel to an edge of the board.

The Play

(a) 12 tokens are placed on the 12 stations of the hexagonal board.

(b) The players play alternately. Each player at his turn makes one move.

The Winner

The winner is the opponent of the player who removes the last token(s) from the board.

Remark

The game was marketed several years ago by Piet Hein. An interesting variation is the case when the outcome is reversed: the player making the last move wins. Both versions have been fully analyzed, but the game on a board with n stations for general n may still be hard.

References

1. P. Hein, Nimbi (game brochure), Denmark, 1967.

2. A. S. Fraenkel & H. Herda, Never rush to be first in playing Nimbi, *Math. Mag.* **53** (1980) 21–26.

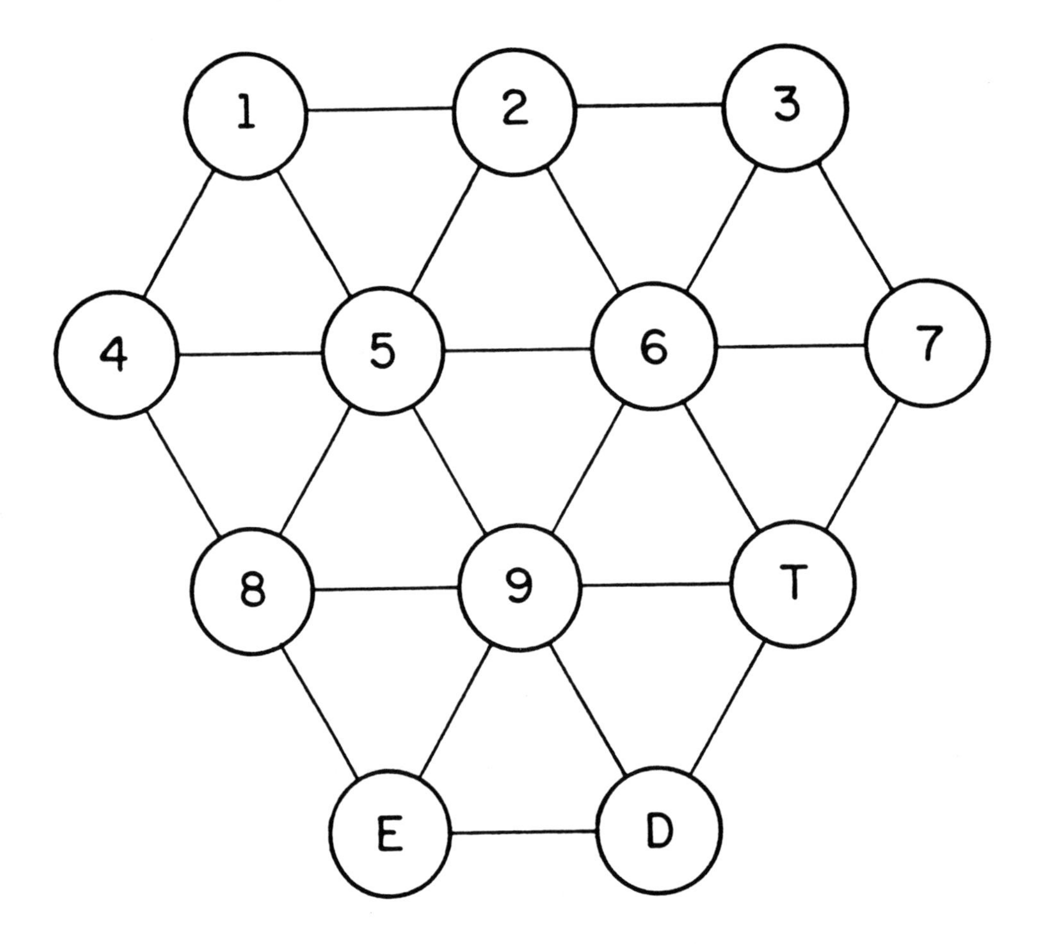

Figure 1: The game of Nimbi

Particles and Antiparticles*

A two-person game.

Aim of Game

To make the last move while moving electrons and positrons along directed orbits.

The Moves

An electron or positron is selected and launched to an adjacent station along a directed orbit, provided that the adjacent station does not harbor a particle of the same type. (As is well-known, particles of the same type repel each other.) If the adjacent station harbors a particle of the opposite type, the two particles annihilate each other and vanish.

The Play

(a) An equal number of electrons and positrons is placed on the stations of each of the five "atoms". The permissible "energy levels" are thus i electrons and i positrons per atom ($i = 0$, 1 or 2).

(b) The players play alternately. Each player at his turn makes one move.

The Winner

The winner is the player who annihilates the last pair of particles. If there is no last move, the outcome is a draw (and then some atoms continue to exist)!

Remark

We do not know whether the game played on an arbitrary directed graph with n vertices has a strategy which is polynomial in n. Such a strategy, if it exists, may not be all that simple: a special case of the game (when all particles are electrons, say) is a generalization of *Welter's* game, which has an interesting intricate theory.

Reference

1. A. S. Fraenkel, The particles and antiparticles game, *Comp. Math. Appl.* **3** (1977) 327–328.

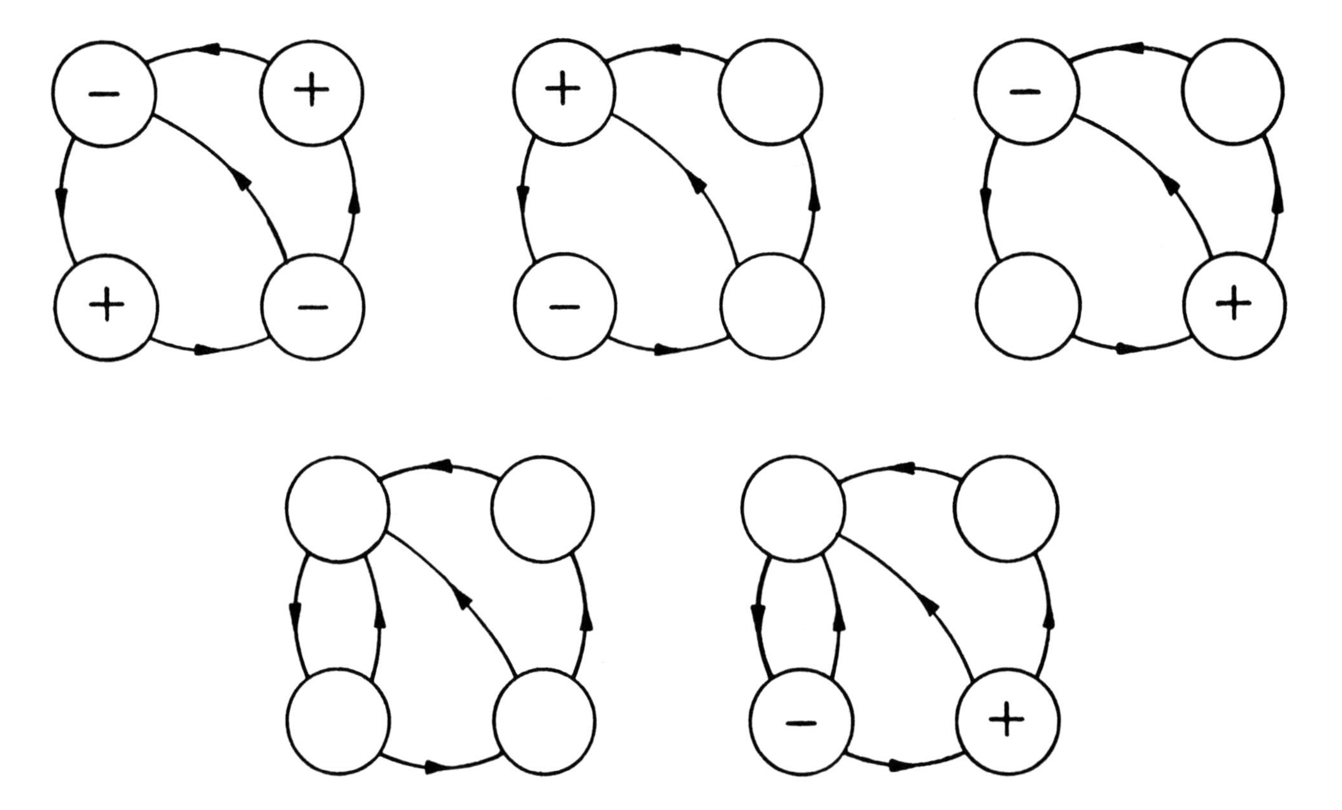

Figure 2: The Particles and Antiparticles Game

Learn to Beat the Grand Masters*

A two-person game.

Aim of Game

To make the last move while moving tokens on the board.

The Moves

Moving a token from a position to an adjacent position along a directed line. Each position may harbor several tokens.

The Play

(a) Several tokens are placed on arbitrary positions. Each position may contain more than one token.

(b) The players play alternately. Each player at his turn makes one move.

The Winner

The winner is the player making the last move. If there is no last move, the outcome is a draw.

Remark

If you find a method enabling you to decide quickly whether any given position is a winning, losing or draw position, you can beat the Grand Masters—at least in this game! You can try it out on either one of the two given boards.

References

1. C. A. B. Smith, Graphs and composite games, *J. Combin. Theory* **1** (1966) 51–81.

2. J. H. Conway, *On Numbers and Games*, Academic Press, 1976 (Ch. 11).

3. A. S. Fraenkel & Y. Perl, Constructions in combinatorial games with cycles, *Colloq. Math. Soc. János Bolyai* **10**, North Holland, 1975 (Vol. 2, pp. 667-699).

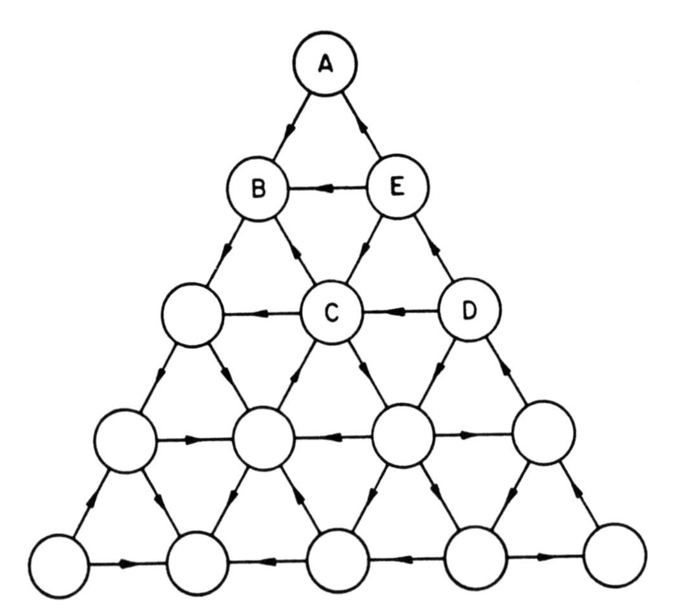

Figure 3: Learn to beat Bobby Fisher.

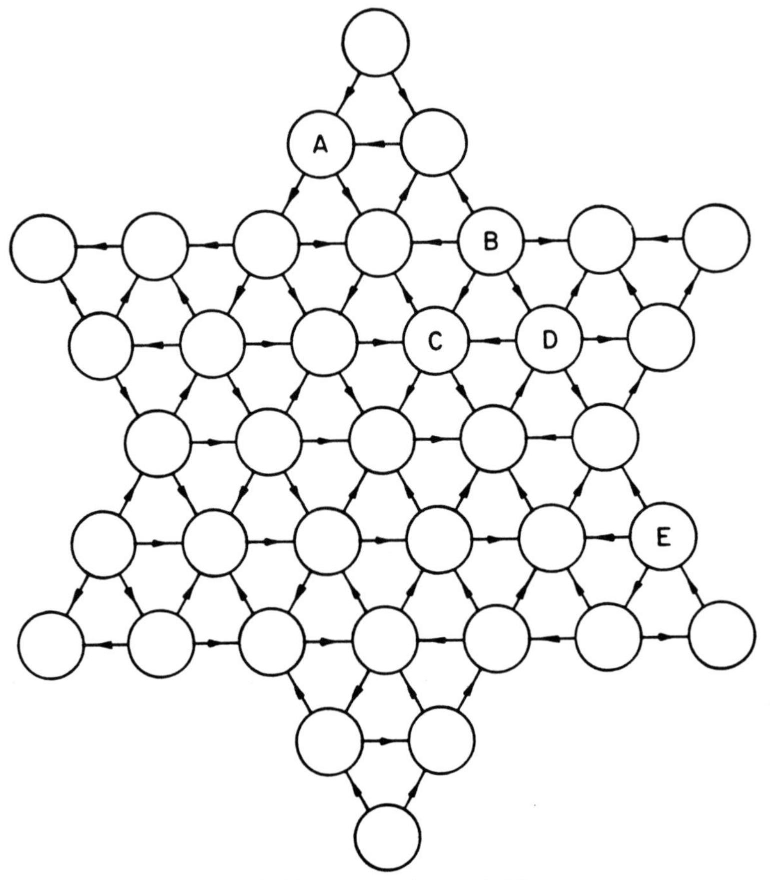

Figure 4: Now beat Anatoly Karpov.

Worlds in Collision *

A two-person game.

Aim of Game

To make the last move while launching worlds from space stations to neighboring space stations along directed orbits, with annihilation.

The Moves

A "world" or "star" is selected and launched to an adjacent "space station" along a directed orbit. If the space station it streaks into is occupied, the colliding stars explode and vanish in a cloud of interstellar dust. If a "star" moves into a space station without exit (there are six such stations in the universe), it phases out of the universe ("falling star").

The Play

(a) A number of stars is placed in space stations (for example: the nine planets in the solar system), not more than one star per station.

(b) The players play alternately. Each player at his turn makes one move.

The Winner

The winner is the player who makes the last move. If there is no last move, the outcome is a draw (and then the universe survives)!

References

1. A. S. Fraenkel, Combinatorial games with an annihilation rule, *Proc. Symp. Appl. Math.*, Vol. 20, pp. 87-91, AMS, 1974.

2. A. S. Fraenkel & Y. Yesha, Theory of annihilation games, *Bull. Amer. Math. Soc.* **82** (1976) 775–777; *J. Combin. Theory* (B) **33** (1982) 60–86.

3. A. S. Fraenkel, U. Tassa & Y. Yesha, Three annihilation games, *Math. Mag.* **51** (1978) 13–17.

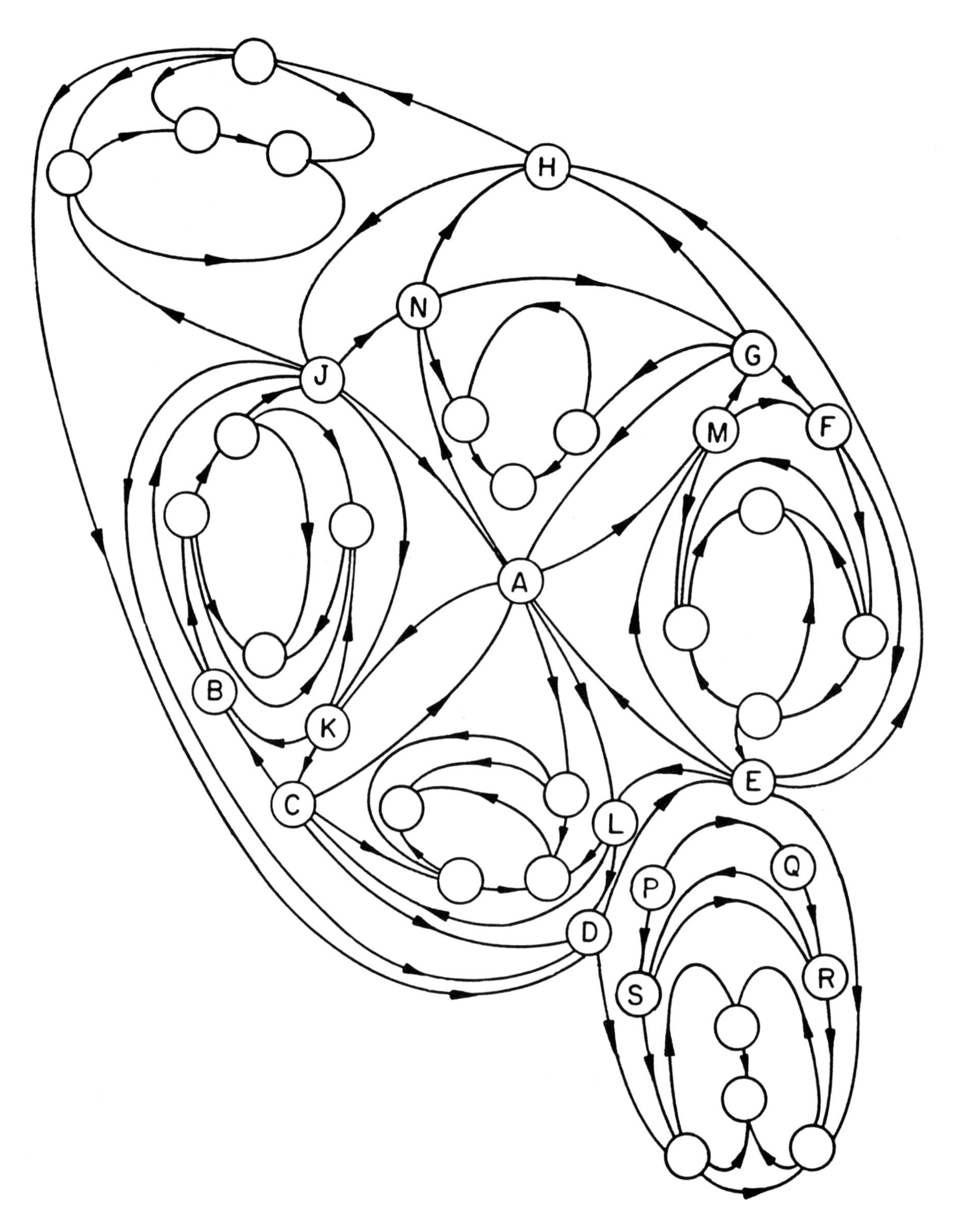

Figure 5: Worlds in Collision

Cannibals *

A two-person game.

Aim of Game

To make the last move while moving and "feasting" on a linear board.

The Moves

Moving a "cannibal" from a square to the neighboring square on its right. If the latter is an occupied "feasting square", then the ingoing cannibal eats up all the occupants (*greedy cannibals*) or only a single cannibal (*polite cannibals*). It turns out that it is much more difficult to be polite than greedy! The particular game suggested here is polite cannibals, where only square no. 1 is a feasting square. A cannibal stepping on a non-feasting square does not eat up any cannibal. A cannibal on square no. 0 "dies" and fades out of the game.

The Play

(a) Several cannibals are placed on arbitrary squares. Each square may contain more than one cannibal.

(b) The players play alternately. Each player at his turn makes one move. Cannibalism (polite) takes place on square no. 1 only.

The Winner

The winner is the player who moves the last cannibal to square no. 0.

Remark

We do not know whether cannibals on an arbitrary directed graph [or a linear board] with n vertices has a strategy which is polynomial in n [or $\log n$].

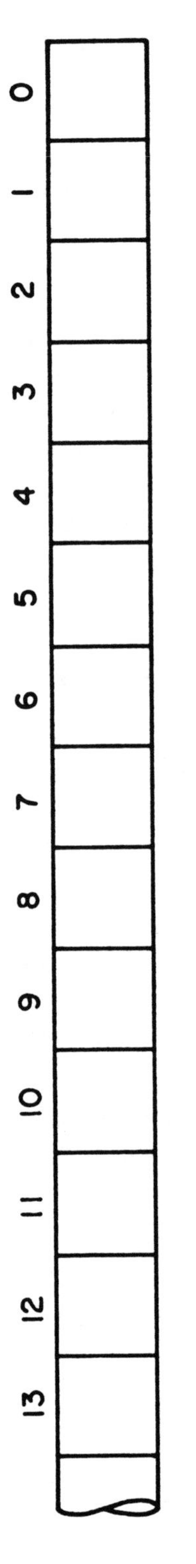

Figure 6: Cannibals

Battle of Numbers *

A two-person game.

Aim of Game

To make the last move while moving tokens on a board with squares numbered 0 through 47.

The Moves

A token on square no. n can move to one of the squares numbered $n-1, n-2, n-3, n-8, n-9, n-10$. (These numbers are written on the sides of each square.) From a red square no. n, a token can move, in addition to the above, to square no. $n+3$. Each square may contain more than one token. A token arriving at square no. 0 is removed from the game.

The Play

(a) Several tokens are placed on arbitrary squares of the board. Each square may contain a multitude of tokens.

(b) The players play alternately. Each player at his turn makes one move.

The Winner

The winner is the player moving the last token to square 0. If there is no last move, the outcome is a draw.

Remark

An interesting variation is obtained when two tokens meeting on a square are both annihilated. In this case no square may contain more than one token initially.

References

1. A. S. Fraenkel & Y. Perl, Constructions in combinatorial games with cycles, *Colloq. Math. Soc. János Bolyai* **10**, North Holland, 1975 (Vol. 2, pp. 667–699).

2. A. S. Fraenkel & U. Tassa, Strategy for a class of games with dynamic ties, *Comput. Math. Appl.* **1** (1975) 237–254.

3. A. S. Fraenkel, U. Tassa & Y. Yesha, Three annihilation games, *Math. Mag.* **51** (1978) 13–17.

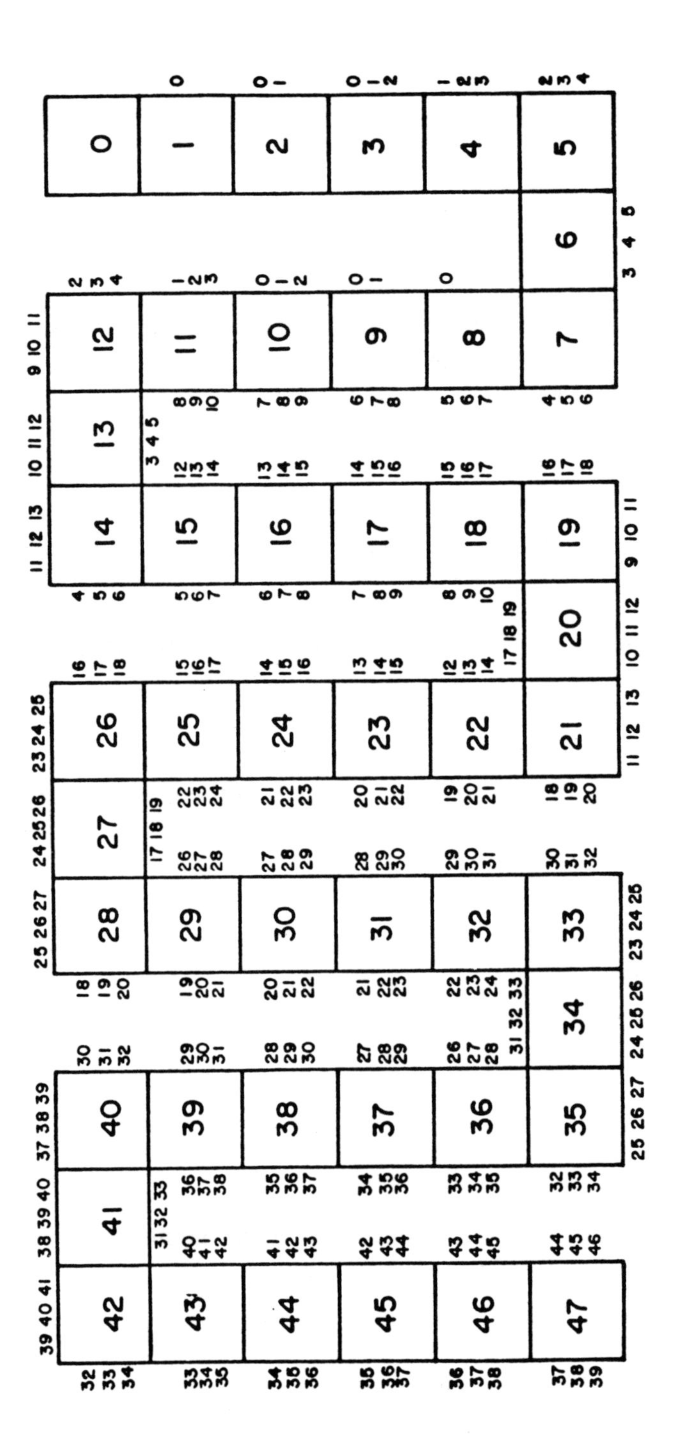

Figure 7: Battle of Numbers

Innocent Marble Game *

A two-person game.

Aim of Game

To make the last move while moving marbles from holes to neighboring holes along directed edges, with annihilation.

The Moves

A "marble" is selected and moved to an adjacent hole in the direction of an arrow. If the adjacent hole already contained a marble, both marbles get annihilated and are removed from the game.

The Play

(a) Zero or two or four marbles are distributed among the holes in each of the five components of the given graph, not more than one marble per hole.

(b) The players play alternately. Each player at his turn makes one move.

The Winner

The winner is the player who removes the last pair of marbles from the game. If there is no last move, the outcome is a draw.

Remark

Note that the five components are only of two types: the top three components are identical, and so are the bottom two. Moreover, the two types, top and bottom, differ only in the direction of the upper horizontal line.

References

1. A. S. Fraenkel, Combinatorial games with an annihilation rule, *Proc. Symp. Appl. Math.,* Vol. 20, pp. 87–91, AMS, 1974.

2. A. S. Fraenkel & Y. Yesha, Theory of annihilation games, *Bull. Amer. Math. Soc.* **82** (1976) 775–777; *J. Combin. Theory Ser. B* **33** (1982) 60–86.

3. A. S. Fraenkel, U. Tassa & Y. Yesha, Three annihilation games, *Math. Mag.* **51** (1978) 13–17.

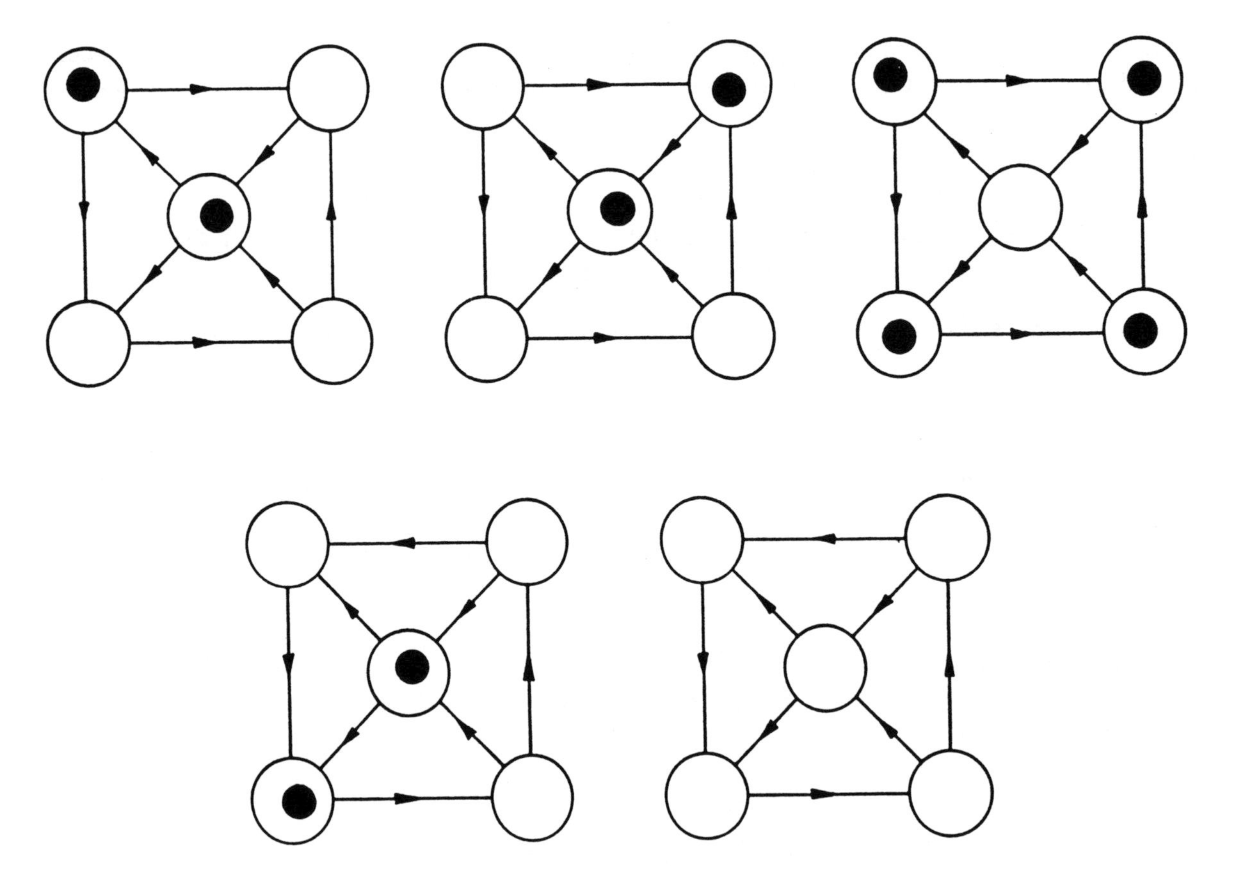

Figure 8: The Innocent Marble Game

Arrows*

A game for 2 players aged 7 and up.

Game includes

Playing board with 24 lettered stations connected by red arrows (one-way) and blue tracks (two-way), and 14 playing pieces (7 of each color).

Object of game

To remove your opponent's playing pieces from the board.

How to play

Each player arranges his 7 playing pieces on the stations nearest to him—4 on the outer row, 3 on the inner row. A turn consists of moving a playing piece along an arrow or a track to an adjacent station. On the arrows, playing pieces may move one way only (in the direction of the arrow) but may move in both directions on the blue tracks.

- First player moves one of his pieces to an adjacent station. Play alternates.
- When a player is able to move into a station occupied by his opponent, his opponent's playing piece is removed from the board. A station **cannot** be occupied by more than one playing piece at any time.
- The winner is the first player to eliminate all his opponent's pieces from the board.

Notes

The arrow pattern was carefully designed to produce a challenging game. Specifically:

(a) The second player cannot win by copying the moves of the first player.

(b) No two-piece position is a draw: one of the two players can always force a win.

---

*Created by Roger B. Eggleton & Aviezri S. Fraenkel.

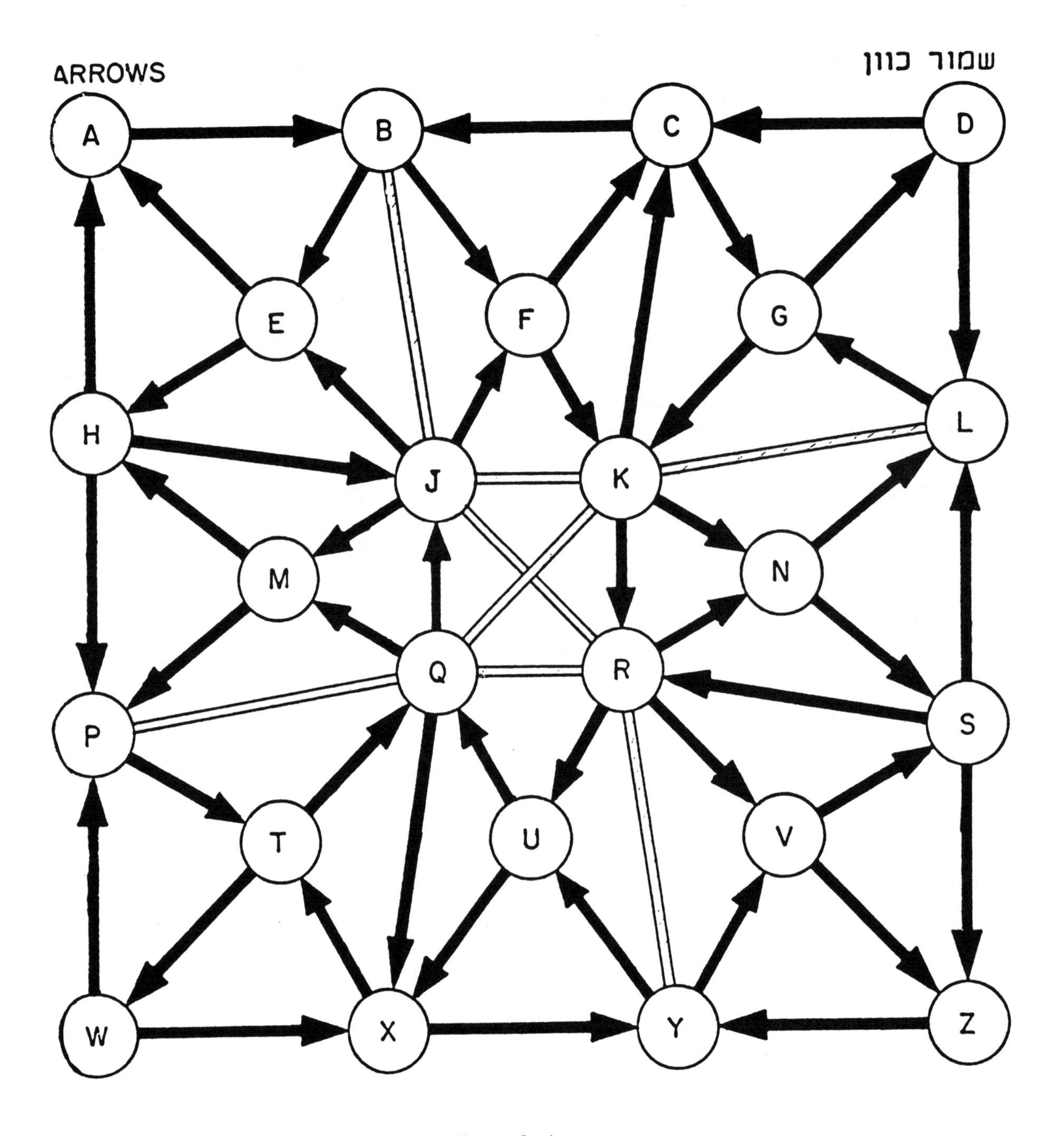

Figure 9: Arrows

Combinatorial Toys

Kathy Jones

At the invitation of the organizers of the Strens Conference, I set up an exhibit of mathematical games and puzzles in one of the classrooms of the University of Calgary, for the amusement and recreation of the conference attendees. Most of these games are based on combinatorial sets and were designed and produced by my company.

It was a great personal pleasure for me to spend several days observing the wonderful intelligence and perseverance the esteemed members of the conference brought to bear on my many little challenges.

Here are brief descriptions of the various sets exhibited, with the tradenames under which they are sold. Canadian Customs had graciously permitted the taking of orders from those who wished to acquire sets of their own, and many attendees happily availed themselves of that opportunity. The games are not available through commercial channels, only directly from Kadon Enterprises.

Quintillions®

The 12 solid pentominoes in hardwood, with a handbook full of two- and three- dimensional puzzle constructions and rules for games.

Super Quintillions®

The non-planar pentacubes (17 distinct shapes) as a companion set to Quintillions, size-compatible for the most wondrous constructions in 3-D. See Figure 1.

Hexacube

The 166 distinct hexacubes, both planar and non-planar, plus four unit cubes, arranged in a $10 \times 10 \times 10$ cube inside a hinged wood box that unfolds to a hexomino shape. Beautifully finished wood, pieces sized to the Quintillions set. Handbook of nomenclature with one solution.

Sextillions™

The 35 hexominoes arranged in a 15×15 pattern with center hole, in laser-cut acrylic, sized $\frac{1}{2}''$ unit square edges. Has a handbook of games and puzzle problems. See Figure 2.

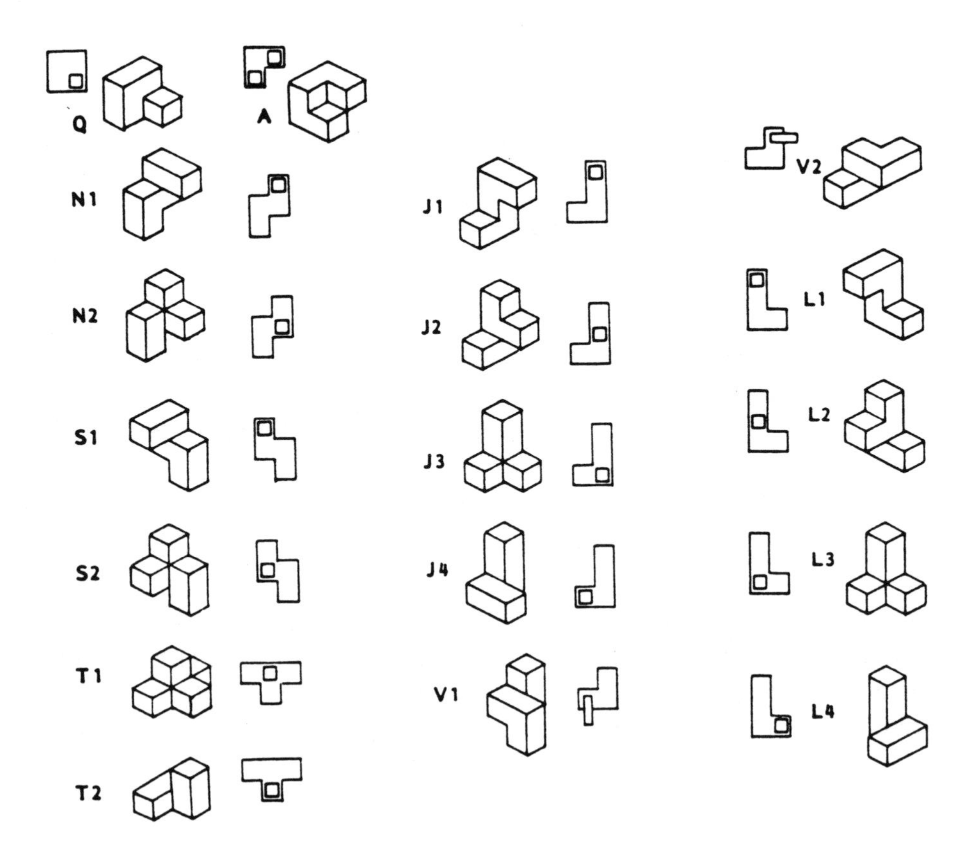

Figure 1: Super Quintillions®

Figure 2: Sextillions™

Poly-5™

The polyominoes orders 1 through 5, in acrylic sized to Sextillions™.

The Heptominoes

The 108 distinct shapes of seven squares joined, arranged as a rectangle in a wooden tray, size-compatible with Sextillions™ and Poly-5.

Quintachex® Chessboard

The 12 pentominoes plus a 2×2 square, checkerboarded on both sides for reversible use, in a framed tray with built-tin game grid. A unique design that allows forming of a maximum number of checkerboarded figures. Handcrafted in wood, $16 \times 16''$ and $20 \times 20''$ overall sizes.

Multimatch®

A cardboard set of MacMahon's Three-Color Squares, originally produced by Wade Philpott and accompanied by his construction problems of figures Percy never thought of. This set is also produced by special order in an extravagant jumbo size in acrylic, in a nice wood tray.

Stockdale Super Square™

A topological cousin of Multimatch® in laser cut acrylic, with the 24 distinct shapes a square becomes when its edges are transformed into all combinations of straight, convex, and concave. Looks a lot like Escher's work. Arranged as a 6×6 square in a clear tray. See Figure 3.

Roundominoes®

Tiling set based on connected circles of sizes 1, 2, and 3, with connecting bridges sizes 1, 2, and 3. Colorful acrylic pieces combine to solve numerous permutational problems, on a 7×7 grid.

Super Roundominoes®

A 10×10 grid expands the Roundominoes set to include order 4 shapes. The handbook lists over 200 adjacency problems. See Figure 4.

Game of Solomon™

Invented by Martin Gardner. Handpainted fabric game board serves for four different two-player strategy games and numerous puzzles.

Figure 3: Stockdale Super Square™

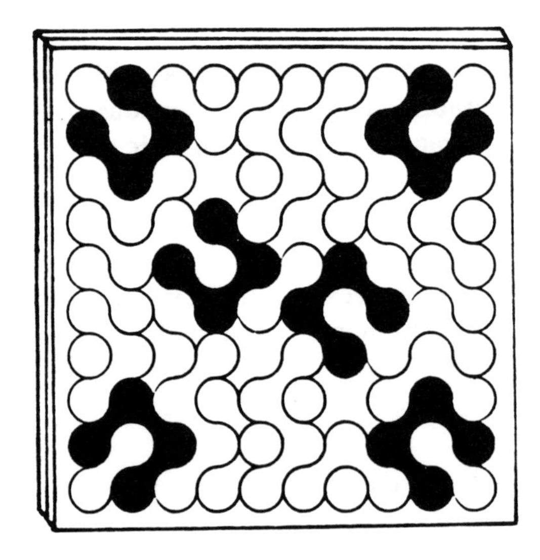

Figure 4: Super Roundominoes®

Tiny Tans™

Three mini-puzzles of polyobolo shapes (baby tangrams) contain four smooth wood pieces each and afford dozens of symmetrical figures to solve.

Grand Tans®

A historical puzzle set from Europe (one of the famous Richter anchorstone puzzles) has several large wood pieces based on multiples of 45° angles but not tangrams. Over 100 silhouette figures to solve.

Octiles™

The 18 path patterns, joining pairs of sides on an octagon, are finely crafted wood tiles that form a game board with square inserts. Each pattern is distinct, so there are a wealth of construction problems with the paths, besides the intricate strategy games. See Figure 5.

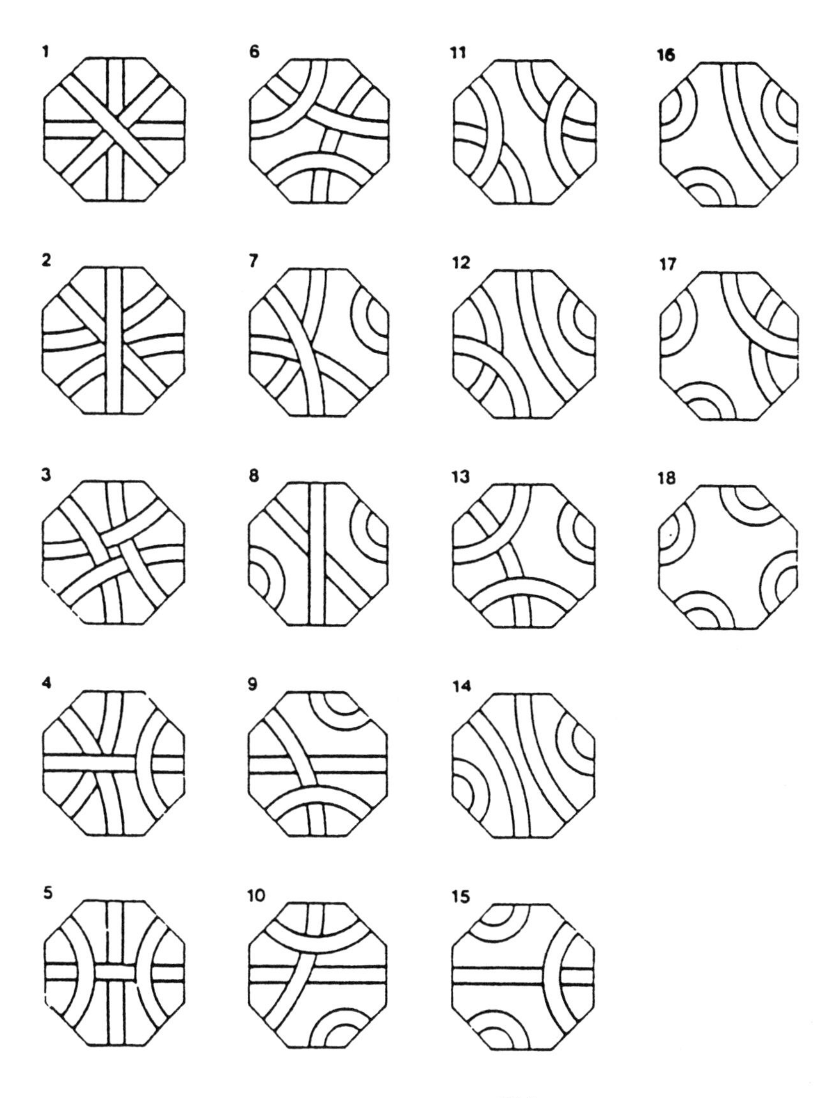

Figure 5: Octiles™

Leap®

A 6×6 grid with a variety of mathematical puzzles, such as magic squares, Wade Philpott's famous peg solitaire, chessboard problems derived from the 8-queens question, and knight's moves activities.

Quantum™

A sophisticated strategy with a random beginning that uses combinations of checkers, king and queen moves.

Proteus®

A meta-game in which the rules change for winning, moving, and trading of the tiles on which the rules are written. This one is an intense game for chess-type thinkers.

Lemma™

A meta-game of evolutionary logic wherein the objective is to keep inventing non-contradictory rules for the game. A major feature of the set is its collection of over 300 topological or positional puzzles on the matrix of grids.

Colormaze™

The 8×8 grid serves for two different groups of color-based problems: 24 tiles (four each of 6 colors) form figures where no color appears twice in any row; and progressively larger arrays of color quadrants are to be untangled in a minimum number of moves. Includes polyomino figures.

Len Gordon's Ball Pyramid Puzzles

These were among the most popular on display. Three sizes—order 4 and order 5 tetrahedrons and an order 4 square-base pyramid—contain various numbers of pieces made of two, three, or four spheres joined. Some are joined orthogonally, some isometrically. Fiendishly tricky to solve because both kinds of pieces must find a place within the overall pyramid shapes. All have unique solutions.

Len Gordon's "Warp-30"

A prototype of this puzzle, under the name of "Life Begins at Thirty," was displayed in an adjacent room. A formal production edition by Kadon was introduced in 1988, on a handsome wooden platform with crystal clear pieces. The object is to fill four different shapes of space with the same eight pieces, in spherical close packing. A square-base order-4 pyramid, a stretched or rectangular pyramid, and two differently proportioned boxes demonstrate different aspects of the coexistence of orthogonal and triangular planes. Solutions are not unique, but are plenty hard anyhow. See Figure 6.

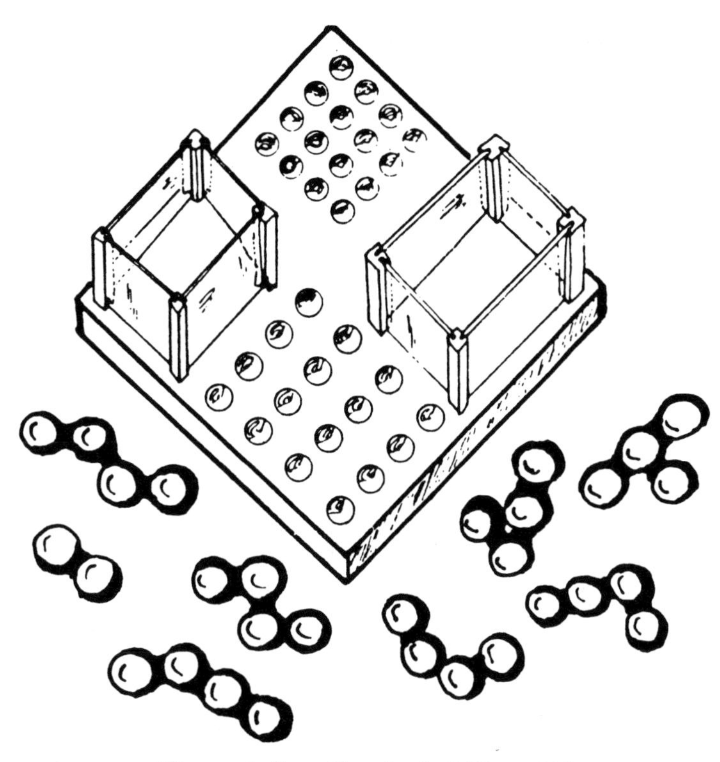

Figure 6: Len Gordon's "Warp-30"

For those wishing to know more about the combinatorial sets produced by Kadon and for those who may wish to acquire some, a catalog is available from Kadon Enterprises, Inc., 1227 Lorene Drive, Suite 16, Pasadena, MD 21122. Within the U.S., a stamped, self-addressed business envelope is appreciated.

By popular demand, combinatorial sets other than those described above are in various stages of development, such as polyhexes, polyiamonds, polytans, octominoes (all 369 of them), MacMahon triangles, and both edge-colored and vertex-colored rhombic MacMahon squares. This list is not exhaustive. We are also open to suggestions, requests, and special orders.

Kadon Enterprises, Inc.
Pasadena, MD 21122-4645

Rubik's Cube—application or illumination of group theory?

Mogens Esrom Larsen

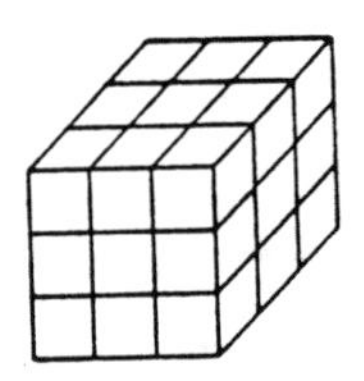

'Εάν μή 'έλπεαι, 'ανέλπιστον 'ονκ 'εξευρήσει.
'Ηράκλειτος.

My favorite advice through the maze of life is the wisdom of Heraclitus, (5th century B.C.): "Only by aiming at the impossible may we reach the unhoped-for!"

As an example allow me to quote *Mathematical Reviews.* Once upon a time someone tried the impossible, i.e., to mark a contribution as *outstanding.* Unfortunately, every possible variety of praise has been abused already. But he found a new way: "It fills a most needed gap in the literature."

The unhoped-for in this case is the hidden quotation we now can take advantage of. E.g., "Rubik's cube fills a most needed gap in group theory."

Some of you might not know the close relation between the cube and the group. It can be said theoretically very briefly. The set of operations is an infinite group with the natural composition. The set of those operations which leaves the initial pattern unaltered makes *accidentally* a normal subgroup. (This is not the case for Rubik's Revenge.) The factor group is the *Hungarian* group, $\mathfrak{U}$, to be identified with a permutation group of the faces by applying the operations to the initial pattern. This correspondence happens to be bijective. (Again, this is not the case for the Revenge.)

So, we can formulate the problem of the cube: Given a pattern, you shall identify the permutation as a group element of $\mathfrak{U}$, i.e., find a representation of this group element in the form of an operation.

To solve this problem in practice was a habit of many youngsters years ago. So, they have the group in hand,—we might want to transfer this group to their minds too!

Because, in the teaching of group theory we need very much some experience to furnish our theory with flesh and blood. And all the other groups we use are so small and trivial that they are not really necessary for a satisfying understanding of the phenomena. Even in crystallography, the groups are hardly used in the names of the classification.

But the Hungarian group offers splendid illumination of crucial points of group theory. E.g., the difference between a *subgroup* and a *factor group.* Suppose we have a complicated group, $\mathfrak{G}$, and want to describe some essentials of it by means of a smaller group, $\mathfrak{H}$. Now, the question is, shall we look for a homomorphism one way or the other:

$$\mathfrak{H} \xrightarrow{\phi} \mathfrak{G} \qquad \text{or} \qquad \mathfrak{G} \xrightarrow{\psi} \mathfrak{H}?$$

Rubik's cube makes the distinction perfectly clear and intuitive. In the group of operations on the cube, as it is illustrated by the patterns or rather permutations, we might simplify matters by looking at the corners only. To simplify further, let's consider their locations only and forget about their twists. Then we can identify the two possible groups easily enough:

The corner subgroup, $\mathfrak{K}$, consists of those operations which leave the edges fixed.

The corner factor group, $\mathcal{K}$, consists of all operations, identifying those making the same permutations of the corners. This becomes even clearer, if we take a cube, paint all edges black and look at the patterns as they appear! The factor group is made *visible* by making the corresponding normal subgroup *invisible.*

To underline the difference between the two groups we are so lucky that they are different.

$$\mathcal{K} \simeq \mathfrak{S}_8$$
$$\mathfrak{K} \simeq \mathfrak{U}_8$$

The difference is understood fairly easily by considering an element of $\mathcal{K}$ without representatives in $\mathfrak{K}$:

$$F^- U^+ B^+ U^- F^+ U^2 B^- U^+ B^+ U^2 B^-$$

(F means *Front, U Upside,* and B *Backside.* + means *a quarter turn to the right,* – *a quarter turn to the left,* and 2 *a half turn* (either way).) See Figure 1.

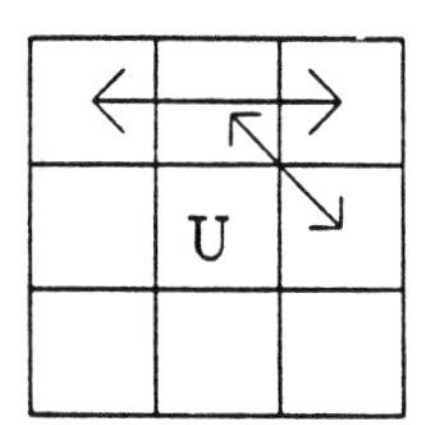

Figure 1:

This time the difference is quite obvious—the difference between odd and even (permutations). But other times the difference is hidden and must be dug out. E.g., in Rubik's Revenge, the four faces on one side are equal and may be invisibly permuted. But as they are painted with the same color in groups of four, this subgroup is not normal. Anyway, it is possible to make an odd permutation of corners at the expense of an odd permutation of equal side-faces, the latter being invisible.

So, as a side effect we also learn to look for hidden variables!

The next step we want to make is to understand what we are doing when we solve the cube. The fun is that it leads us to understand the concept of a *solvable* group! Though, the Hungarian group is not solvable, the solution illustrates the idea behind solvability just as well.

To make the idea as simple and intuitive as possible, let us define in analogy to the corner groups above the edge subgroup of operations having the corner-pattern fixed, $\mathfrak{M}$. This is exactly

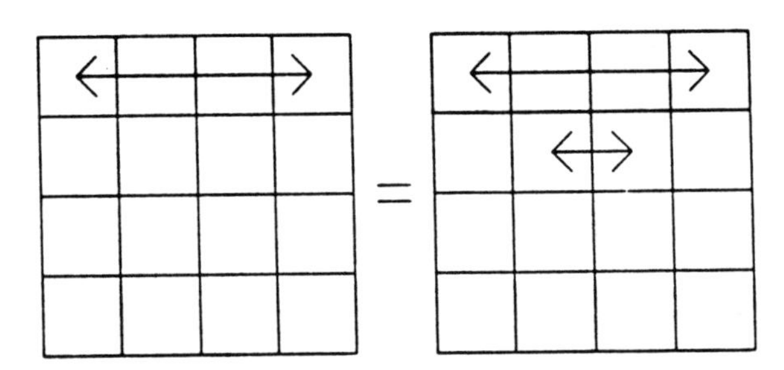

Figure 2:

the kernel of the homomorphism from $\mathfrak{U}$ to $\mathcal{K}$. So we get a short exact sequence of homomorphisms

$$0 \to \mathfrak{M} \to \mathfrak{U} \to \mathcal{K} \to 0$$

If we forget about edge-flips and only consider the permutations involved, we find $\mathfrak{M} \simeq \mathfrak{U}_{12}$. This suggests the following solution of the permutation problem: First we work with the generators of $\mathcal{K}$ to reduce the pattern to one with the corners in place. This means that we have obtained a pattern in $\mathfrak{M}$. Then we use—more sophisticated—generators of $\mathfrak{M}$ to reduce the pattern to the original one. The solution is a path through the subgroup $\mathfrak{M}$.

$$\mathfrak{U} \triangleright \mathfrak{M} \triangleright \mathfrak{E}$$

with factors

$$\mathfrak{G}_8 \qquad \mathfrak{U}_{12}$$

To deserve the name *solvable* the factors must be abelian. But to solve a scrambled cube it is enough that we can generate the groups conveniently, and our main aid is the generators of the alternating groups called 3-cycles.

A real solution to the cube problem is—e.g., the solution suggested by David Singmaster—a 5-step *normal series* analogous to the 2-step *normal series* above.

The first step is to make both permutations even. This is to generate the factor group $\mathbb{Z}_2$. Then we replace the edges by using 3-cycles as generators of the factor group $\mathfrak{U}_{12}$, etc.

The normal series in Singmaster's solution is

$$\mathfrak{U} \triangleright \mathfrak{H}K \triangleright \mathfrak{H}V \triangleright \mathfrak{D}V \triangleright \mathfrak{D} \triangleright \mathfrak{E}$$

with factors

$$\mathbb{Z}_2 \qquad \mathfrak{U}_{12} \qquad \mathfrak{U}_8 \qquad \mathbb{Z}_2^{11} \qquad \mathbb{Z}_3^7$$

So, Singmaster's recipe should be understood as the following 5-step advice.

1. Make both permutations even by a possible twist of one face.

2. Replace the edges by the given 3-cycles.

3. Replace the corners by the given 3-cycles.

4. Flip the edges by the given pair-flippers.

5. Twist the corners by the given pair-twisters.

Hence, the important thing is to understand what you are doing rather than to get the right answer!

References

1. Christoph Bandelow, *Inside Rubik's Cube and Beyond,* Birkhäuser, Basel, 1982.

2. J. A. Eidswick, Cubelike puzzles—what are they and how do you solve them?, *Amer. Math. Monthly* 93 (1986) 157–176.

3. Mogens Esrom Larsen, *Rubiks terning,* Arnold Busck, København, 1981.

4. Mogens Esrom Larsen, *Gruppeteori,* Arnold Busck, København, 1981.

5. Mogens Esrom Larsen, An undergraduate proof of the main theorem on permutations, *Internat. J. Math. Ed. Sci. Tech.* 14 (1983) 67–68.

6. Mogens Esrom Larsen, Rubik's Revenge: the group theoretical solution, *Amer. Math. Monthly* 92 (1985) 381–390.

7. David Singmaster, *Notes on Rubik's 'Magic Cube',* 5th ed., London, 1980.

8. Edward C. Turner & Karen F. Gold, Rubik's groups, *Amer. Math. Monthly* 92 (1985) 617–629.

København universitets matematiske institut
Universitetsparken 5
DK-2100 København Ø

Golomb's Twelve Pentomino* Problems

Andy Liu†

1 A Brief Survey

The polyominoes are geometric shapes formed of unit squares joined edge-to-edge. There are one monomino, one domino (from which the "family name" is derived), two trominoes and four tetrominoes. The most attractive are the twelve pentominoes, shown in Figure 1 with their "letter names."

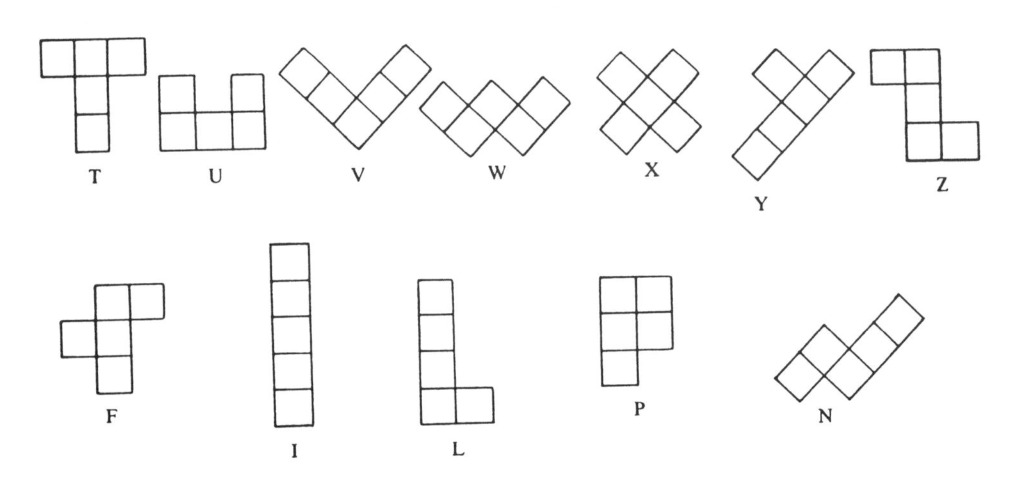

Figure 1: Golomb's names for the twelve pentominoes.

Their number is small enough to be manageable and large enough to offer diversity. They were first featured in a classic problem of Dudeney [3], then discussed by Dawson and Lester [2], and popularized by Golomb [10] and especially Martin Gardner [7, 8, 9].

In his masterpiece *Polyominoes,* Golomb [11] discussed many fascinating aspects of these combinatorial pieces. To whet the appetite of enthusiasts, he posed twelve "readers' research problems." More than twenty years after the publication of this classic, one of these problems is still unsolved.

In each of Problems 84 to 87, the readers are asked to construct a pair of rectangles using a complete set of pentominoes, or prove that no such constructions are possible. The rectangles are

*"Pentomino" is a registered trademark of S.W. Golomb.

†Research partially supported by NSERC grant A5137.

4 by 5 and 5 by 8 in Problem 84, 3 by 5 and 5 by 9 in Problem 85, 4 by 5 and 4 by 10 in Problem 86, and 2 by 10 and 5 by 8 in Problem 87. In each of Problems 81, 82, 83, 89 and 90, a specific shape is to be constructed. These are shown in Figure 2.

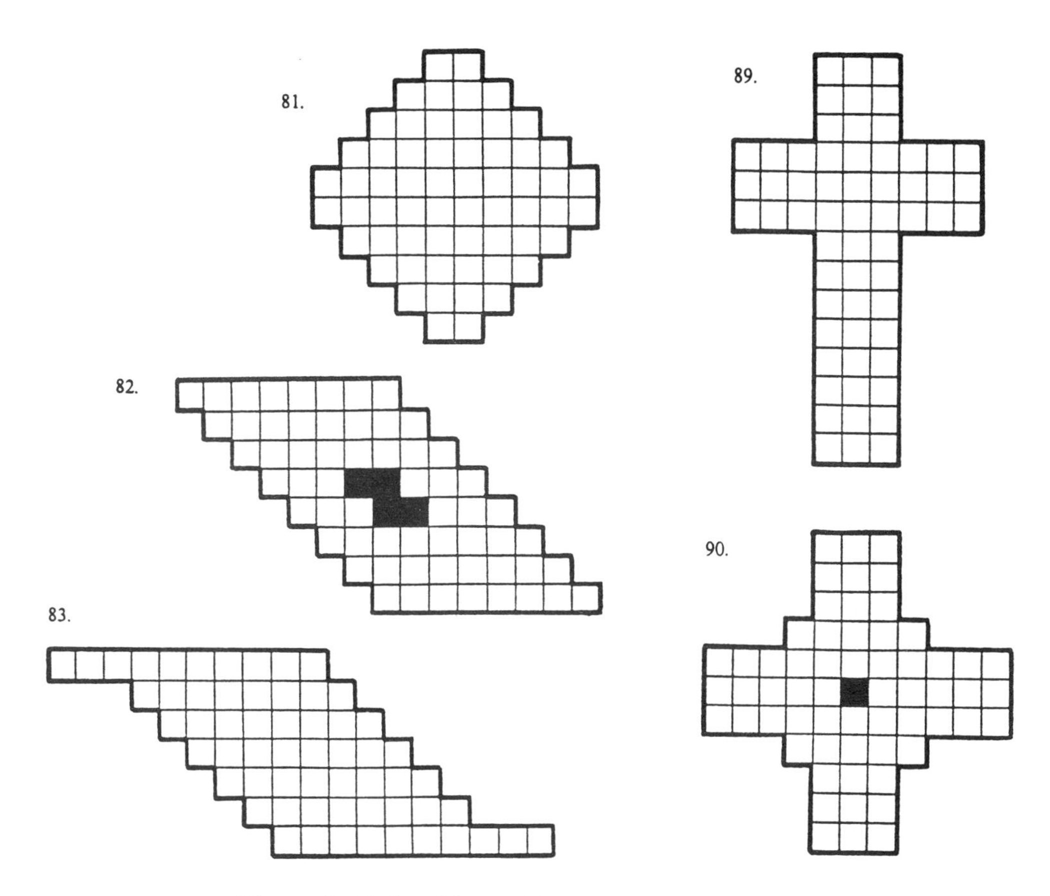

Figure 2: Golomb's problems 81, 82, 83, 89 & 90.

Problems 82 and 83 were solved by Walsh [**17**], Problems 89 and 90 by Garcia [**5**] and Problem 85 by Fairbairn [**4**]. These solutions are all in the affirmative, and unpublished [**6**].

Fairbairn's solution shows up in an exhaustive list he had compiled of 1010 ways of constructing a 5 by 12 rectangle. Haselgrove & Haselgrove [**12**] had an earlier computer program which generated an exhaustive list of 2339 ways of constructing a 6 by 10 rectangle. Miller [**15**] pointed out that it could have been run for a 5 by 12 rectangle as well, but this was not done.

Fairbairn's list contains no construction which would solve Problem 84. It may be desirable to obtain a proof of impossibility without reference to such lists.

Published solutions to Problems 82, 83, 85, 89, and 90, found independently, appeared in [**14**], which also contained a proof that the simultaneous rectangles in Problem 87 cannot be constructed. An affirmative solution to Problem 86 appeared in [**1**]. In this paper, we shall prove that the construction required in Problem 81 is not possible.

These nine "standard" pentomino problems share the following characteristics:

1. The construction of a specific shape is sought.
2. Only one complete set of pentominoes is used.
3. The pentominoes are considered planar figures.

In each of the remaining three problems, exactly one of the above is suppressed.

A problem in which 1. is suppressed simply asks the readers to use a complete planar set of pentominoes to construct an unspecified shape. Clearly a modification is in order. In Problem 88, the only unsolved of the twelve, the readers are asked to choose a shape consisting of fifteen unit squares and construct four copies of it.

In Problem 91, the readers are asked to use nine complete sets of pentominoes to construct a set of models of the pentominoes three times as long and as wide. Implicit are the requirements that the nine pieces used to construct each model must not contain duplicates, and must not contain the piece corresponding to the model.

Figure 3 shows one such construction, independent of prior unpublished solutions obtained by Wade Philpott **[13]** and by several people in Japan **[18]**, some by hand and some with the assistance of computers. Figure 3 was obtained by hand.

In Problem 92, the pentominoes are considered solid figures with unit thickness, and the construction of a model of the F-pentomino is sought, where the model is twice as long and as wide but three times as thick. A solution was published in **[16]**.

2 Solution to Problem 81

We shall prove that the desired shape, featured in Figures 4 and 5, cannot be constructed with a complete set of pentominoes.

There are twenty cells on the boundary of the shape. The maximum number of these cells that can be covered by a particular pentomino is as follows:

F	I	L	N	P	T	U	V	W	X	Y	Z
3	1	2	2	3	1	2	1	3	2	2	1

The total is 23, so there is no immediate contradiction. However, these boundary cells will play an important role in our argument. If the X-pentomino is to occupy its maximum quota of boundary cells, or in fact any boundary cells, it must be placed, up to rotation and reflexion, as shown in the upper part of Figure 4. Note that the top part of the shape must now be occupied by the P-pentomino.

Consider cell number 1. It cannot be occupied by the I-pentomino. If it is occupied by any of the T, U, V or Y-pentominoes, regions in which no pentomino can fit will be isolated. If it is occupied by any of the F, L, N or Z-pentominoes without isolating impossible regions, then the leftmost part of the shape must be occupied by the P-pentomino, which however has been committed.

It now follows that cell number 1 must be occupied by the W-pentomino. The orientation shown is the only possible one since alternative placements of the W-pentomino either isolate an impossible region or necessitate the use of another copy of the P-pentomino.

Figure 3: The pentominoes formed with nine sets of pentominoes.

Consider the cells numbered 2 and 3. We claim that they cannot be occupied by the same pentomino. Suppose the contrary is true. Then such a pentomino cannot also occupy cell number 4. Otherwise, it must occupy cell number 5 as well, but the P-pentomino is the only one that can cover a 2 by 2 square. Now the bottom part of the shape must be occupied by the P-pentomino, which is no longer available.

It follows that the pentomino occupying cell number 2 cannot occupy cell number 3, nor can it, by symmetry, occupy cell number 6. Hence it must be one of the L, N, U, W and Z-pentominoes. However, all except the W-pentomino will isolate impossible regions, and the W-pentomino has been committed. Hence the X-pentomino cannot occupy any boundary cell.

This means that exactly one other pentomino may occupy one less than its maximum quota of boundary cells while the others must occupy their maximum quota. A pentomino other than the X-pentomino is said to be **deficient** if it occupies less than its maximum quota.

Suppose the U-pentomino is not deficient. Then it must be placed, again up to rotation and reflexion, as shown in the rightmost part of Figure 4. Now cell number 7 can only be occupied by one of the F, L and W-pentominoes which will then be deficient. It follows that either the U-pentomino is deficient or one of the F, L and W-pentominoes is deficient, and in the manner described.

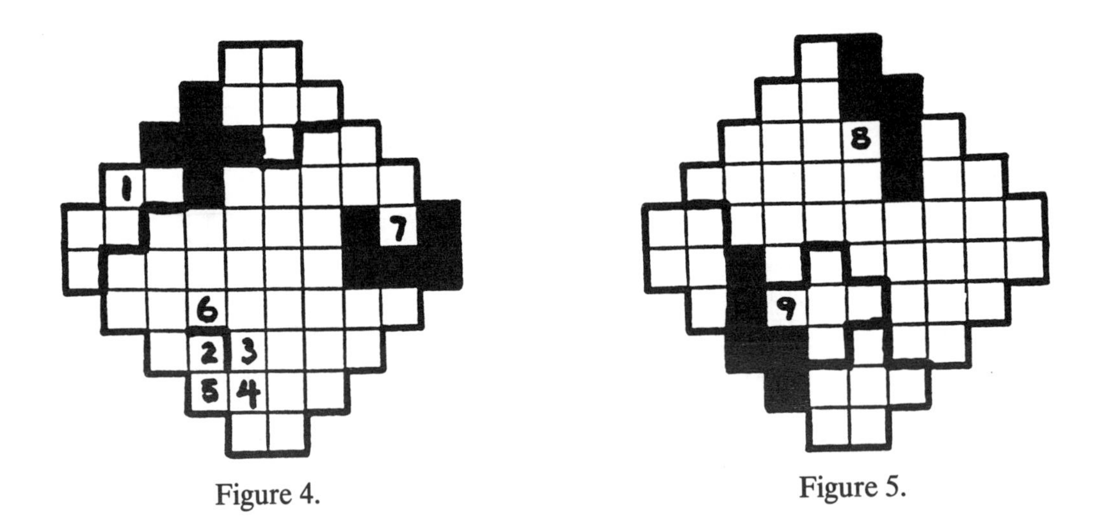

Figure 4. Figure 5.

In any case, the N-pentomino is not deficient. It has two non-equivalent non-deficient placements, as shown in the upper and the lower parts of Figure 5. In the former case, cell number 8 cannot be occupied without creating multiple deficiency when the U-pentomino is taken into consideration.

Finally, in the latter case, cell number 9 can only be occupied non-deficiently by the X-pentomino as shown, but then two copies of the P-pentomino would be required to complete the construction. This completes the proof that the construction required in Problem 81 is not possible.

Acknowledgement: The author thanks Richard Guy for references [**2, 12, 15**], Kathy Jones [**13**] and Nob Yoshigahara [**18**] for their oral communications and Martin Gardner [**6**] for his informative letter.

References

1. C.J. Bouwkamp, Simultaneous 4 by 5 and 4 by 10 pentomino rectangles, *J. Recreational Math.*, **3**(1970) 125.

2. T.R. Dawson & W.E. Lester, A notation for dissection problems (III), *Fairy Chess Review*, **5**(1937) 46–47.

3. H.E. Dudeney, The Canterbury Puzzles, Dover, New York, 1958, pp. 119–121.

4. R.A. Fairbairn, letter to Martin Gardner, August 4, 1967.

5. A.A. Garcia, letter to Martin Gardner, July 9, 1971.

6. Martin Gardner, letter to Andy Liu, May 8, 1982.

7. Martin Gardner, The Scientific American Book of Mathematical Puzzles and Diversions, Fireside Books, New York, 1959, pp. 124–140.

8. Martin Gardner, New Mathematical Diversions from Scientific American, Fireside Books, New York, 1966, pp. 150–161.

9. Martin Gardner, Mathematical Magic Show, Math. Assoc. Amer., Washington, 1990, pp. 172–187.

10. S.W. Golomb, Checkerboards and polyominoes, *Amer. Math. Monthly*, **61**(1954) 675–682.

11. S.W. Golomb, Polyominoes, Charles Scribners' Sons, New York, 1965, pp. 165–167.

12. C.B. Haselgrove & J. Haselgrove, A computer program for pentominoes, *Eureka*, **23**(1960) 16–18.

13. Kathy Jones, oral communication, July 30, 1986.

14. Andy Liu, Pentomino problems, *J. Recreational Math.*, **15**(1982) 8–13.

15. J.C.P. Miller, Pentominoes, *Eureka*, **23**(1960) 13–16.

16. J.M.M. Verbakel, The F-pentacube problem, *J. Recreational Math.*, **5**(1972) 20–21.

17. M.R. Walsh, letter to Martin Gardner, March 13, 1975.

18. Nob Yoshigahara, oral communication, July 30, 1986.

Department of Mathematics,
The University of Alberta,
Edmonton, T6G 2G1, Canada

A New Take-Away Game

Jim Propp*

1 Introduction

Two children take turns stealing cookies from a larder, each removing a single cookie every other day. Some of the cookies may go bad while on the shelf and become unsafe to eat, but fortunately each cookie has an expiration date written on it in frosting. The goal of each child is *not* to maximize the number of cookies he or she eats, but rather to have the spiteful pleasure of getting the last cookie.

If none of the cookies goes bad during the course of play, then of course the game will be a dull one, as the outcome depends only on whether the number of cookies present at the start was even or odd. But if the initial provisioning is such that some of the cookies might go bad during the course of play, then who gets the last cookie (thereby winning the game) may depend on who eats which cookie when.

We may represent each cookie by a heap of n counters, where n is the number of days remaining before the cookie goes bad. A heap of size 0 represents a cookie whose time has expired; we assume that the two children are unwilling to risk food poisoning and will not eat a possibly spoiled cookie. On any given turn, a player removes *one* of the non-empty heaps and diminishes all the others by 1 counter; this corresponds to eating one of the cookies and noting that the rest have aged by a single day. When no counters remain, the player who just moved is deemed the winner. Let us represent each position schematically by a list of boldface numbers arranged in non-descending order (called the **components** of the position), each of which gives the size of a corresponding heap. Thus **1 2 4** signifies a three-heap position with heaps of size 1, 2 and 4. In this representation, a sample game would appear as:

1 2 4	
↓ (under 2)	1st player removes the 2-heap;
0 3	
↓ (under 3)	2nd player removes the 3-heap;
0	
	2nd player wins.

Though the heap-notation is probably the most convenient form in which to display cookie-game positions, there is another interesting representation: one can represent a heap of n counters

*Partially supported by an NSF Postdoctoral Fellowship during the writing of this article.

graphically by a stack of n squares, and stick all the stacks together in a row, with the shorter stacks on the left and the taller stacks on the right. Thus, the position **1 2 4** would be depicted

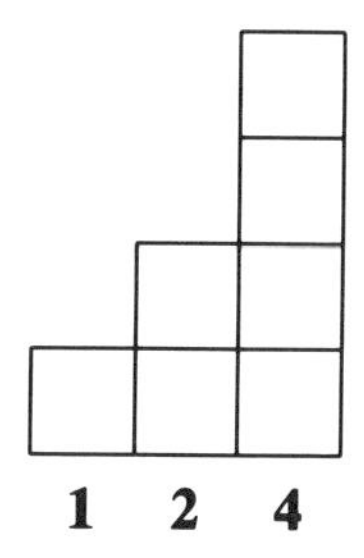

A legal move in this box-diagram representation is to cross out one of the columns and then to cross out the lowest row. Players continue in alternation until all the squares have been crossed out.

Jonathan King has made the interesting suggestion of regarding the *rows* of this diagram as the "heaps", rather than the columns. Then we are effectively playing a heap-game in which a legal move consists of first removing a single counter from every heap that equals or exceeds a certain size (which may be chosen by the player at will, provided there is at least one heap of that size) and then removing the largest heap in its entirety (unless no heaps remain).

Lastly, John Conway has a way of viewing the game as a game of pure passing, in which there are no game pieces and no legal moves! Under normal play such a game would end as soon as it started, if not sooner, but in our modified scenario the players jointly own a supply of permits, each of which entitles the bearer to skip one turn. A permit can be used only once, and what is more each has an expiration date, so that if it is not used on or before that date it cannot be used at all. (Conway calls a permit that entitles the bearer to skip a turn anywhere up until the nth move of the game an "n-day pass".) Players take turns using these passes until no usable passes remain on the table. At this point, the player whose turn it is to move, no longer having the option of passing, loses the game.

2 The Reduction Theorems

Since a game that starts with n heaps must terminate in finitely many (indeed, at most n) moves, one of the two players must have a winning strategy. For instance, the position **1** is a clear win for the first player. On the other hand, note that from the initial position **1 2 4** the second player can win regardless of the first player's opening move:

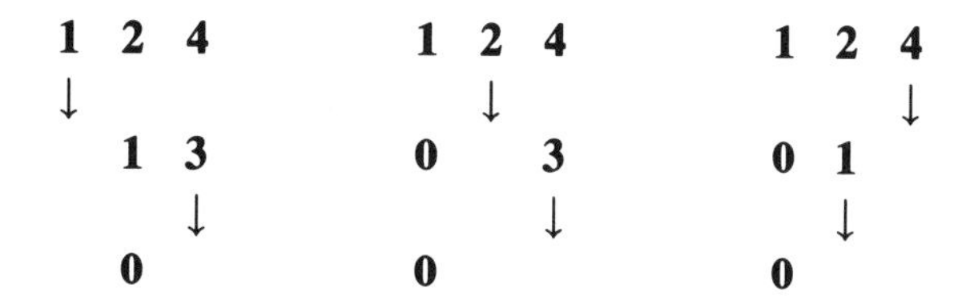

Following tradition [1], we call positions like **1**, in which the player about to move has a winning strategy, $\mathcal{N}$-positions (wins for the Next player). If a game is not an $\mathcal{N}$-position, then it is a win for

the second player; such positions are called $\mathcal{P}$-positions (wins for the Previous player). We will call the $\mathcal{N}$-versus-$\mathcal{P}$ character of the position the *type* of the position. Thus, **1 2 4** has type $\mathcal{P}$. Note that a position is of type $\mathcal{N}$ if and only if there is a move from it to a position of type $\mathcal{P}$. If one could efficiently determine whether an arbitrary position was of type $\mathcal{N}$ or type $\mathcal{P}$, one would have an effective winning strategy for one of the two players, since the rule "Always move to a $\mathcal{P}$-position if possible" can never lead a player astray.

If two positions G and H are of the same type (that is, if both are $\mathcal{P}$-positions or both are $\mathcal{N}$-positions), then we will write "$G \sim H$". For instance, it is easy to see in advance that

$$\mathbf{1\,5\,6\,7} \sim \mathbf{1\,4\,4\,4} \tag{1}$$

without even knowing whether both are $\mathcal{P}$-positions or both are $\mathcal{N}$-positions, because the exact size of a heap cannot matter if it exceeds the maximum possible length of the game, which is at most the number of heaps present at the start of the game. As a second example, one can see that

$$\mathbf{1\,1\,2} \sim \mathbf{1\,2}\,, \tag{2}$$

because only k heaps of size k can be relevant strategically—extra heaps of that size may be ignored, since they will vanish before they can be used. Relations (1) and (2) are special cases of two general theorems which greatly reduce the set of positions requiring analysis.

Depletion Theorem: Let G be the position $\mathbf{a_1\,a_2\,a_3\,\cdots\,a_n}$. Any component of size $> n$ may be depleted down to size n without affecting the type of the position.

Deletion Theorem: Let G be the position $\mathbf{a_1\,a_2\,a_3\,\cdots\,a_n}$. If $a_k < k$ for some k, then the component $\mathbf{a_1}$ may be deleted without affecting the type of the position. Indeed, if we define the **surplus** of G as the smallest non-negative integer s such that $k - a_k \leq s$ for all $1 \leq k \leq n$, then whenever $s > 0$, the components $\mathbf{a_1}, \mathbf{a_2}, \cdots, \mathbf{a_s}$ may all be deleted.

Proofs of the two theorems will appear in a forthcoming article with Dan Ullman.

It's worth emphasizing that heaps may have size 0; in practice, one ignores 0-components, since they can never be selected, but the theoretical analysis of the game is simpler if one permits them. A position without 0-components will be called **proper**.

We call a proper position $\mathbf{a_1 \cdots a_n}$ **reduced** if $1 \leq a_k \leq n$ for all k and if the surplus of the position is 0, that is, if $a_k \geq k$ for all k.

As an application of these theorems, consider the position

$$\mathbf{1\,2\,3\,3\,6}.$$

By the depletion theorem, the position **1 2 3 3 5** has the same type. This position has surplus $4 - a_4 = 4 - 3 = 1$, and so is tantamount to **2 3 3 5**, by deletion. We may apply depletion once more to obtain the reduced position **2 3 3 4**.

The operations of deletion and depletion allow every position to be converted into a unique reduced form. To understand why, it is helpful to have an alternative way of conceiving of depletion: instead of replacing heaps of size $> n$ by heaps of size n (where n is the number of heaps), replace all heaps of size $\geq n$ by heaps that are effectively infinite, to be denoted by "$*$"; these correspond to cookies with an infinite shelf-life that remain edible indefinitely. This formal device makes it much easier to see what happens when one alternates the operations of deletion and depletion, because the process of "depletion" now leaves the surplus of a position alone. It becomes clear how one can determine the reduced form of a position: Perform deletion as many times as possible, and then depletion. For instance, given the position **1 2 3 3 6** $\sim$ **1 2 3 3** $*$, we should first delete the smallest component, yielding **2 3 3 6** $\sim$ **2 3 3** $*$; we should then deplete the largest component, yielding **2 3 3 4**. If the position we started with was proper, then "delete as many components as possible" means simply delete the s smallest components, where s is the surplus of the position.

The reduced positions with n heaps correspond to box-diagrams of width n and height n in which the bounding lattice-path that runs along the top of the diagram to connect the lower-left corner with the upper-right corner never goes below the diagonal line joining those two corners. Thus, the five reduced 3-heap positions may be represented as:

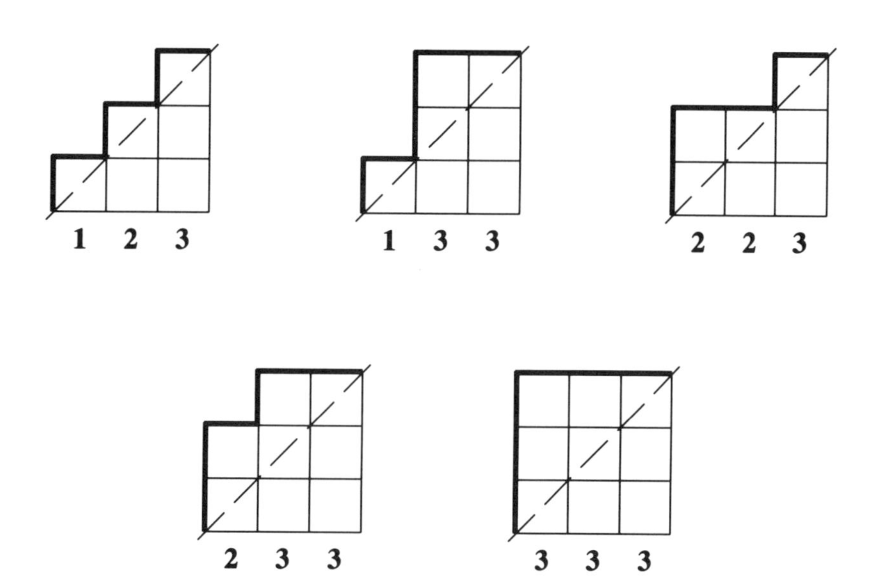

Such positions are in one-to-one correspondence with the lattice-paths that bound them, and these in turn are well known to be enumerated by the Catalan numbers (see for example **[2]**, **[3]**, and **[4]**). Thus the number of reduced positions with n components is

$$\binom{2n}{n} \Big/ (n+1).$$

Table 1 shows all of the reduced positions involving 4 or fewer components; in the $\mathcal{N}$-positions, the winning components are circled. (We call a component *winning* if selecting that component leads to a $\mathcal{P}$-position.)

①				$\mathcal{N}$
1	②			$\mathcal{N}$
2	2			$\mathcal{P}$
1	2	3		$\mathcal{P}$
①	3	3		$\mathcal{N}$
2	2	3		$\mathcal{P}$
②	3	3		$\mathcal{N}$
③	③	③		$\mathcal{N}$
①	②	3	4	$\mathcal{N}$
1	②	4	4	$\mathcal{N}$
①	③	③	④	$\mathcal{N}$
1	③	④	④	$\mathcal{N}$
1	④	④	④	$\mathcal{N}$
②	②	3	4	$\mathcal{N}$
2	2	4	4	$\mathcal{P}$
②	③	③	④	$\mathcal{N}$
2	3	④	④	$\mathcal{N}$
2	4	4	4	$\mathcal{P}$
③	③	③	④	$\mathcal{N}$
3	3	④	④	$\mathcal{N}$
3	4	4	4	$\mathcal{P}$
4	4	4	4	$\mathcal{P}$

Table 1

3 An Intriguing Pattern

If we tabulate positions of the form **1 2** $\cdots$ **n**, we notice a striking regularity:

①												$\mathcal{N}$
1	②											$\mathcal{N}$
1	2	3										$\mathcal{P}$
①	②	3	4									$\mathcal{N}$
1	2	3	④	⑤								$\mathcal{N}$
1	2	3	4	5	6							$\mathcal{P}$
①	②	③	④	5	6	7						$\mathcal{N}$
1	2	3	4	5	⑥	⑦	⑧					$\mathcal{N}$
1	2	3	4	5	6	7	8	9				$\mathcal{P}$
①	②	③	④	⑤	⑥	7	8	9	10			$\mathcal{N}$
1	2	3	4	5	6	7	⑧	⑨	⑩	⑪		$\mathcal{N}$
1	2	3	4	5	6	7	8	9	10	11	12	$\mathcal{P}$

Table 2

That is, with the exception of $n = 1$, we have:

For $n \equiv 0 \pmod 3$, the position has type $\mathcal{P}$.

For $n \equiv 1 \pmod 3$, the winning components are $\mathbf{1}, \cdots, \frac{2}{3}(\mathbf{n}-\mathbf{1})$.

For $n \equiv 2 \pmod 3$, the winning components are $\frac{2}{3}(\mathbf{n}+\mathbf{1}), \cdots, \mathbf{n}$.

This pattern has been verified up to $n = 26$, but it has not been proved to hold for all n.

More generally, we can consider the two-dimensional family of positions of the form

$$\mathbf{a} \;\; \mathbf{a+1} \;\; \mathbf{a+2} \;\; \cdots \;\; \mathbf{a+n-1}\,.$$

Call them arithmetic progression positions, or AP-positions for short. Table 3 gives, for each value of a and n, the winning components of the corresponding AP-position. A blank entry means that the position has no winning components, i.e., no winning move is possible because the position has type $\mathcal{P}$.

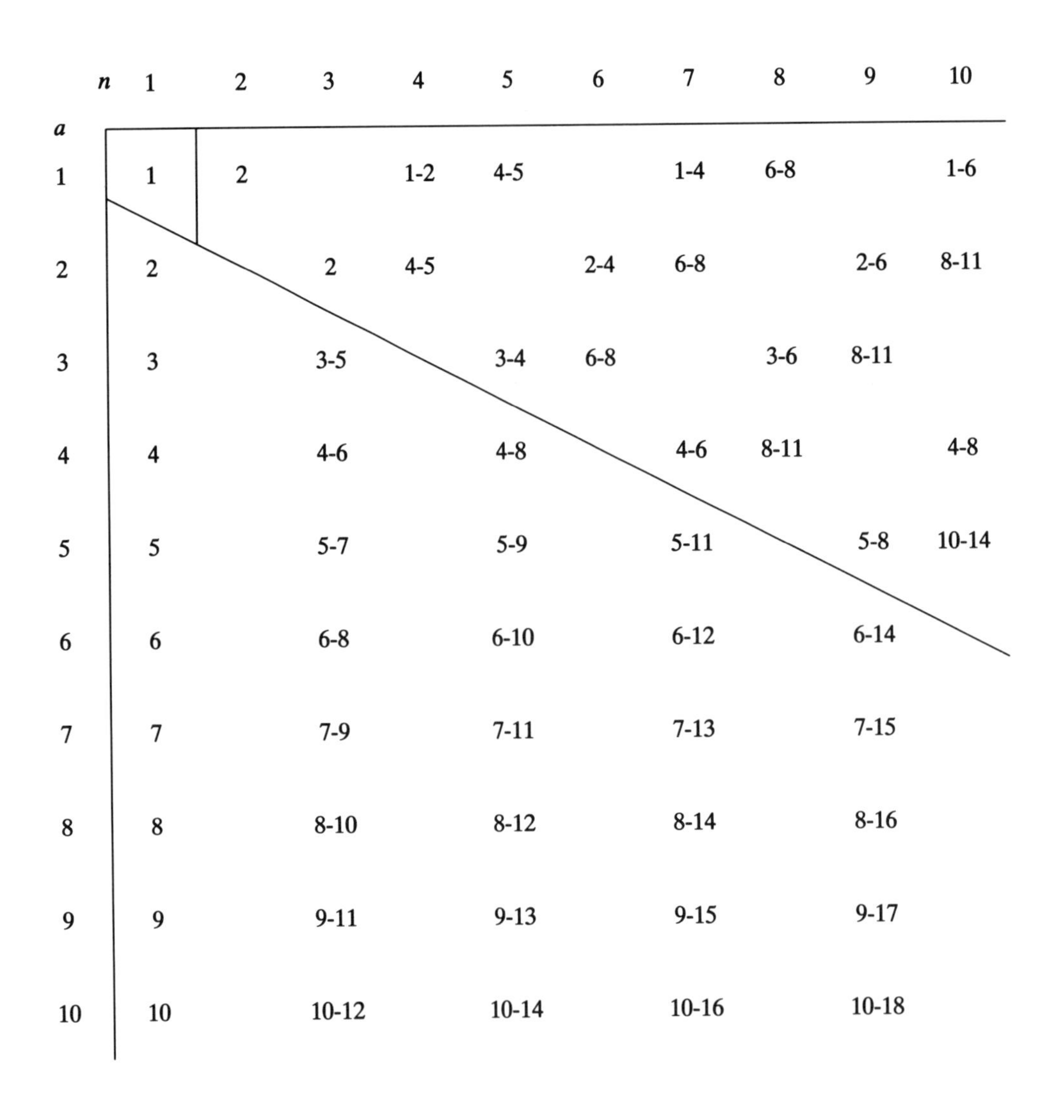

a \ n	1	2	3	4	5	6	7	8	9	10
1	1	2		1-2	4-5		1-4	6-8		1-6
2	2		2	4-5		2-4	6-8		2-6	8-11
3	3		3-5		3-4	6-8		3-6	8-11	
4	4		4-6		4-8		4-6	8-11		4-8
5	5		5-7		5-9		5-11		5-8	10-14
6	6		6-8		6-10		6-12		6-14	
7	7		7-9		7-11		7-13		7-15	
8	8		8-10		8-12		8-14		8-16	
9	9		9-11		9-13		9-15		9-17	
10	10		10-12		10-14		10-16		10-18	

Table 3

Lines are used to divide the behavior of the family of AP-positions into three distinct regimes. One regime (the "zeroth") consists of the single position **1** (with $a = n = 1$); the other two are infinite, and are determined by whether $n \leq 2a - 2$ or $n \geq 2a - 2$ (the boundary case $n = 2a - 2$ is properly considered as belonging to both).

In the regime $n \leq 2a - 2$, the type of the position is $\mathcal{P}$ or $\mathcal{N}$ according to whether n is even or odd. In fact, the same is true under the weaker condition $n \leq 2a - 1$. We prove this by considering the even case and odd case separately, and showing that one player can control the game by making sure that no heap ever goes down to zero. Suppose first that n is even, so that $n \leq 2a - 2$. At the start of the game there are n heaps, each of size at least a. If the second player always selects the smallest possible component, then after two turns (one turn by each player) there will be $n - 2$ heaps, each of size at least $a - 1$; after two more turns there will be $n - 4$ heaps, each of size at least $a - 2$; and so on. Finally, after $n - 2$ turns, there will be 2 heaps, each of size at least $a - (n - 2)/2 \geq 2$, which is a $\mathcal{P}$ position. Hence the strategy of always removing the smallest heap guarantees a win for the second player. The argument for n odd is very similar, except that we conclude our analysis by looking at what happens after $n - 1$ turns: there will be 1 heap of size at least $a - (n - 1)/2 \geq 1$, which is an $\mathcal{N}$ position.

The second infinite regime ($n \geq 2a - 2$) is more interesting: it contains among other things the special AP-positions **1 2** $\cdots$ **n**, and exhibits the same sort of period-3 behavior shown in Table 2. The patterns manifested in Table 3 have not been proved to hold in general, but they have been checked as far as $a = n = 20$. Also, the pattern can be shown to hold for the infinite diagonals $n = 2a - 1$ and $n = 2a$; the key idea is that games that begin in the second regime but very close to its boundary with the first regime can only get closer to the (already-understood) first regime, or move into it altogether, as the game progresses. The same method of analysis should in principle allow one to prove the pattern for any particular diagonal $n = 2a + k$, but as k gets large and the starting position gets further from the boundary between the two regimes, the amount of detailed analysis required seems to grow without bound.

A simple rule for the opening move, apparently applicable to every AP-position (though for quite different reasons in the two regimes) is:

> If the largest component is congruent to 0, 1, or 2 (mod 3), then select indifferently, select the smallest component, or select the largest component, respectively.

J.C. Kenyon has looked at the Grundy values of various positions in the cookie game. Following his lead, we can tabulate the Grundy values of the AP-positions. Once again, the regularity is striking (see Table 4). Notice that within the $n \geq 2a - 2$ regime, the Grundy value is constant on each northeast-to-southwest diagonal—that is, it depends only on $a + n$. Indeed, the Grundy value appears to be a periodic function of $a + n$ with period 6, given by the repeating sequence $0, 1, 2, 0, 1, 3, \ldots$; this pattern has been observed as far out as $a = n = 20$. Can we somehow prove it in general by induction? If we try to devise such a proof, we almost immediately run afoul of the fact that the AP-positions do not form an isolated subset of "position-space". Indeed, starting from an AP-position, either player can force his opponent into positions that look nothing at all like AP-positions. It follows that a winning strategy for a game whose first position is an AP-position must involve an understanding of many non-AP-positions, and it is correspondingly unlikely that one can prove facts about the class of AP-positions without having to consider the general behavior of the game. We can either find this fact of life vexing or encouraging: it is vexing because we are prevented from obtaining partial results about special classes of positions which by all evidence are orderly and understandable, yet it is encouraging because we are led to suspect that there is an over-arching pattern that governs not just the AP-positions but all positions, or at least all those in which the heap-sizes are distinct.

a \ n	1	2	3	4	5	6	7	8	9	10
1	1	2	0	1	3	0	1	2	0	1
2	1	0	1	3	0	1	2	0	1	3
3	1	0	1	0	1	2	0	1	3	0
4	1	0	1	0	1	0	1	3	0	1
5	1	0	1	0	1	0	1	0	1	2
6	1	0	1	0	1	0	1	0	1	0
7	1	0	1	0	1	0	1	0	1	0
8	1	0	1	0	1	0	1	0	1	0
9	1	0	1	0	1	0	1	0	1	0
10	1	0	1	0	1	0	1	0	1	0

Table 4

4 Strict Positions

Let us call a position $\mathbf{a_1} \cdots \mathbf{a_n}$ **strict** if it is proper and if all its components are distinct, with the possible exception of components of size n, which may be duplicated. (The motivation for this exception comes from the Depletion Theorem: If the position **2 3 4** deserves to be called strict, then so does its reduced form **2 3 3**.) Clearly all AP-positions are strict, and it is left to the reader to check that the number of strict reduced positions with n heaps is $2n - 1$.

The strict reduced positions seem to have some special characteristics.

Grundy Value Conjecture: Every strict position has Grundy value 0, 1, 2 or 3.

The evidence in favor of this conjecture is quite good. Strict positions with Grundy value 3 appear fairly early (starting with **2 3 4 4**), but exhaustive computer search reveals that there are no strict positions with Grundy value 4 having 20 or fewer components. In contrast, for non-strict positions, Grundy value 4 first appears for positions of size $n = 7$, and Grundy value 5 first appears for positions of size $n = 11$ (which is as far as the search went); so, for general positions, it would be rash indeed to conjecture that the Grundy values remain bounded.

Winning Interval Conjecture: If G is a strict position of type $\mathcal{N}$, then the winning components of G are exactly those whose sizes lie in a certain "winning interval" of natural numbers; that is, the winning components form a single uninterrupted subsequence $\mathbf{a_i}\ \mathbf{a_{i+1}} \cdots \mathbf{a_j}$.

This conjecture is *not* true for non-strict positions. For example, the position **1 3 3 5 5** does not have the winning interval property, since its winning components are the **1** and the **5**'s. On the other hand, the Winning Interval Conjecture has been checked for all strict reduced positions of length up to 20.

Shortly before this article went to press, Dan Ullman made the shrewd suggestion that every strict reduced position G has a certain innate duration, and that the way a player wins such a game is by controlling not merely the parity of the length of the game but the actual length of the game itself. This is indeed the case, and the observation turns out to be the key to the structure of the game's position-space. By the time this article appears in print, we hope to have a full theory for strict positions, and to have proofs of all the conjectures in this article.

REFERENCES

1. Elwyn R. Berlekamp, John H. Conway & Richard K. Guy, *Winning Ways for Your Mathematical Plays,* Academic Press, 1982.

2. Martin Gardner, Catalan numbers: an integer sequence that materializes in unexpected places, *Scientific American* 234, No. 6 (June, 1976), *Mathematical Games,* pp. 120–125 and 132.

3. Fred S. Roberts, *Applied Combinatorics,* Prentice-Hall, 1984, pp. 234–243.

4. I. M. Yaglom & A. M. Yaglom, *Challenging Mathematical Problems with Elementary Solutions,* Holden-Day, 1964, volume 1, problems 53, 83.

Department of Mathematics,
Massachusetts Institute of Technology,
Cambridge, MA 02139-4307

Confessions of a Puzzlesmith

Michael Stueben

Mathematical curiosities have always been a passion of mine. Perhaps that is why I became a high school mathematics teacher. From 1979 to 1984 I edited a puzzle column in the Washington DC area Mensa newsletter. From 1984 to 1987 I edited the "Brain Bogglers" puzzle column in *Discover* magazine. My experiences, the opinions I formed and the insights gained from these activities are the subject of this chapter.

Algebraic drill problems were sometimes fun in our school days because of the competition with other students and the satisfaction of a job well done, but now most adults desire entertainment and stimulation, not exercises of endurance. To be worthy of a reader's time and effort, a popular puzzle should possess some special quality of merit. There are two simple criteria I use to judge a puzzle as entertaining.

- First, the puzzle should be short in statement and easy to comprehend. As Sam Loyd said, it is largely the first glance at a problem that challenges the solver to attempt it.
- Second, the solution should be easy to understand, but psychologically difficult to discover. This type of solution is a reward even to the solver who has failed; a complex solution makes the solver wish he or she had never started.

Beyond these criteria, perhaps the greatest quality a puzzle may have is an answer that runs counter to our intuition. This quality is very rare in puzzles. Another valuable characteristic is the hidden trap for the careless worker. This is the sole basis for many enjoyable puzzles. To perceive and sidestep a trap is pleasing to the ego of a solver. (Trick questions, of course, must be stated as such to avoid wasting the reader's time.)

It is sometimes possible to associate a puzzle with a historically well known person or event. This definitely adds to the solver's interest. I recall a puzzle dealing with an application of Kepler's Second Law. I know next to nothing about physics, but the realization that I could understand a clever, albeit elementary, application of a famous law was very satisfying.

Occasionally a puzzle can be presented with the success rate of past solvers. This allows the isolated solver to compete in private and rate himself against others. Even a dull puzzle becomes more interesting if the challenge ends with a declaration that, "...this puzzle was published with an answer which was not discovered to be wrong for ten years" or "your puzzle editor spent 20 minutes on this conundrum before the obvious dawned."

Finally, a puzzle that teaches us something curious about our world is usually worth the effort. If such puzzles were plentiful, school books would be vastly more interesting. No puzzle will have all of these desirable qualities, but the majority of the pastimes found in books and magazines have none.

The most common complaints about puzzles pertain to ambiguity in wording and unstated assumptions. This is something that is usually a matter of taste. The best puzzlesmiths keep ambiguity to a minimum, but usually cannot remove it entirely without incurring other difficulties. Some very pretty problems can be stated in a few words that would otherwise lose their charm if exact interpretation were required. Unstated assumptions must be made for the same reason. The most precise books of puzzles are mathematical textbooks which are rarely singled out as enjoyable or interesting and that is what puzzling is all about: entertainment.

> If you shut out these tricks and quibbles in a puzzle, you spoil it by overloading the conditions. It is better (except in the case of competitions) to leave certain things to be understood.
>
> H.E. Dudeney (1857-1930),
> Found in *536 Puzzles & Curious Problems*,
> Edited by Martin Gardner,
> (Scribners, 1967) page 290.

There is one very real danger with this philosophy: it can be used to justify poor craftsmanship. A puzzle author's standards for accuracy of solution and rigor in presentation must remain very high. If a puzzle is not meant to be serious or an approximate answer is acceptable this fact should be communicated to the solver, possibly by disclaimer. Writing an interesting, simple to understand puzzle with an accurate answer is one of the more difficult puzzles.

The popularizer has two audiences to please: his readership and the editor who usually has a feel for what the readership wants. The popularizer must win the favor of both to keep his job. When an editor makes a suggestion with which I disagree, I give him a short quick argument. If he still wants the change, I incorporate his ideas into my puzzles. It is vital to keep a good working relationship with an editor. I always have my work edited locally before I send it off to a magazine editor who will judge a writer by his style as much as by the content.

> A surprising number of those who write for print disclaim any grasp of punctuation or taste for its subtleties. They leave the commas to be put in by copyists or secretaries, sometimes implying that their own minds are intent on higher things. But so much of punctuation is inseparable from meaning, it is so integral with the truth or falsity of what we write, that anyone can write himself down a fool through wrong punctuation as readily as through wrong words. It is a frivolous abdication of responsibility for a writer not to know more than a copyist does about what he is trying to say, or not to respect the means of making it clear.
>
> Wilson Follett,
> *Modern American Usage*,
> (Hill & Wang, 1966) pp. 395-396.

I have discovered through bitter experience to beware of last minute changes in wording, which have been requested over the phone by an editor. These changes always seem acceptable, but there isn't enough time to examine them and consider their implications. And, of course, once the approval for a change is made, it is clear to whom the blame for an error belongs.

> Woe betide the puzzle editor who fathers deformed brain children. Puzzle solvers take their pastime mighty seriously and will not overlook even the occasional *faux pas* permitted other departments of a publication. The puzzleman is expected to sail along pretty close to 100 percent with respect to accuracy, if he hopes to preserve his job and good name.... An unsolvable puzzle set adrift will, like the flying Dutchman, sail the seven seas of puzzledom eternally.

Sam Loyd II (1874-1934),
Sam Loyd and His Puzzles,
An Autobiographical Review,
(New York: Barse & Co., 1928), p. 7.

Two of the more serious errors in writing expository material are flogging an idea to death and letting the ego show through. The first bores a reader; the second annoys him. For psychological reasons, both are difficult for an author to see. A good editor can save a writer from these mistakes if the writer will listen.

I have always tried to write thematic puzzle columns. In doing so, I discovered another trap. The columns that were too thematic were the least interesting and received the fewest letters. Evidently, purely thematic columns fail because they appeal only to a small number of readers. For example, gambling is a reasonable theme for a set of puzzles. But five gambling puzzles about poker hands is too narrow a theme. Better is a collection of puzzles about a carnival game, a poker hand, a bar bet, a dice game and a roulette wager. Such a column will appeal to three times as many readers as a column of poker puzzles.

Story lines present another pitfall. It is true that a clever story can do much to make a puzzle more interesting, but it can also obscure the essence of a puzzle. A story must not be too involved.

A puzzle column like my "Brain Bogglers" in *Discover* is not at all the same thing as an expository column like Martin Gardner's "Mathematical Games." The worst columns I've ever written were the ones that waxed eloquently about the implications of a few puzzles. The two just don't mix. I used to think that it was harder to come up with five or six really good puzzles than write a page or two of mathematical exposition, but I eventually surrendered that illusion. For me it is actually easier to fill a page with puzzles rather than exposition; for others, the reverse is probably true.

Many fans of Martin Gardner wonder where he finds his material. The sources for all recreational material (listed in order of importance) are (1) the back issues of mathematical magazines; (2) books on recreational mathematics; (3) the author's own imagination; (4) puzzle books and (5) readers' submissions.

Whenever I find a mathematical curiosity or puzzle, the page number is copied down in the front of the book or back of the magazine. Later, the material is photocopied and filed away for future reference. All of the books and magazines in my personal library are so marked. If I wish to recall something in a certain book, I only need to refer to my files or the notes in the book.

The result is that I have an enormous collection of materials to consult in my work. A fire would probably end my career. When I visited Martin Gardner I noticed that he had used a razor blade to cut out parts of the pages in various books and magazines. (This was before the invention of the photocopy machine.) His collection of the *Mathematical Gazette* is so cut up that it is now worthless to others, but his files have become an invaluable treasury of ideas – and that is what matters.

> Procure, in lots of twenty thousand or more, slips of stiff paper of the size of post cards, made up into pads of 50 or so. Have a pad always about you, and note upon one of them anything worthy of note, the subject being stated at the top and reference being made below to available books or to your own notebooks. If your mind is active, a day will seldom pass when you do not find a dozen items worth such recording; and at the end of twenty years, the slips having been classified and arranged and rearranged, from time to time, you will find yourself in possession of an encyclopedia adapted to your own special wants. It is especially the small points that are thus to be noted; for the large ideas you will carry in your head.

Charles Sanders Peirce (1908),
Found in Kenneth O. May,
Bibliography and Research Manual of the History of Mathematics,
(University of Toronto, 1973) p. 18.

If someone were to ask me for advice in writing a puzzle column I would tell them: Don't be afraid to borrow ideas. You must, of course, not reproduce the writing of another author, but you may borrow and hopefully improve upon another person's ideas. The great composers of music often did this. Martin Gardner, who has always been my greatest source of good ideas, once modestly wrote me that, with few exceptions, all of his ideas came from others. Ideas are universal; they do not belong to individuals. The idea behind a puzzle, like a literary device, belongs to the world. Perhaps 95 percent, maybe more, of my published work is based on the ideas of others that I find in magazines and books.

The three qualities that seem to be important for day-to-day puzzlesmithing are:

1. A great attention to minor details. (In other words, puzzle writing requires a lot of clerical work: checking the wording, looking for alternate interpretations, adjusting the level of difficulty, drawing the illustrations, etc.)

2. An unflagging enthusiasm for math, wordplay, curiosities and puzzles. (This is required to keep the files well stocked with good ideas.)

3. The ability to recognize (and then borrow) the good ideas of others. (A good idea should not die in its original puzzle, but should return to the reading public in varied form so that its maximum potential for entertainment can be realized.)

I have two secrets for improving the quality of my work. First, I pretest my puzzles and use the comments and opinions of the testers to improve my work. Second, after finishing a set of puzzles I put them aside for a week or more and then rework them. This often generates new insights. Remember the Italian proverb, "Never make a good move without first looking for a better one."

Hint to solvers: Check your solutions for correctness and clarity by reading them a few days after you have written them, or get a friend to read them.

Rubric to the Problems Department,
Mathematics Magazine, **58**, #2
(March 1985) page 114.

It is not easy to create a clever name for a town or character used in a puzzle. Sam Loyd often set his posers in "Puzzleville." A.K. Dewdney coined the words "Solutionville" and "Problemtown" in the June 1987 issue of *Scientific American*. Around 1975, Scot Morris used the terms "A lphaville" and "Betaberg" in a puzzle in *Omni*. The late David Silverman created the mythical country of "Puevigi" ("I give up" spelled backwards). Spell Dewdney backwards (changing the "w" to an "r") and we get Yendred the name of the main character in his 1984 book, The Planiverse. Hugh ApSimon created the nonexistent towns of Ayling, Beeling and Ceiling in his puzzle book "Mathematical Byways". Paul Foerster's algebra text book contains a problem on the road from Tedium to Ennui. (Quite appropriate for a textbook.)

I have always liked the wordplay in the university names "Camford" and "Oxbridge". Tom Hood's book on wordplay, "Excursions into Puzzledom" (1879), referred to its readers as "Pilgrims of Puzzledom." Some of the more clever names used by other puzzlists are the following: Arthur Bungle, Dinny Dimwit, Mr. Nozitawl, Gherkin, Gunther and Gooner Gesundheit, the League Against Restrictive Diets (LARD) and the National Association of Universities, Schools and Educational Authorities (NAUSEA).

Certain sets of names in my files seem to be waiting for a puzzle to bring them together: three sets of sisters: Paula, Pauleen and Paulette; Marlene, Arlene and Darlene; and Dolly, Holly, Molly and Polly; four brothers: Edward, Edmund, Edgar and Edwin; three animals: a turtle, a tortoise and a terrapin; four dwarfs or elves: Grumpy, Grouchy, Crabby and Cranky; six children: Scooter, Scamper, Skipper, Skeeter, Scotty and Skippy; and four bread foods: a biscuit, a roll, a bun and a croissant.

The late Polish mathematician Hugo Steinhaus in his book "One Hundred Problems in Elementary Mathematics" has a chapter entitled "Practical and Non-practical Problems." This must be a joke; what other kind is there?

When I wrote a Mensa puzzle column, I received about 22 letters a month. The most ever for one month was 66. But these letters were submissions of answers for the purpose of publishing the names of the solvers in the newsletter. The Mensa letters rarely required a reply.

My former *Discover* column received about 30 letters a month, (92 was the most ever) all of which the editors required me to answer. We had a circulation of about 900,000. A.K. Dewdney once wrote me that he averaged 300 letters a month. (He answers many of his letters by hand on post cards.)

Martin Gardner said that he averaged between 50 and 400 letters a month. My experience leads me to believe that I could not answer more than 100 letters a month on a regular basis.

THE TWO GOLDEN RULES FOR ANSWERING MAIL

1. Answer all letters you are going to answer within seven days.
2. Admit your mistakes.

There are two penalties for letting the mail pile up beyond the seven day limit: 1) Your correspondent loses interest in the subject and 2) you must reread all your letters a second time before responding because you can't remember what your correspondents wrote about.

> One of the major chores of academic life is correspondence. I have developed a few basic rules to live by. (1) Answer all *bona fide* letters without delay [of more than 7 days]. (2) Do not use flamboyant rhetoric; be plain. (3) If you have said what is needed saying, stop; be brief.
>
> Paul Halmos,
> *I Want To Be A Mathematician*
> (Springer-Verlag, 1985), pp. 333, 336.

Beth Adler, an editor at Time-Life, gave me a wonderful time-saving idea for answering the mail. I simply combined the replies to many letters into one long discussion about that month's column. The discussion contained nearly all criticisms, complaints, observations and generalizations along with my responses. This gigantic reply was then sent as a letter to most of the correspondents.

Believe it or not, even a puzzlesmith gets hate mail. I never answered these kinds of letters. Possibly, the letters editor answered them with form letters.

Some letters require a personal answer. I have always given in to letters from children or prisoners. Readers who submit puzzle ideas require a thank-you note in reply. When I correspond with an individual, I will enclose a handout of mathematical curiosities to make our correspondence more interesting. A correspondent's name and the date of his letter go on the outside of a file that contains a handout. That way I never send the same handout twice to the same individual even if we correspond several years later.

It is only too easy for an expert to patronize a correspondent and belittle his observations. Success clouds the mind to others' desires and feelings, and breeds arrogance. This is a pitfall because arrogance is the enemy of excellence; it will generate and rationalize poor workmanship.

What is the value of recreational mathematics in general and a puzzle column in particular? On this matter the past puzzlists and expositors of recreational material are strangely silent. A search for the positive benefits derived from noncompetitive puzzle solving has turned up little beyond anecdotal data. My own feeling as a junior high teacher (three years), a high school teacher (nine years) and a puzzle writer for almost a decade is that the value of a puzzle is entertainment, period.

The goals of most classroom instruction are to build skills and to teach concepts. This is best done through practice (seat work), interactive lecture and teaching the same concept in different ways in the same lesson. If recreational mathematics were as valuable in the classroom as some would have us believe, then our textbooks and lectures would be full of clever demonstrations, odd and sundry observations, cute stories and mathematical curiosities. They aren't. Good textbooks will drive bad textbooks out of circulation. Mathematical recreations and textbooks have developed side by side for over 2000 years. It is unrealistic to think that the pedagogic value of mathematical recreations and puzzles has been overlooked for two millennia.

Nevertheless, mathematical entertainment fills a need that cannot be denied and has its place both in and out of the classroom. The puzzle and the recreation need not feel embarrassed about their worth. Entertainment is also a noble endeavor.

4651 Brentleigh Court,
Annandale, VA 22003

Puzzles Old & New: Some Historical Notes

Jerry Slocum

The history and development of mechanical puzzles is often interwoven with that of mathematical recreations. David Singmaster's research has traced the problem of the wolf, goat and cabbages to Alcuin in the 9th century. Mechanical puzzles representing the same problem (shown in Figure 1), date from the late 1800's.

Figure 1: Wolf, Goat and Cabbages Puzzle.

Magic squares, a well-known mathematical recreation, have also appeared as mechanical puzzles for at least a century. Rubik's cube stimulated recreational mathematics in group theory. One only has to review W.W. Rouse Ball & H.S.M. Coxeter's classic book *Mathematical Recreations & Essays* or Martin Gardner's extensive writings to appreciate the close link between mechanical puzzles and recreational mathematics.

Professor Hoffman, in his book *Puzzles Old and New* published in 1893, included many arithmetic recreations and, in addition, described at length for the first time the rich variety of mechanical puzzles available in England in the late 19th century.

Some examples of the wide diversity of mechanical puzzles are included in the following. The earliest known example of the "put-together" class of puzzles is the Loculus of Archimedes (also called the Stomachon—"The Problem That Drives One Mad") which dates from the 3rd century B.C. and was a 14-piece dissection of a square (see Figure 2).

A well-known dissection puzzle, the Tangram (see Figure 3), uses seven pieces. It was developed in China about 1800 and is still used today to teach geometry and as a stimulating puzzle.

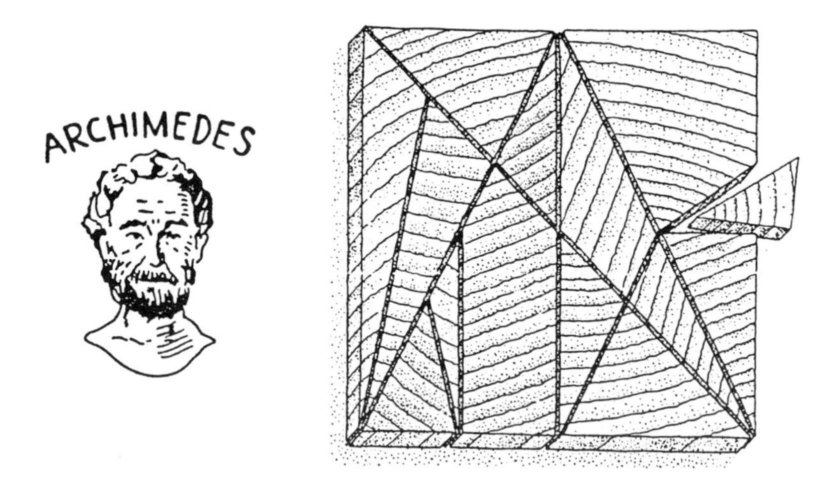

Figure 2: The Loculus of Archimedes.

Figure 3: The Tangram Puzzle.

More than 40 examples of "take-apart", or the "secret opening" class of puzzles, were included in Hoffman's *Puzzles Old and New.* Most were made of English boxwood or brass and some were used to carry matches, rings, snuff and even to hide drugs such as cocaine (see Figure 4). Secret opening or "trick" puzzle locks have been made in India for many centuries. In some locks the keyholes were hidden and in others no key was needed, because the lock was opened by pressing a part of the body or decoration (see Figure 4).

Figure 4: Hoffmann Puzzles & Puzzle Locks.

Some toy catalogs as early as 1803 included 6- and 24-piece interlocking puzzles (burrs). An extraordinary variety of beautiful and challenging new polyhedral interlocking puzzles have been designed and built by Stewart Coffin (see Figure 5).

Figure 5: Polyhedral Interlocking Puzzles. Designed & made by Stewart Coffin.

Comprehensive computer analyses of a 6-piece burr have been done by Bill Cutler who has also designed and built many new interlocking puzzles (see Figure 6). Bill is using the computer to help him design even more interesting and difficult 6-piece burrs. One of his designs, Bill's Baffling Burr, requires five moves before the first piece can be removed. He has stimulated 6-piece burrs by other puzzle designers with as many as ten moves to remove the first piece.

Figure 6: Interlocking Puzzle – designed & made by Bill Cutler.

The Chinese Rings puzzle (Figure 7) is the oldest example of the "disentanglement" class of puzzles. According to legend, Hung-Ming, a Chinese hero of the 2nd century, invented the puzzle to keep his wife occupied while he was away at war. Its solution is equivalent to the binary based Gray Code and several more recent puzzles use the same principle. L. Gros published the theory of the solution in 1872 and calculated that if a 10-ring puzzle can be solved (the rings removed) in eight minutes, a 60-ring puzzle would take 55 billion years to solve.

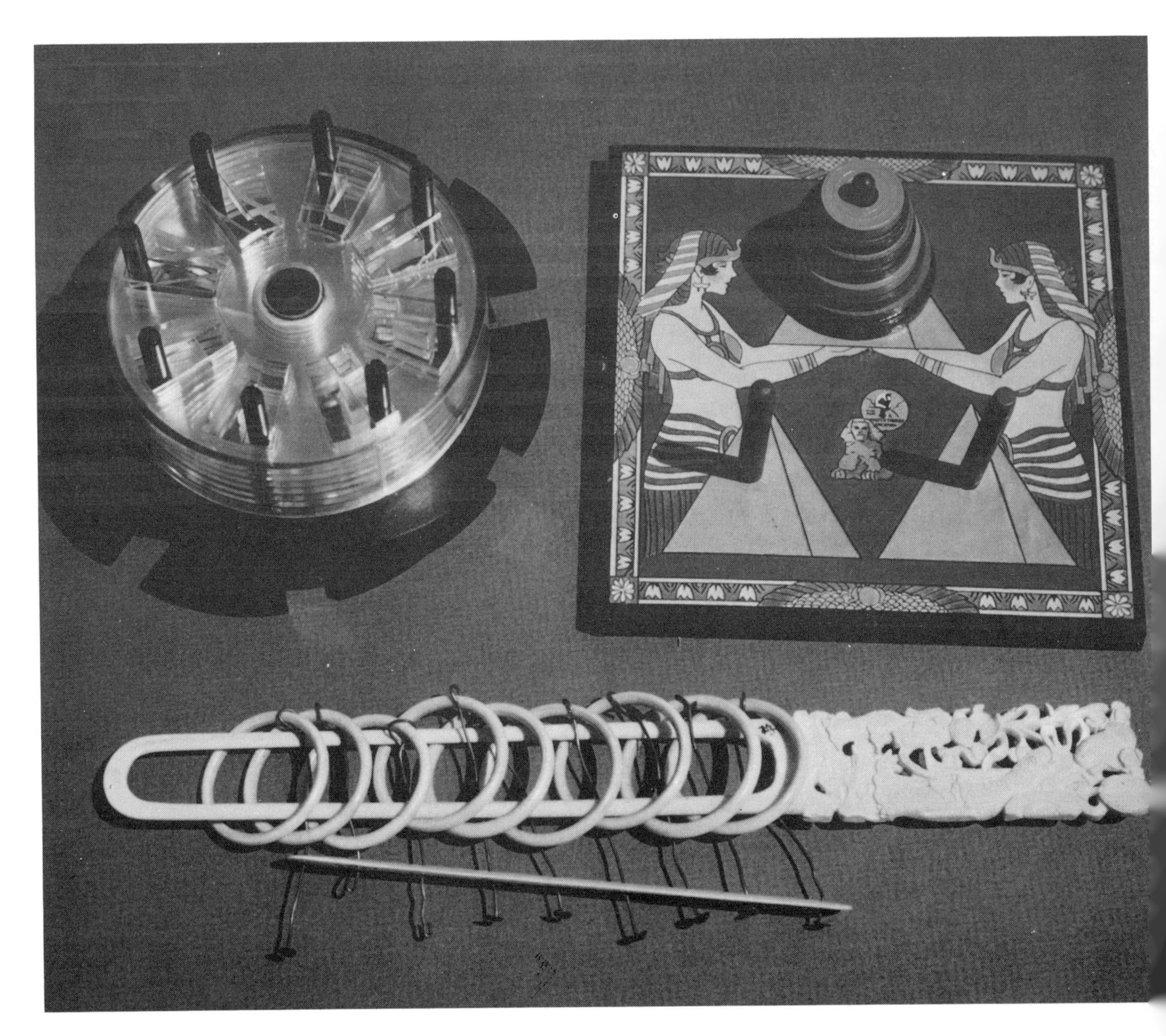

Figure 7: Chinese Rings Puzzle—made in China of ivory (c. 1850).

Many string disentanglement puzzles were used for advertising in the late 19th century (see Figure 8).

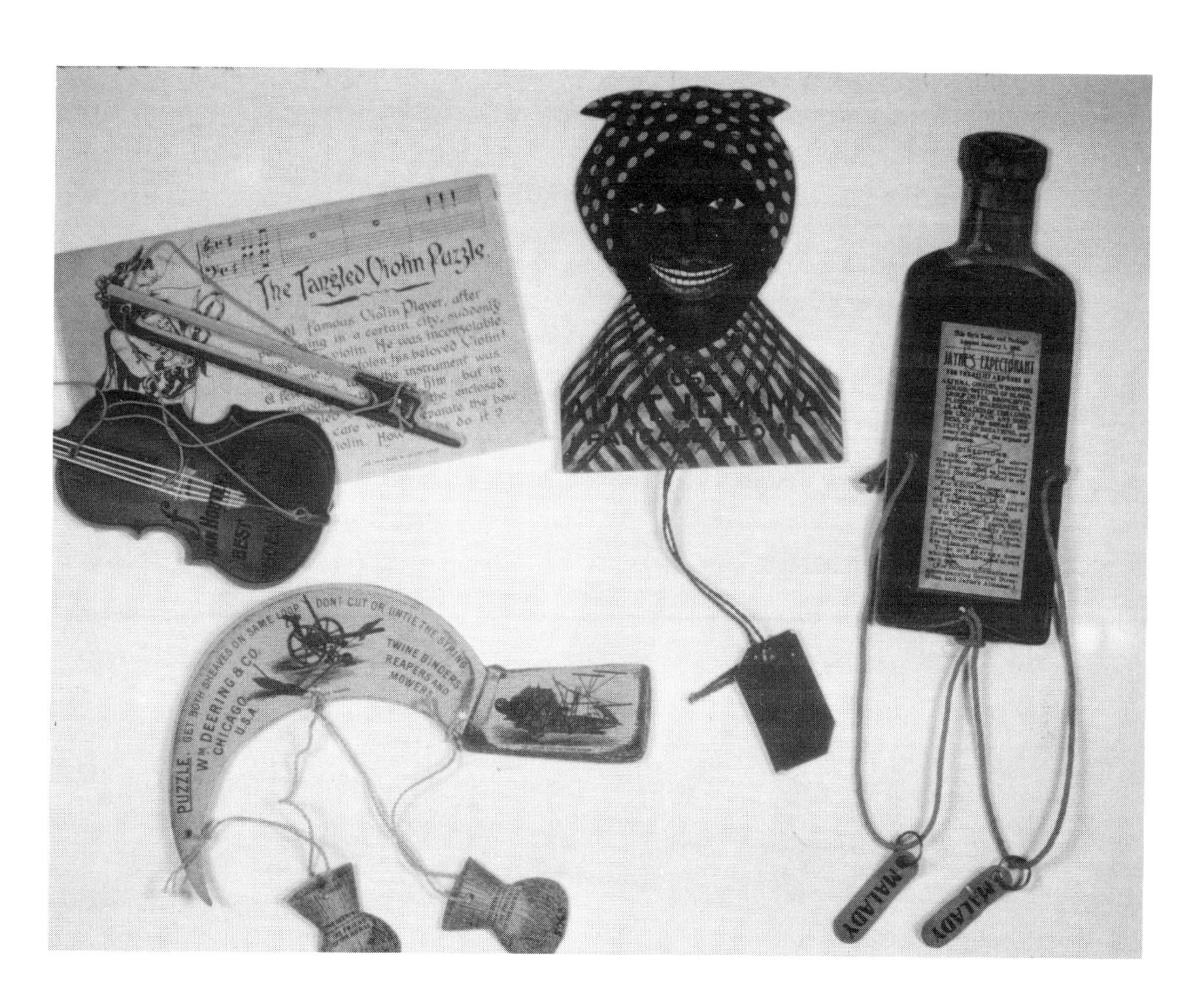

Figure 8: Advertising Disentanglement Puzzles.

Sequential movement puzzles involve a series of moves, following a set of rules to reach a specified goal. Peg Solitaire (Figure 9), described by Gottfried Leibniz in 1710, sliding block puzzles (Figure 10), popular since the 1870's, and Rubik's Cube (Figure 11), the rage of the early 1980's, are all examples of sequential movement puzzles.

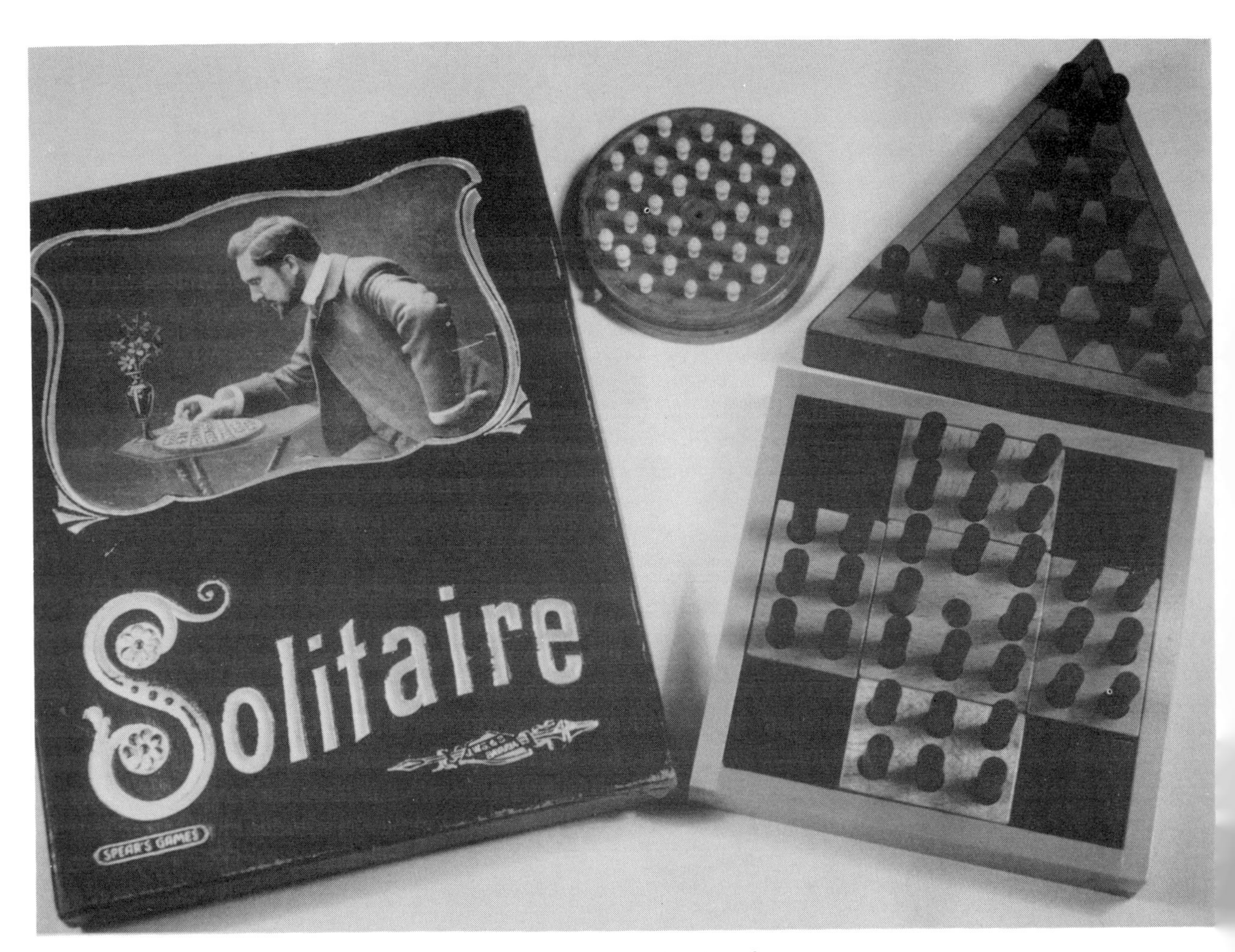

Figure 9: Peg Solitaire Puzzles.

Figure 10: "14–15" and "RATE YOUR MIND, PAL" Sliding Block Puzzles.

Sam Loyd, a famous American puzzle inventor, offered a prize of $1000 to anyone that found a solution to the "14–15" puzzle (Figure 10). It required reversing the order of the 14 and 15 blocks, a task shown as impossible by articles in *The American Journal of Mathematics* in 1879. Another sliding block puzzle (also shown in Figure 10) is similar to the 14–15. It requires the letters LA to be reversed and it can be solved to read "rate your mind, pal".

Figure 11: Rubik's ("Nichols") Cubes & Relatives.

Figure 12: Cup & Ball (Bilbouquet) Dexterity Puzzle. A rage in 16th century France.

Dexterity puzzles have been popular in many cultures for centuries. The cup and ball (bilboquet) was the rage in 16th century France (Figure 12). In 1889, in the United States, Charles Crandall started selling "Pigs in Clover" (Figure 13). Within weeks it was a "rage". It was so well-known it was used in political cartoons of the day. Plastic versions of "Pigs in Clover" are still very popular.

Figure 13: "Pigs in Clover" was very popular & well-known in 1889.

RECREATION CVI.

The geometric money.

DRAW on paſteboard the following rectangle ABCD, whoſe ſide AC is three inches, and AB ten inches.

Divide the longeſt ſide into ten equal parts, and the ſhorteſt into three equal parts, and draw the perpendicular lines, as in the figure, which will divide it into thirty equal ſquares.

From A to D draw the diagonal A D, and cut the figure, by that line, into two equal triangles, and cut thoſe triangles into two parts, in the direction of the lines E F and G H. You will then have two triangles, and two four-ſided irregular figures, which you are to place together, in the manner they ſtood at firſt, and in each ſquare you are to draw the figure of a piece of money; obſerving to make thoſe in the ſquares, through which the line A D paſſes, ſomething imperfect.

As the pieces ſtand together in the foregoing figure, you will count thirty pieces of money only; but if the two triangles and the two irregular figures be joined together, as in the following figures, there will be thirty-two pieces.

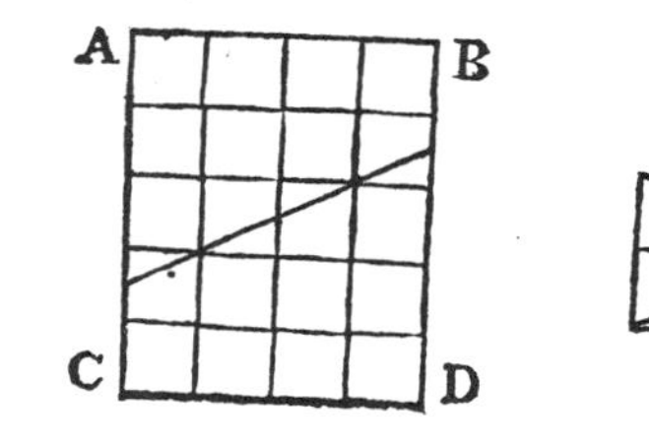

RECRE-

Figure 14: The Geometric Money Paradox (William Hooper's "Rational Recreations").

William Hooper, in his book *Rational Recreations* published in 1794, included a paradox entitled *The Geometric Money* (Figure 14). Mechanical puzzles based on the same paradox appeared in the late 19th century in several forms. Sam Loyd's version, published in 1896 and called "Get Off The Earth", was one of his best puzzles (Figure 15). When the circular centre section of the puzzle is rotated, 13 figures of Chinese men in the original position are reduced to twelve. Sam Loyd's question, "Which one has vanished; where does he go?" is still perplexing to puzzlers today.

Figure 15: Sam Loyd's "Get Off The Earth" Puzzle.

The existence of puzzle vessels, the only ancient mechanical puzzles made of durable material that have survived many centuries, indicate that puzzles and tricks existed at least 3000 years ago. Puzzle jugs (Figure 16) were quite popular in Europe in the 17th & 18th centuries. The puzzle is to drink the contents without spilling. It can be done by using a built-in "straw" and sucking the liquid out while closing off unused holes. The same principle was used in ceremonial bowls during the 11th to 10th century B.C. in the Middle East.

Figure 16: The puzzle of this Puzzle Jug was to drink without spilling.

Reference

Jerry Slocum & Jack Botermans, *Puzzles Old and New*, University of Washington Press, 1986, 160pp.

257 South Palm Drive,
Beverly Hills, CA 90212.

Part 3

People & Pursuits

The Marvelous Arbelos

Leon Bankoff

The figure of three mutually tangent semicircles with their centers on the same straight line was known among the ancient Greeks as the arbelos or the shoemaker's knife, so called because of its resemblance to a tool used by leather workers. The figure was first described by Archimedes in his Book of Lemmas, Propositions IV, V and VI. Select a point C anywhere on a line AB and draw semicircles on AC, CB and AB as diameters. The space bounded by the three arcs is an arbelos. (See Figure 1.)

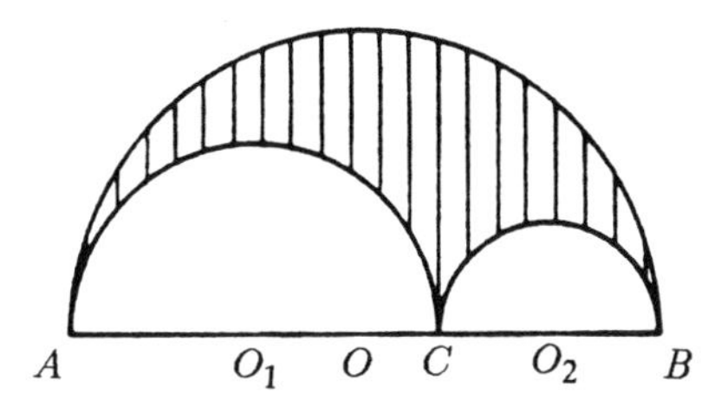

Figure 1: The Arbelos

The figure has a seemingly endless array of amusing properties and amazing coincidences most of which are decidedly counterintuitive. Nevertheless each surprising property can be demonstrated mathematically. Perhaps the pleasure experienced by the triumph of reason over intuition accounts for the fascination of this simple yet deceptive geometrical figure.

But the shoemaker's knife is more than a mere mathematical plaything. As with many other popular topics in the field of so-called recreational mathematics, the arbelos is a fruitful source of mathematical knowledge. The figure lends itself admirably to the study of geometrical concepts that now come under the heading of modern geometry, college geometry or advanced Euclidean geometry. Investigation of the properties of the arbelos brings us in contact with topics such as inversion, the radical axis, the golden ratio, loci of conics, similitude and antisimilitude, circles inscribed in mixtilinear and curvilinear triangles, chains of successively tangent circles, rational right triangles and power theorems, to name a few. Particularly striking is a recent development revealing properties of exceptional beauty when the golden ratio determines the location of C on the line AB.

It is small wonder then that the shoemaker's knife managed to attract the attention of mathematical luminaries such as Pappus, Vieta, Descartes, Fermat, Newton, Steiner and, in our own time, Victor Thébault. The esthetic appeal of the figure makes the arbelos an important source of enrichment material, stimulating and instructive for tyro and sophisticate alike.

In the years between 1953 and 1959 I collaborated with Victor Thébault in the preparation of a 10-chapter manuscript for a book embodying a comprehensive collection of old and newly discovered properties of the figure. The excerpts given here cover essentially the material presented in a talk delivered at the Eugene Strens Memorial Conference on Intuitive and Recreational Mathematics and Its History, given at the University of Calgary from July 27 to August 2, 1986.

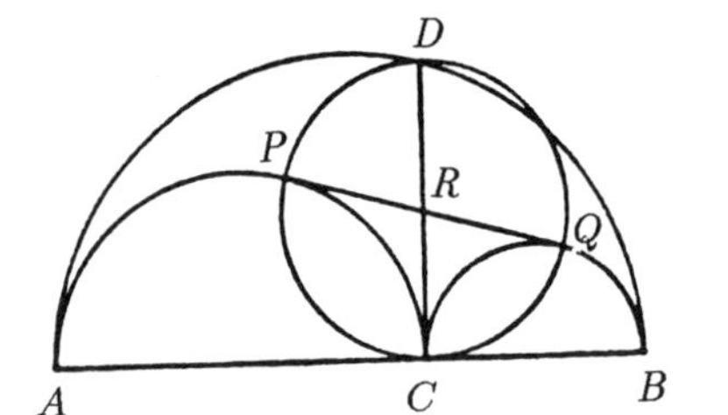

Figure 2: Circle on CD

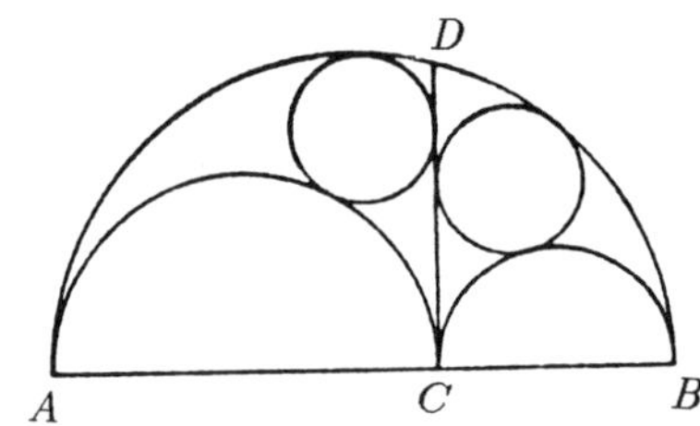

Figure 3: Equal inscribed circles

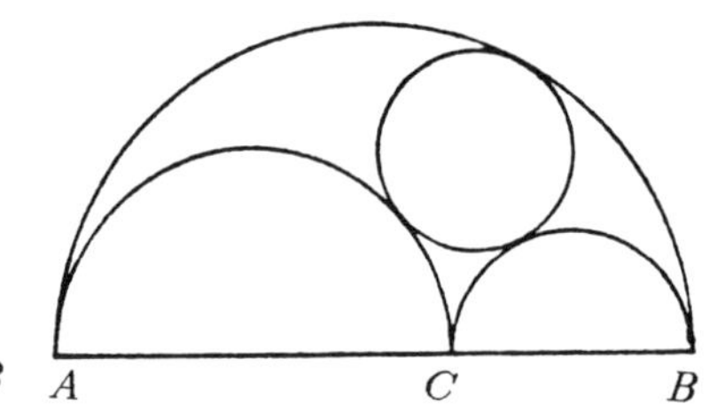

Figure 4: Inscribed circle

In Proposition IV, Archimedes, after forging his tools with three preliminary lemmas, erects a half-chord perpendicular to AB at C and cutting the outer circle at D. It is then easily shown that the circle whose diameter is CD is equal in area to the arbelos. (See Figure 2.) Proposition V considers the circles described on each side of CD, each touching two arcs of the knife. The two inscribed circles, each having a diameter equal to $(AC \times CB)/AB$, remain equal regardless of the location of C on the base AB. (See Figure 3.) In Proposition VI Archimedes indicates how to calculate the diameter of the circle inscribed in the arbelos itself. He assigns a ratio of $3/2$ to AC/CB and shows that the diameter of the inscribed circle is equal to $6/19$ times AB. He mentions that the same method could be used for any other initial ratio, but does not offer any generalization. (See Figure 4.)

After its first appearance in the Book of Lemmas the arbelos lay dormant, at least in the literature, for 500 years until Pappus resurrected it in his mathematical collection. There he established the startling properties of the famous circles of Pappus thus ensuring the survival of the arbelos to this day.

Pappus starts with the inscribed circle of Proposition VI, $(\omega_1)\rho_1$, and continues with a chain of consecutively tangent circles $(\omega_i)\rho_i$, contained between the arcs AB and AC and radiating in the horn angle toward A, where the subscripts denote the order number of the circles in the chain. The centers of the circles so described follow the path of an ellipse while the points of contact between the circles of the chain lie on a circle. (See Figure 5.) Pappus discovered that if h_n denotes the distance from the center of circle $(\omega_n)\rho_n$ to the baseline AB, then $h_n = 2n\rho_n$. On the other hand, if the semicircle on CB is omitted and the chain starts with the circle $(\omega_1)\rho_1$, tangent to the base AB, then $h_n = (2n-1)\rho_n$. (See Figure 6.) Several other elementary properties are worth noting:

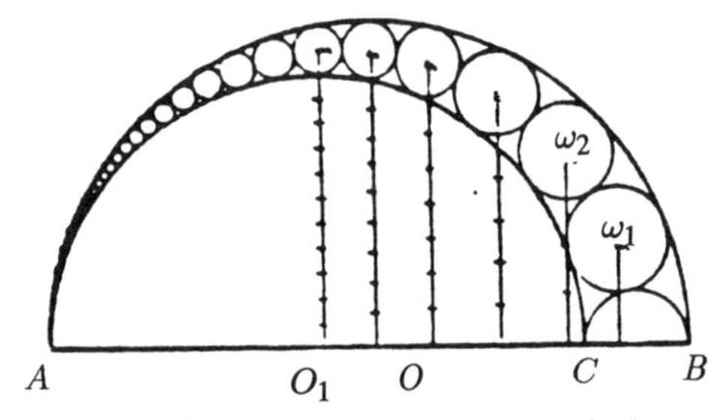

Figure 5: Distance of ω_n to AB is $2n\rho_n$

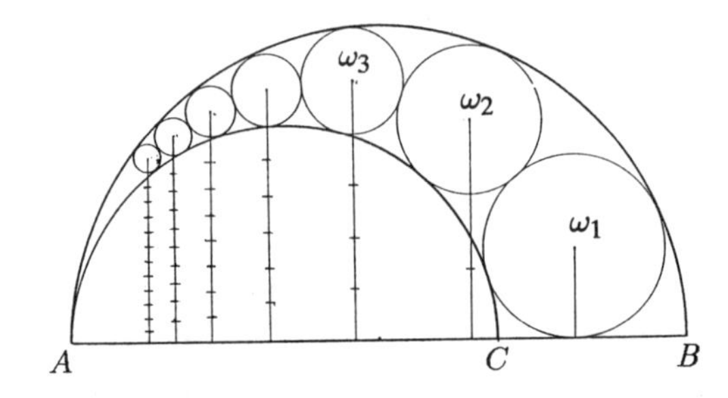

Figure 6: Distance of ω_n to $AB = \rho(2n-1)$

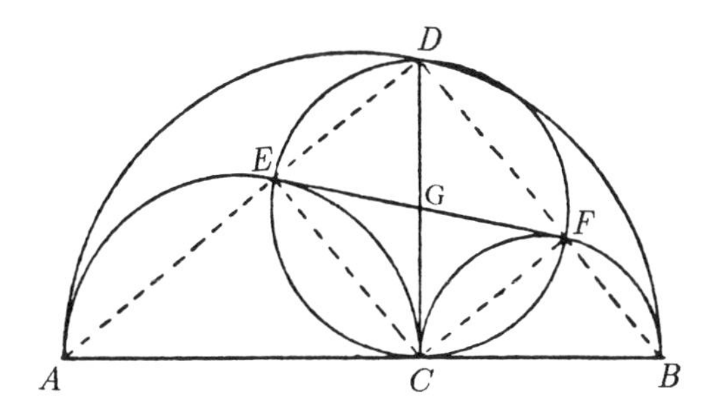

Figure 7: Rectangle in circle

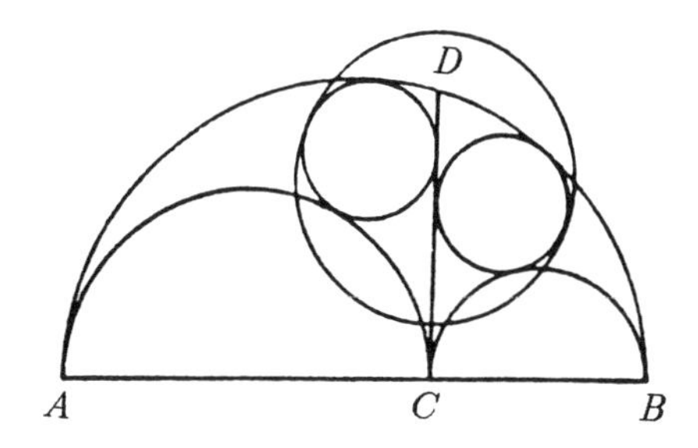

Figure 8: Circle equal to that on CD

1. Arc AB = arc AC + arc CB.

2. If EF is the common external tangent to arcs AC and CB, then CD and EF bisect each other and are the diagonals of the rectangle inscribed in the circle on CD. Note that E lies on DA and F on DB. (See Figure 7.)

3. The smallest circle that circumscribes the twin circles of Archimedes is equal to the circle on diameter CD and is therefore equal to the area of the arbelos. (See Figure 8.)

4. Consider the arbelos having only one inscribed circle touching arcs AC, CB, and AB in P, Q, R, respectively. Let M be the midpoint of arc AC and N the midpoint of arc CB. With M as center and MA as radius, describe a circle. (See Figure 9.) The circumference of this circle will pass through A, C, Q, and R. Similarly the circle with radius NB centered at N will pass through the points B, C, P, and R.

5. If an arbelos is reflected in AB, we obtain two congruent figures. (See Figure 10.) In the lower Arbelos draw CD and two inscribed twin circles as shown. In the upper figure inscribe a single circle touching arcs AC and CB in P and Q respectively. It has been shown that the circle through P, Q, and C is also equal to each of the twin circles of Archimedes. The proof is contained in my article "Are the Twin Circles of Archimedes Really Twins?" (*Mathematics Magazine,* Vol.47, No.4, September 1974, pages 214–218.)

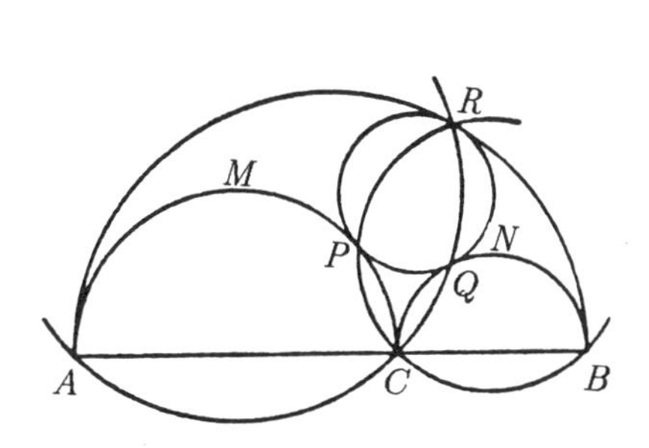

Figure 9: Circles $ACQR$, $BCPR$

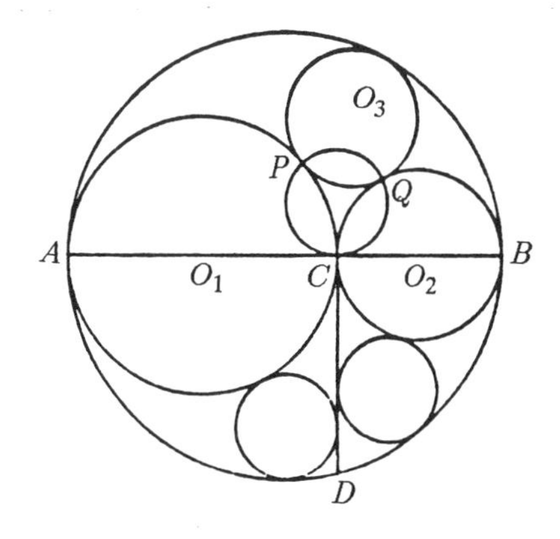

Figure 10: The Arbelos reflected

We now find that the twin circles are in reality only members of a set of triplets. Since the publication of that paper, more developments have arisen. Consider the segment of the arc AB cut off by the chord tangent to arcs AC and CB. The maximum circle inscribed in this segment also happens to be equal to each of the three triplets. Consequently the triplets have now become quadruplets. (See Figure 11.)

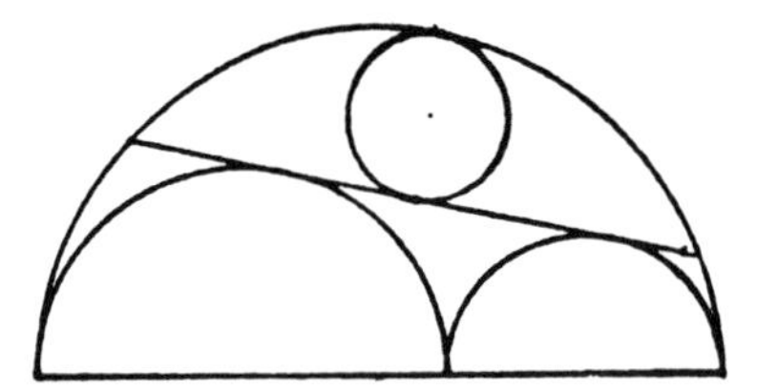

Figure 11: A fourth equal circle

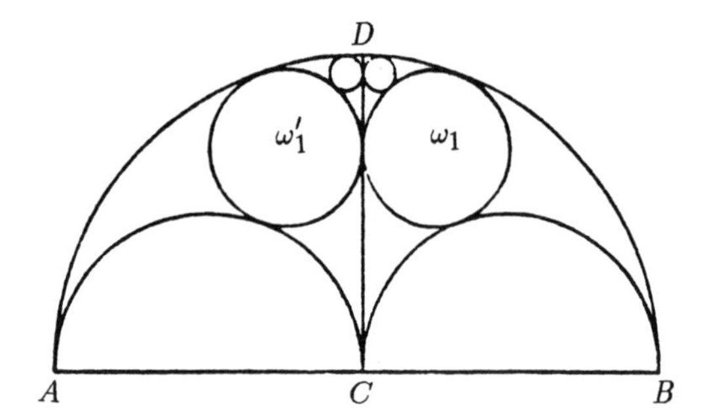

Figure 12: A symmetrical Arbelos

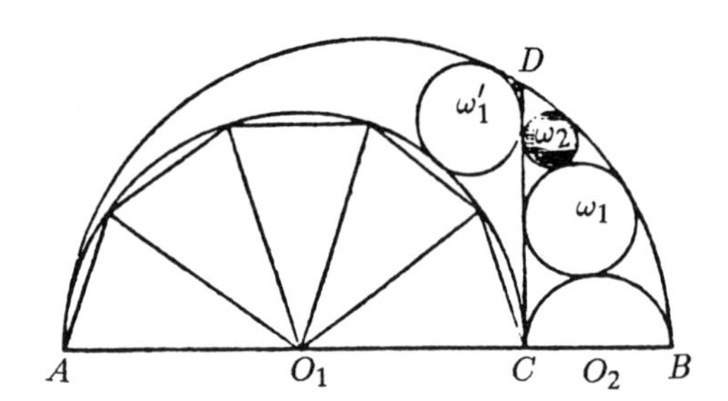

Figure 13: The decagonal Arbelos

6. Figure 12 illustrates a symmetrical shoemaker's knife where C bisects AB. Two sets of circles are inscribed as shown. If C is permitted to move toward B, the set of circles touching arcs AC and CB will, of course, remain equal to each other. But what happens to the second set above them? The one at the left diminishes continuously in size and finally approaches zero as C approaches B. But the one on the right increases up to a certain point and then decreases. In my effort to display this circle to the best advantage in a diagram I determined to locate the point C so as to obtain the variable circle of maximum diameter. It came as a great surprise to me to learn that the desired maximum is achieved when CB is equal to the side of a regular decagon inscribed in the circle whose diameter is AC. See the January 1956 issue of the *American Mathematical Monthly,* Problem E 1166. (See Figure 13.)

7. With the unexpected intrusion of the golden ratio into the study of the shoemaker's knife, it occurred to me to investigate what happens when C divides AB in the golden ratio, that is, when AC is constructed so that $AC = CB \times AB$. Figures 14 and 15 portray only a small fraction of the elegant and exquisite collinearities and tangencies that arise following a haphazard exploration of sequences of tangent circles. Some of the results were published in *Scripta Mathematica,* v. 21, 1955 under the title of "The Golden Arbelos." Two of the resulting configurations are shown in Figures 14 and 15.

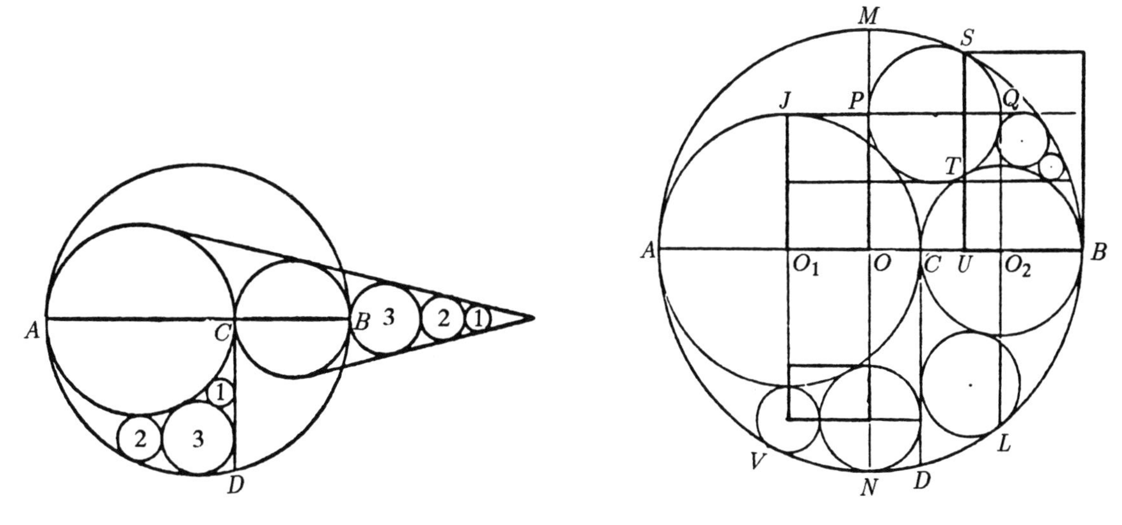

Figure 14: The Golden Arbelos

Figure 15: The Golden Arbelos

8. Many of the calculations were carried out with the help of an ingenious set of formulae derived by inversion and sent to me by Thébault in response to my urgent cry for help. The venerable French mathematician was a virtual wizard in solving the most complicated geometrical problems. My visit to him in the little hamlet of Tennie, France, a suburb of Le Mans, in 1958, was one of the most memorable and rewarding experiences in my life. Figure 16 illustrates, without comment, the inversion used by Thébault in solving a most intricate problem concerning tangent circles.

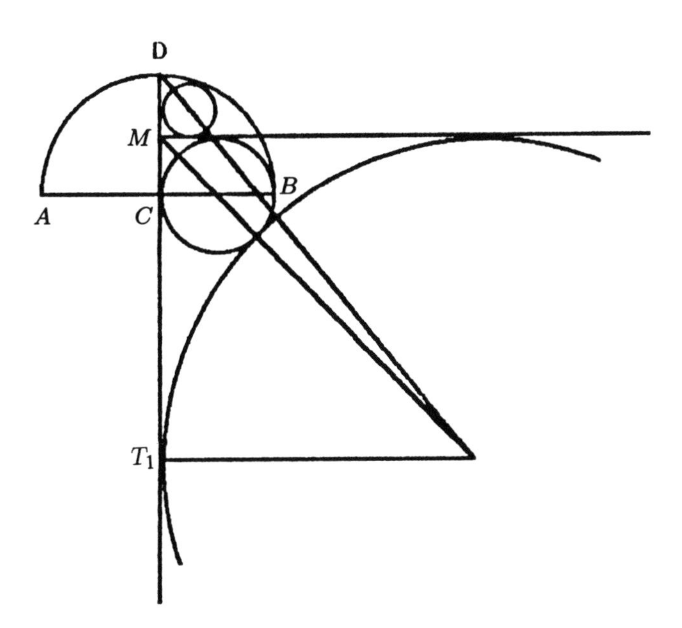

Figure 16: Thébault's construction

9. The proof of the Pappus circle theorem is clear and simple when arrived at by inversion. But Pappus lived sixteen centuries before the discovery of that wonderful mathematical tool, a method taken for granted in our century. So the natural question arises, "How Did Pappus Do It?" To learn the answer to that question consult my article published in the *Mathematical Gardner,* a collection of assorted mathematical papers edited by David A. Klarner and published by Wadsworth International in commemoration of Martin Gardner's 65th birthday. Figure 17, one of the many different inversion procedures for the solution of the Pappus circle theorem, shows how the complicated deployment of the Pappus circles yield to a simple analysis when inverted to a vertical chain of successively tangent circles.

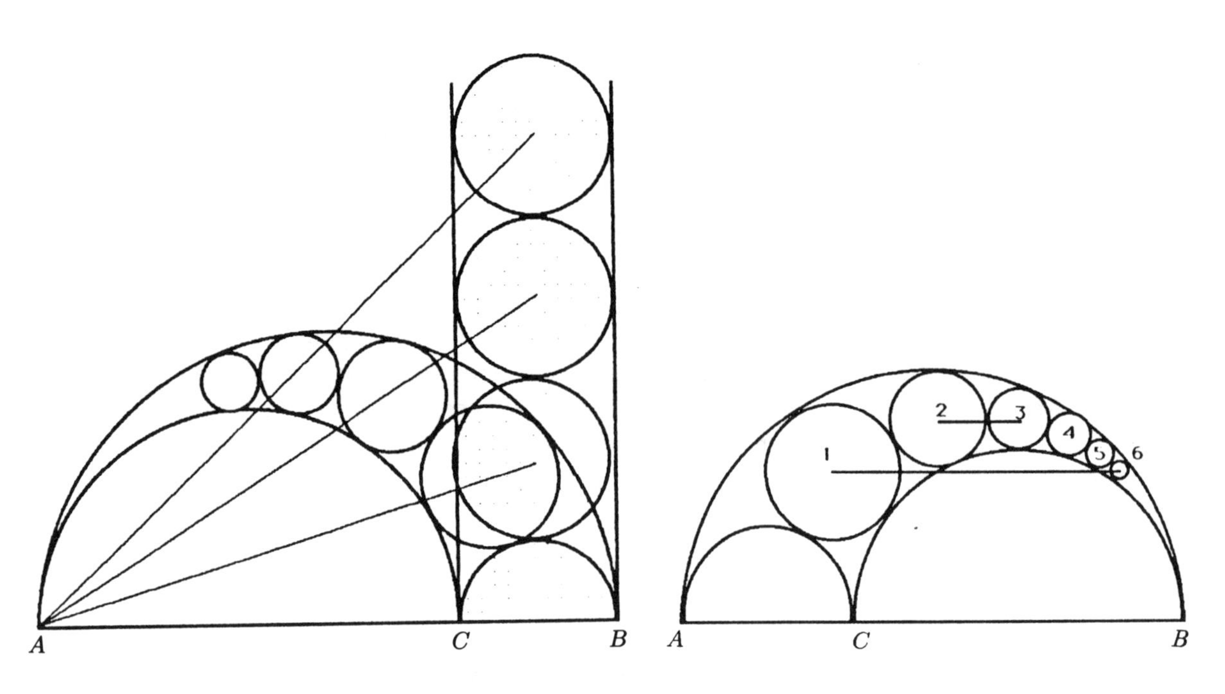

Figure 17: Inverting the Pappus chain

Figure 18: Gaba's 3 to 2 arbelos

10. Another surprising feature of the shoemaker's knife was discovered by M.G. Gaba and was published in *The American Mathematical Monthly,* Vol. 47, January 1940, pages 19–24. If $AB/CB = k/(k-1)$, the centers of at least one pair of circles of Pappus will lie on a line parallel to AB. The circles paired in this manner are those whose order numbers in the chain are factors of $k(k-1)$. In the example shown in Figure 18, the integer k is equal to 3, that is, $CB = 2$ and $AB = 3$. Then $k(k-1) = 6$. The integer 6 can be factored in two ways, 1×6 and 3×2. So the centers (ω_1, ω_6) and (ω_2, ω_3) are the only pairs lying on lines parallel to AB. This curiosity was further elaborated on by Victor Thébault (*American Mathematical Monthly,* Nov. 1940, pages 640–642.)

11. Figures 19 and 20 exhibit a variation of the property regarding the equality of the twin circles. If the semicircles AC' and CB intersect or have no point in common, circles ω_1 and ω_2 are equal provided TT' is the radical axis of circles AC' and CB.

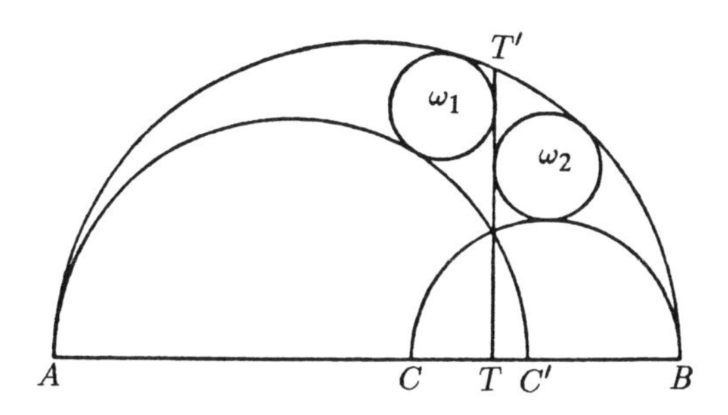

Figure 19: Twin circles

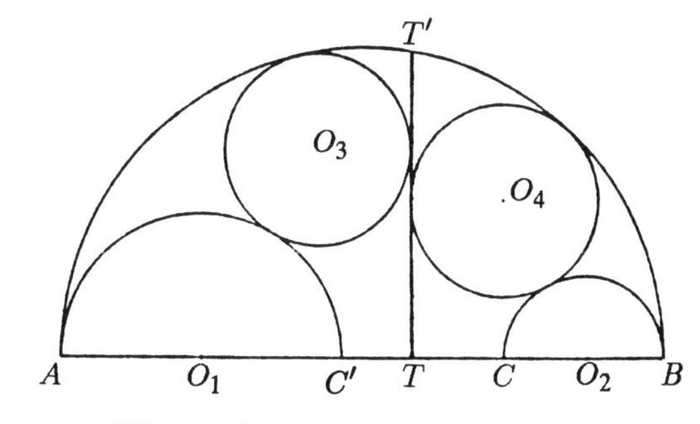

Figure 20: Another equal pair

12. As mentioned earlier in this paper, there exists enough material on the arbelos to more than fill a 10-chapter book on the subject. Consequently this brief presentation merely suggests the kinds of inquiries that have been undertaken since the time of Archimedes and makes no pretense of being thorough. A list of references is offered for those interested in pursuing the subject further.

References

1. R. A. Johnson, *Modern Geometry,* Houghton Mifflin, 1929, p. 113. (Reissued by Dover as *Advanced Euclidean Geometry.*)

2. J.S. Mackay, *Proceedings of the Edinburgh Math. Soc.,* 1884, p. 2.

3. J.S. Shively, *Modern Geometry,* Wiley, 1939, p. 151.

4. T.L. Heath, *Works of Archimedes,* Dover, NY, 1953, pp. 304–308.

5. *Mathematics Magazine,* Vol.26, No.2, Nov.–Dec. 1952, pp. 111–115.

6. V. Thébault, *Amer. Math. Monthly,* Vol.47, 1940, p. 640.

7. L. Bankoff, *Scripta Mathematica,* Vol.XXI, No. 1, March 1955, p.70.

8. Martin Gardner, *Scientific American,* Jan. 1979, p. 118.

9. *Mathematics Magazine,* Vol. 47, No.4, Sept. 1974, pp. 214–218.

10. *The Mathematics Teacher,* Vol.34, No.3, March 1961, pp. 134–137.

11. J.H. Caldwell, *Topics in Recreational Math.,* Cambridge Univ. Press, 1966.

12. M.G. Gaba, *Amer. Math. Monthly,* Vol 47, Jan. 1940, pp. 19–24.

13. S. Stanley Ogilvy, *Excursions in Geometry,* Oxford Univ. Press, 1969, p.54.

14. Howard Eves, *A Survey of Geometry,* Allyn and Bacon, 1965, p. 34.

471 Rodeo Drive
Beverly Hills, CA 90212

Cluster Pairs of an n-Dimensional Cube of Edge Length Two

I.Z. Bouwer & W.W. Chernoff

1 Introduction

D.A. Engel [1] asked if it is possible to color the 2^n unit cells of an n-dimensional cube of edge length 2 with two colors, black and white, so that

(a) half of the cells are colored black and half white, and

(b) the cluster formed by the black cells is not congruent to the cluster formed by the white cells.

He showed that for $n \leq 3$ no such coloring exists. B.L. Schwartz [4] illustrated that such a coloring exists for $n = 4$. In this paper we show that such colorings exist for each $n \geq 4$ and enumerate them.

2 Clusters in Cell 2-Colorings of K

Let K denote an n-dimensional cube of edge length k, with k a positive integer. Thus K consists of k^n cells. A **cluster** of K is the geometric configuration formed by a subset of cells of K. When $k = 2$, each cluster is **connected** (as point sets in E^n), since any two (closed) cells have at least a point in common. Two clusters A and B of K are **congruent**, written $A \sim B$, if they can be superposed by rigid motions (including reflexions) in E^n.

Suppose that each cell of K is colored black or white. Let B (or W) denote the cluster of black (or white) colored cells of K. The coloring will be called **equicardinal** if B and W have the same number of cells, and **symmetric** if $B \sim W$. Two cell 2-colorings of K are **equivalent** when there is a symmetry operation of K (including reflexions) that transforms the one coloring to the other.

Figure 1: Congruent black but incongruent white clusters, $n = 2$, $k = 4$.

If two cell 2-colorings of K have congruent black clusters, then their white clusters may not be congruent, as is illustrated in Figure 1. An isolated cell is free to move independently. Even when the white clusters also happen to be congruent, the colorings may not be equivalent. This is illustrated in Figure 2 where the colorings happen to be symmetric.

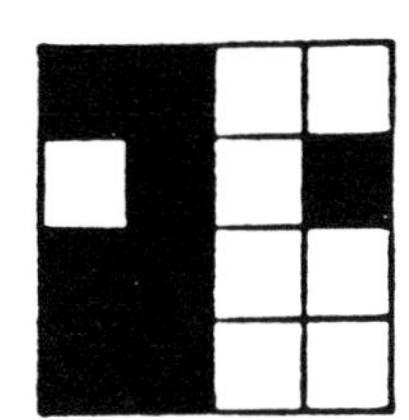
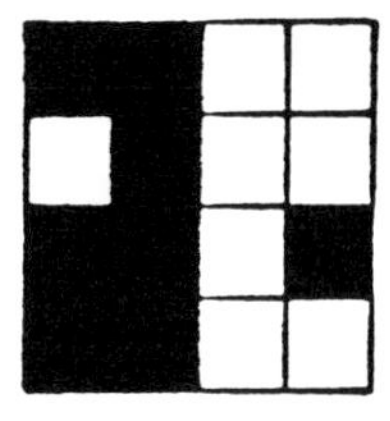

Figure 2: Symmetric but inequivalent 2-colorings, $n = 2$, $k = 4$.

Here, the one black cluster can be superposed on the other, and the white cluster on the other, but these two individual superpositions cannot be effected simultaneously by a rigid motion of the square as a whole. The following theorem states that the situations illustrated in Figures 1 and 2 cannot occur when $k = 2$.

Theorem 1 *Let $k = 2$. If two cell 2-colorings of K have congruent black clusters (say), then their white clusters are also congruent, and the colorings are equivalent.*

Proof: Let (B, W) and (B', W') be the cluster pairs associated with the two colorings, with $B \sim B'$. We show that any rigid motion γ superposing B onto B' can be extended to a symmetry of K (which thus maps W to W'). For any cluster C of K, define C^* to be the minimal rectangular n-solid containing C. Then C^* is a cluster of K of size $k_1 \times k_2 \times \ldots \times k_n$, where $1 \leq k_i \leq 2$ for each i. Consider B^*. If $k_i = 2$ for each i, then $B^* = K$, and γ is a symmetry of K. If $k_i = 1$ for some i, then B^* may be translated, within K, through a distance of one unit, in the direction parallel to the ith axis. Since $k = 2$, the same effect is achieved by a reflexion of K in the bisecting hyperplane perpendicular to the edge. It follows that any rigid motion superposing B on B' within K can be effected by a symmetry of K. □

3 Subgraphs in 2-colorings of Q

We now restrict ourselves to the case $k = 2$. If the cells of K are represented by their centre points and two centre points are joined by an edge just when the corresponding cells are adjacent (in the sense that they share an $(n-1)$-dimensional face), then K is represented by the n-dimensional unit cube Q. The cell 2-colorings of K are in one-to-one correspondence with the vertex 2-colorings of Q or, equivalently, of the graph of Q. The clusters B and W then correspond to the respective induced subgraphs on the black and white vertices, which we shall call the **black** and **white subgraphs** associated with the given coloring.

We note that if two vertices in an induced subgraph are not joined by an edge, then the corresponding cells in the cluster intersect in an m-dimensional face, where m can take any of the values $0, 1, \ldots, n-2$. It should not therefore come as a surprise that 2-colorings of Q are not uniquely determined by their black and white subgraphs. Figure 3 illustrates the occurrence (first arising for the case $n = 5$) of a 2-coloring of Q which is not symmetric, although the black and white subgraphs are isomorphic.

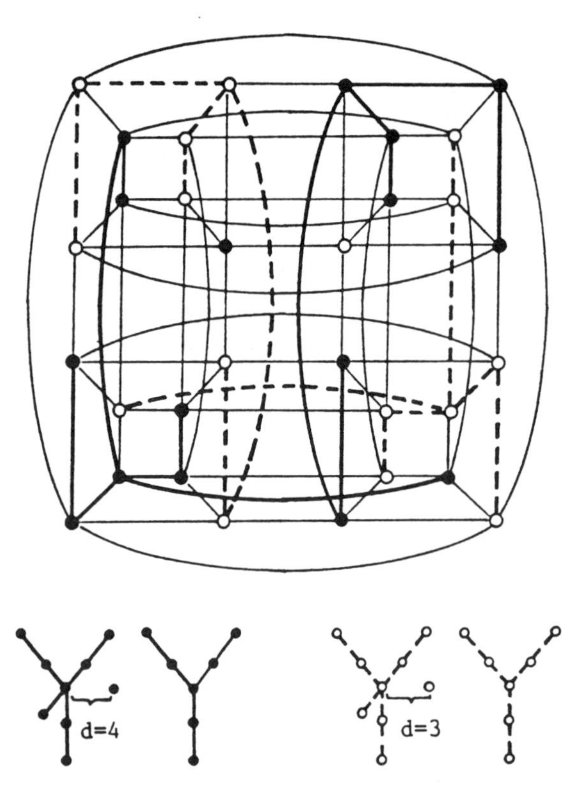

Figure 3: An unsymmetric 2-coloring with isomorphic black and white subgraphs.

In the two subgraphs, the isolated vertex and the vertex of valence four are at distances $d = 4$ and $d = 3$ in the graph of Q and therefore represent cells that respectively intersect in 1- and 2-dimensional faces. Thus the black cluster cannot be congruent to the white.

Figure 4 shows two symmetric 2-colorings of Q with isomorphic black (and white) subgraphs, but the colorings are not equivalent. This situation first arises in the case $n = 4$.

Figure 5 illustrates the same situation (which first arises in the case $n = 5$) for unsymmetric 2-colorings of Q. Thus, we cannot reduce the problem of enumerating the 2-colorings of Q to that of enumerating the subgraphs of the graph of Q. However, knowledge of the subgraphs is still useful in that if two 2-colorings of Q have non-isomorphic black subgraphs (say), then the colorings cannot be equivalent.

Theorem 2 *For each $n \geq 4$ there exists a 2-coloring of Q which is equicardinal but not symmetric.*

Proof: Label the vertices of Q with the binary sequences of length n such that sequences corresponding to adjacent vertices differ in exactly one position. Color black the 2^{n-1} vertices whose

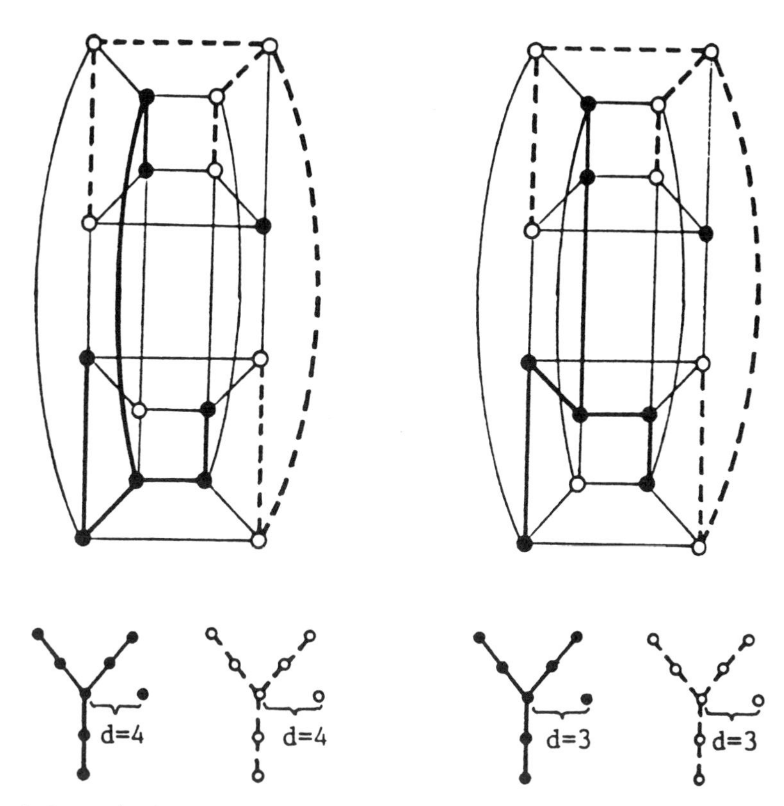

Figure 4: Inequivalent 2-colorings with isomorphic black and white subgraphs.

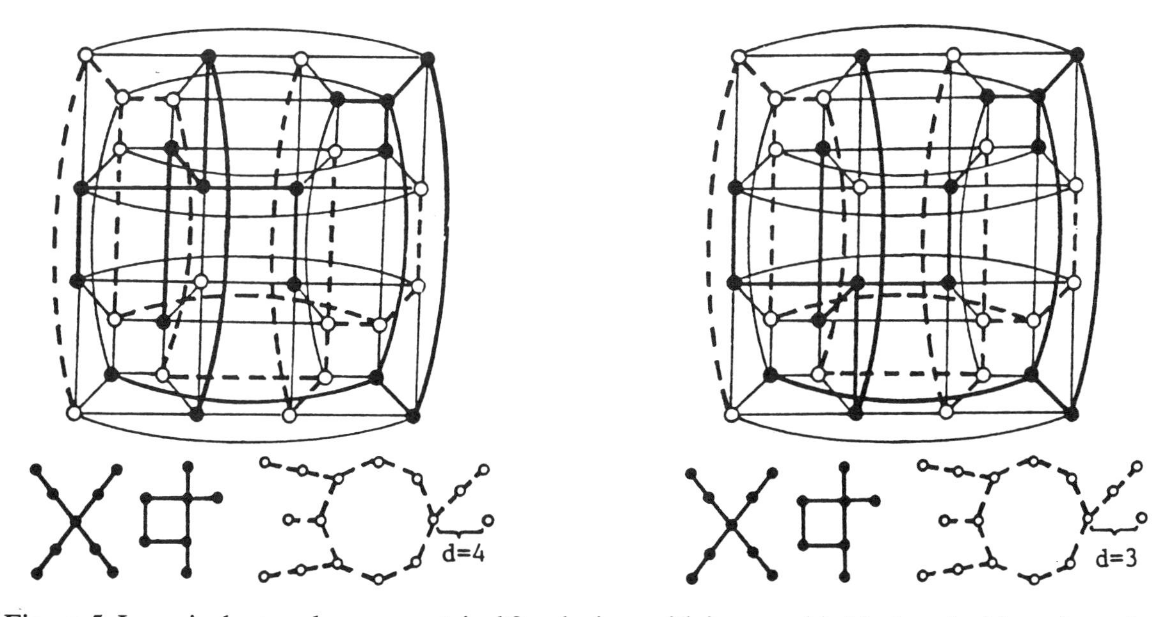

Figure 5: Inequivalent and unsymmetrical 2-colorings with isomorphic black and white subgraphs.

labels start with 1000, 1100, 0100, 0110, 0010, 0011, 0001 or 1001, and color the remaining vertices white. Then the black subgraph is connected, while the white subgraph is not. It has a component (the one on the vertices whose labels start with 0000) isomorphic to the graph of the $(n-4)$-dimensional unit cube. The coloring in the case $n = 4$ is shown in Figure 6.

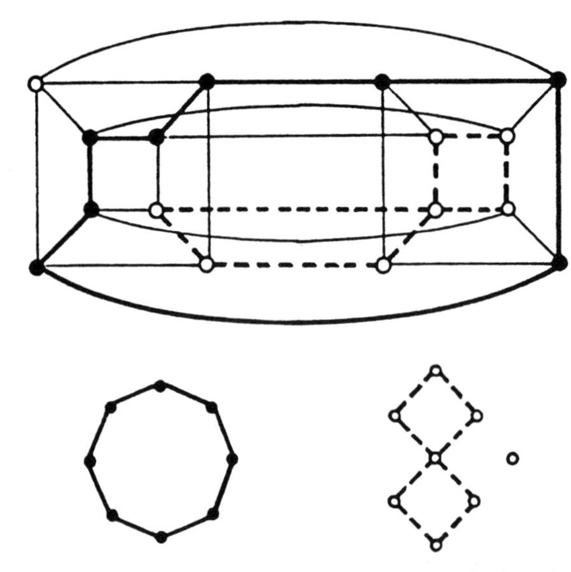

Figure 6: An equicardinal but unsymmetric 2-coloring.

4 Enumeration

We have seen that, for $k = 2, n \geq 1$, the problem of counting the cluster pairs (B, W) is equivalent to that of counting the vertex 2-colorings of Q. This is readily done by use of Pólya's enumeration theorem (see **[3]**).

The group of all symmetries (including reflexions) of the unit cube Q is the exponential group $G = [S_2]^{S_n}$ whose cycle index $Z(G; a_1, a_2, \ldots)$ is computed in **[2]**. If the variables a_i in the cycle index are replaced by $b^i + w^i$, then the resulting polynomial is the enumerating polynomial for the colorings, the coefficient of $b^i w^j$ giving the number of (non-equivalent) 2-colorings with i black and j white vertices. On applying a technique used by Read **[3]** to enumerate self-complementary graphs, we know that the number of symmetric colorings is given by $Z(G; 0, 2, 0, 2, \ldots)$. We have thus proved part (1) of the following theorem.

Theorem 3 *Let K be the n-cube with $k = 2$, $G = [S_2]^{S_n}$, $\alpha =$ the coefficient of $b^s w^s$ in $Z(G; b + w, b^2 + w^2, \ldots)$, where $s = 2^{n-1}$, and $\beta = Z(G; 0, 2, 0, 2, \ldots)$. Then*

(1) The number of equicardinal, non-symmetric cell 2-colorings of K is $\alpha - \beta$, or $\frac{1}{2}(\alpha - \beta)$ if colors may be interchanged.

(2) The number of distinct cluster 2-sets $\{B, W\}$ associated with equicardinal cell 2-colorings of K is $\frac{1}{2}(\alpha + \beta)$.

Part (2) follows readily from the fact that the equicardinal, non-symmetric 2-colorings of K occur in pairs such that their cluster pairs (B, W) and (B', W') satisfy $B \sim W'$ and $B' \sim W$ (with $B \not\sim W$).

The case $n = 4$:

Here $G = [S_2]^{S_4}$ and $Z(G; a_1, a_2, \ldots, a_{16}) =$

$$\frac{1}{384}\left(a_1^{16} + 51a_2^8 + 12a_1^8a_2^4 + 12a_1^4a_2^6 + 84a_4^4 + 32a_1^4a_3^4 + 96a_2^2a_6^2 + 48a_1^2a_2a_4^3 + 48a_8^2\right).$$

If we substitute $b^i + w^i$ for each a_i we get the enumerating polynomial for the 2-colorings of Q:

$$b^{16} + b^{15}w + 4b^{14}w^2 + 6b^{13}w^3 + 19b^{12}w^4 + 27b^{11}w^5 + 50b^{10}w^6 + 56b^9w^7 + 74b^8w^8 + 56b^7w^9 +$$
$$+50b^6w^{10} + 27b^5w^{11} + 19b^4w^{12} + 6b^3w^{13} + 4b^2w^{14} + bw^{15} + w^{16}.$$

The number of equicardinal 2-colorings is given by $\alpha = 74$, the coefficient of b^8w^8. The number of symmetric colorings is found as $\beta = Z(G; 0, 2, 0, 2, \ldots) = 42$. The number of equicardinal but non-symmetric colorings is therefore equal to $\alpha - \beta = 32$, or 16 if we allow interchange of color. Thus, 16 of the equicardinal colorings have non-congruent black and white clusters. There is therefore a total of 58 distinct cluster 2-sets $\{B, W\}$ such that B and W have the same number of cells.

We have also made these calculations for $n = 5$, and summarize our results for $n \leq 5$ in the accompanying table.

Table for Equicardinal 2-Colorings of K, $k = 2$

n	α	β	$\frac{1}{2}(\alpha - \beta)$	$\frac{1}{2}(\alpha + \beta)$
1	1	1	0	1
2	2	2	0	2
3	6	6	0	6
4	74	42	16	58
5	169112	4094	82509	86603

REFERENCES

1. D.A. Engel, An N-dimensional binary coloring problem, *J. Recreational Math.*, **4**(1971) 199–200.

2. E.M. Palmer, The exponential group as the automorphism group of a graph, in F. Harary (editor), *Proof Techniques in Graph Theory*, Academic Press (1969) 125–131.

3. R.C. Read, On the number of self-complementary graphs and digraphs, *J. London Math. Soc.*, **38**(1963) 99–104.

4. B.L. Schwartz, Black and white vertices of a hypercube, *J. Recreational Math.*, **11**(1979) 284–285.

University of New Brunswick,
Fredericton, N.B., Canada E3B 5A3

The Ancient English Art of Change Ringing

Kenneth J. Falconer

The ancient English art of change ringing dates back to 1630 and continues to be practised in some 5,000 churches in England, and a very small number of churches elsewhere in the world.

Bells are mounted on wheels, and strike in the course of being rotated from the mouth upwards position through 360° back to the mouth upwards position. Each bell is controlled by an individual ringer manipulating a rope. A typical church tower houses 5, 6 or 8 bells tuned to consecutive notes of a scale. The bells are numbered 1, 2,..., n going from highest to lowest pitch, with number 1 called the **treble** and number n called the **tenor** bell.

The aim of change ringing is not to ring tunes, but to ring **changes** or **permutations**, which many ringers regard as just as musical. A **change** consists of the n bells rung in some order, for example 2 3 1 5 4. The descending scale 1 2 3 4 5 is called **rounds**. Typically, a sequence of changes will be rung according to some predetermined scheme — **Bob Doubles**, **Stedman** and **Grandsire** are commonly used methods. There are various rules that methods must satisfy. For example, any sequence of changes always starts and ends with *rounds*, no other change being repeated. Perhaps the most important rule is that no bell moves more than one place between two consecutive changes (this is dictated by the mechanics of the bell). Thus 2 3 1 5 4 could be followed by 2 1 3 4 5 but not by 2 5 3 1 4 (the 5 has moved two places). There are various other rules that apply to particular methods. It is regarded as unethical for ringers to have any form of memory aid when ringing changes. They must know precisely when to ring their bell relative to the others; this is made easier by the symmetrical pattern underlying the standard methods.

1	2	**3**	4	5	
					a
2	1	4	**3**	5	
					b
2	4	1	5	**3**	
					a
4	2	5	1	**3**	
					b
4	5	2	**3**	1	
					a
5	4	**3**	2	1	
					b
5	**3**	4	1	2	
					a
3	5	1	4	2	
					b
3	1	5	2	4	
					a
1	**3**	2	5	4	
					b
1	2	**3**	4	5	

Figure 1: Plain Hunt.

An **extent** on n bells consists of a sequence of $n! + 1$ changes (starting and ending with rounds). Extents on 5 bells takes a few minutes to ring, on 6 bells about 20 minutes, and on 7 bells nearly 3 hours. An extent on 8 bells has only ever been rung once, in Loughborough in 1963, taking about 18 hours.

The most basic sequence used in change ringing is the sequence of $2n + 1$ changes on n bells known as **Plain Hunt**. This is shown in figure 1 in the case of 5 bells.

Each change is a permutation of the previous one, consisting either of transposing the bells in places 1 and 2, and also those in places 3 and 4 (marked 'a' and abbreviated to (12)(34)) or transposing those in places 2 and 3 and also those in places 4 and 5 (marked 'b' and written (23)(45)). Thus Plain Hunt may be described as $(ab)^5$. Note that each bell follows the same pattern of places, but starting at a different position — this is how ringers remember sequences of changes.

1	2	3	4	5	a
2	1	4	3	5	b
2	4	1	5	3	a
4	2	5	1	3	b
4	5	2	3	1	a
5	4	3	2	1	b
5	3	4	1	2	a
3	5	1	4	2	b
3	1	5	2	4	a
1	3	2	5	4	c
1	3	5	2	4	a
3	1	2	5	4	b
3	2	1	4	5	a
2	3	4	1	5	b
2	4	3	5	1	a
4	2	5	3	1	b
4	5	2	1	3	a
5	4	1	2	3	b
5	1	4	3	2	a
1	5	3	4	2	c
1	5	4	3	2	a
5	1	3	4	2	b
5	3	1	2	4	a
3	5	2	1	4	b
3	2	5	4	1	a
2	3	4	5	1	b
2	4	3	1	5	a
4	2	1	3	5	b
4	1	2	5	3	a
1	4	5	2	3	c
1	4	2	5	3	a
4	1	5	2	3	b
4	5	1	3	2	a
5	4	3	1	2	b
5	3	4	2	1	a
3	5	2	4	1	b
3	2	5	1	4	a
2	3	1	5	4	b
2	1	3	4	5	a
1	2	4	3	5	c
1	2	3	4	5	

Figure 2:
Plain Bob Doubles.

In order to obtain further changes, one must replace the 'b' at the end of Plain Hunt by a different permutation. Taking c to mean transposing the bells in places 3 and 4 and leaving the rest fixed, that is (34), we may join four blocks of Plain Hunt to get the method called Bob Doubles, see figure 2. This may be denoted by $((ab)^4(ac))^4$. This still gets only 40 out of the 120 possible changes on 5 bells. To get the remainder it is usual to insert the permutation $d = (23)$ (called a '**bob**'), after appropriate changes. For example

$$(((ab)^4ac)^3(ab)^4ad)^3$$

gives an extent on 5 bells.

Bob Doubles, and, indeed, other traditional methods of ringing changes, have several mathematical interpretations. If we think of the set of 120 permutations on 5 bells as the permutation group S_5, then the Plain Hunt block may be regarded as the 10 element subgroup H generated by the permutations $a = (12)(34)$ and $b = (23)(45)$. The four blocks that make up Bob Doubles are cosets of H, that is H, Hbc, $H(bc)^2$ and $H(bc)^3$. Thus, ringing Bob Doubles is just enumerating cosets of H. To ring an extent of 5 bells based on the Bob Doubles method requires inserting extra permutations to enable the other cosets of H to be reached. It is remarkable that change ringing methods such as these were in use a century or more before the introduction of group theory as a branch of mathematics. The ease with which many experienced ringers with little knowledge of mathematics can use properties of permutations, such as evenness or oddness, to analyze complicated sequences of changes, for example to check that a method for generating 5040 changes on 7 bells does not repeat or omit any changes, is impressive.

Another way of representing sequences of changes and displaying the various symmetries involved is by means of a **Cayley graph**. For 5 bells, the 120 vertices of the graph represent the changes, and two vertices are joined if it is allowable to follow one change by the next, i.e., with no bell moving more than one place. For 5 bells the following permutations from one change to another may occur:

a	(12)(34)
b	(23)(45)
c	(34)
d	(23)
e	(12)
f	(12)(45)
g	(45)

To save Figure 3 from becoming unduly complicated, we have only marked the permutations corresponding to a, b and c. A circuit of the central decagon represents the Plain Hunt subgroup H, the other 11

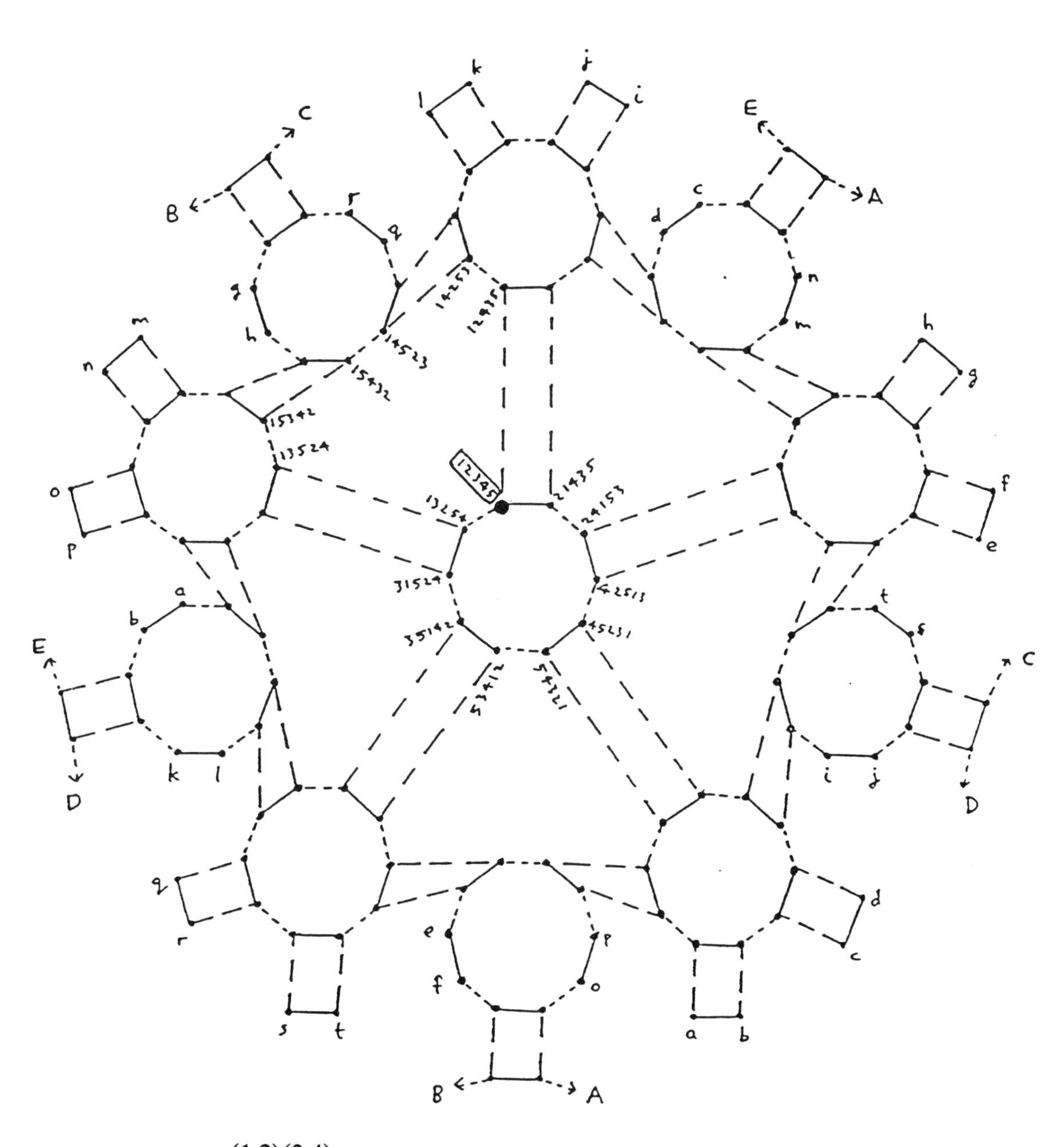

$a = (1\,2)(3\,4)$

$b = (2\,3)(4\,5)$

$c = (3\,4)$

Figure 3: Cayley graph of the changes on five bells.

decagons (10 are obvious and one is found by joining external vertices by edges $ABCDE$) correspond to the cosets of H.

It is suggested that the reader follow round the Plain Bob method on the Cayley graph. Once a decagon is joined, 9 sides are followed, before a 'c' or other permutation is used to transfer to another decagon. (In fact, Figure 3 as it is shown, comprises the 120 vertices and 180 edges of a *truncated icosidodecahedron* (see Rouse Ball and Coxeter, page 138).

Obviously the Cayley graphs become almost unmanageable for 6 or more bells, although study of their local structure can be of interest.

In this briefest of introductions to the fascinating pastime of change ringing we have done no more than point out that sophisticated mathematics underlies this ancient art. Some references are given below for those who wish to pursue the subject further. Change ringing is put into a 'social context' in the detective novel by Dorothy Sayers. The book by Wilson is a good general introduction to change ringing, and mathematical aspects are explored rather more deeply in some of the other articles, in particular the recent papers by White.

Should you find yourself in England and hear changes being rung on church bells, you should make an effort to introduce yourself to the ringers. They are usually only too glad to 'show the ropes' to any visitor, and you may be amazed at the mental agility of ringers, when it comes to thinking through complicated sequences of changes.

The Strens Conference on Recreational Mathematics was held in Calgary, and some of the conference participants took time off to ring the bells of Christ Church, Elbow Park, Calgary, one of the eight rings of bells in Canada.

References

F. J. Budden, *The Fascination of Groups,* Cambridge Univ. Press, 1972.

D. J. Dickinson, On Fletcher's paper 'Campanological groups', *Amer. Math. Monthly*, **64**(1957) 331–332.

T. J. Fletcher, Campanological groups, *Amer. Math. Monthly*, **63**(1956) 619–628.

M. Hodgson, *A Symbolic Treatment of False Course Heads,* Woodbridge Press, Guildford, 1962.

B. D. Price, Mathematical groups in campanology, *Math. Gaz.*, **53**(1969) 129–133.

R. A. Rankin, A campanological problem in group theory, *Proc. Cambridge Phil. Soc.*, **44**(1948) 17–25.

W. W. Rouse Ball & H.S.M. Coxeter, *Mathematical Recreations & Essays,* 12th Ed., University of Toronto, 1974.

D. L. Sayers, *The Nine Tailors,* Victor Gollancz, London, 1934.

J. Snowdon & W. Snowdon, *Diagrams,* C. Groome, Kettering, 1972.

A. T. White, Ringing the changes, *Math. Proc. Cambridge Phil. Soc.*, **94**(1983) 203–214.

A. T. White, Ringing the cosets, *Amer. Math. Monthly*, **94**(1987) 721–746.

A. T. White, Ringing the changes II, *Ars. Combin.*, **20A**(1985) 65–75.

A. T. White, Ringing the cosets II, *Math. Proc. Cambridge Phil. Soc.*, **105**(1989) 53–65.

W. G. Wilson, *Change Ringing,* Faber, 1965.

University of St. Andrews
St. Andrews, Fife, Scotland

The Strong Law of Small Numbers*

Richard K. Guy

This article is in two parts, the first of which is a do-it-yourself operation, in which I'll show you 35 examples of patterns that *seem* to appear when we look at several small values of n, in various problems whose answers depend on n. The question will be, in each case: do you think that the pattern persists for all n, or do you believe that it is a figment of the smallness of the values of n that are worked out in the examples?

Caution: examples of both kinds appear; they are not all figments!

In the second part I'll give you the answers, insofar as I know them, together with references.

Try keeping a scorecard: for each example, enter your opinion as to whether the observed pattern is known to continue, known not to continue, or not known at all.

This first part contains no information; rather it contains a good deal of disinformation. The first part contains one theorem:

You can't tell by looking.

It has wide application, outside mathematics as well as within. It will be proved by intimidation.

Here are some well-known examples to get you started.

Example 1. The numbers $2^{2^0}+1=3$, $2^{2^1}+1=5$, $2^{2^2}+1=17$, $2^{2^3}+1=257$, $2^{2^4}+1=65537$, are primes.

Example 2. The number 2^n-1 can't be prime unless n is prime, but $2^2-1=3$, $2^3-1=7$, $2^5-1=31$, $2^7-1=127$, are primes.

Example 3. Apart from 2, the oddest prime, all primes are either of shape $4k-1$, or of shape $4k+1$. In any interval $[1,n]$, the former are at least as numerous as the latter ($4k-1$ wins the "prime number race"):

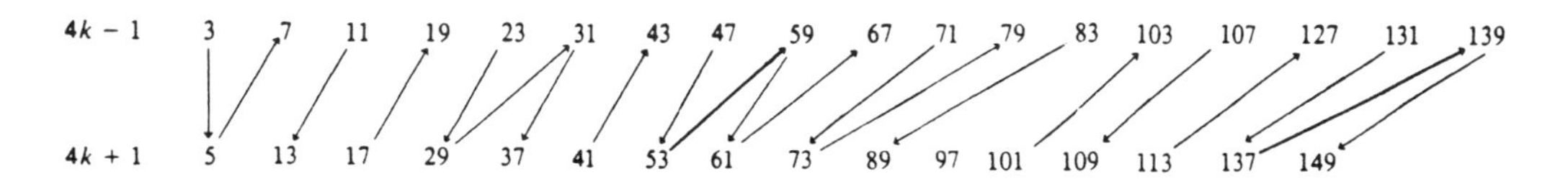

Example 4. Pick several numbers at random (it suffices just to look at odd ones). Estimate the probability that a number has more divisors of shape $4k-1$, than it does of shape $4k+1$. For

*Reprinted with permission from the *American Mathematical Monthly* **95**(1988) 697–712. Winner of the Lester R. Ford award.

example, 21 has two of the first kind (3 &7) and two of the second (1 & 21), while 25 has all three $(1, 5, 25)$ of the second kind.

Example 5. The five circles of Figure 1 have $n = 1, 2, 3, 4, 5$ points on them. These points are in general position, in the sense that no three of the $\binom{n}{2}$ chords joining them are concurrent. Count the numbers of regions into which the chords partition each circle.

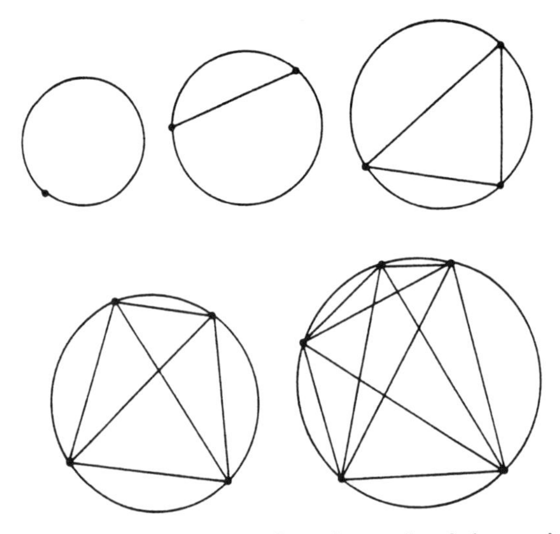

Figure 1: How many regions in each of these circles?

I've been trying to formulate the **Strong Law of Small Numbers** for many years [9]. The best I can do so far is

> There aren't enough
> small numbers to meet
> the many demands made of them.

It is the enemy of mathematical discovery. When you notice a mathematical pattern, how do you know it's for real?

> Superficial similarities
> spawn spurious statements.

> Capricious coincidences
> cause careless conjectures.

On the other hand, The Strong Law often works the other way:

> Early exceptions
> eclipse eventual essentials.

> Initial irregularities
> inhibit incisive intuition.

Here are some misleading facts about small numbers:

Ten per cent of the first hundred are perfect squares.

A quarter of the numbers less than 100 are primes.

Except for 6, all numbers less than 10 are prime powers.

Half the numbers less than 10 are Fibonacci numbers

$$0, 1, 1, 2, 3, 5, 8, \ldots$$

and alternate Fibonacci numbers, $1, 2, 5, \ldots$ are both Bell numbers and Catalan numbers.

Example 6. The numbers 31, 331, 3331, 33331, 333331, 3333331, are each prime.

Example 7. The alternating sums of factorials,

$$3! - 2! + 1! = 5$$
$$4! - 3! + 2! - 1! = 19$$
$$5! - 4! + 3! - 2! + 1! = 101$$
$$6! - 5! + 4! - 3! + 2! - 1! = 619$$
$$7! - 6! + 5! - 4! + 3! - 2! + 1! = 4421$$
$$8! - 7! + 6! - 5! + 4! - 3! + 2! - 1! = 35899$$

are each prime.

Example 8. In the table

row	entries	count
row 1	1 1	2
row 2	1 2 1	3
row 3	1 3 2 3 1	5
row 4	1 4 3 2 3 4 1	7
row 5	1 5 4 3 5 2 5 3 4 5 1	11
row 6	1 6 5 4 3 5 2 5 3 4 5 6 1	13
row 7	1 7 6 5 4 7 3 5 7 2 7 5 3 7 4 5 6 7 1	19
row 8	1 8 7 6 5 4 7 3 8 5 7 2 7 5 8 3 7 4 5 6 7 8 1	23
row 9	1 9 8 7 6 5 9 4 7 3 8 5 7 9 2 9 7 5 8 3 7 4 9 5 6 7 8 9 1	29

row n is obtained from row $n-1$ by inserting n between each pair of consecutive numbers which add to n. The number of numbers in each row is shown on the right. Each is prime.

Example 9. Is there a prime of shape $7013 \times 2^n + 1$?

Example 10. Are all the numbers $78557 \times 2^n + 1$ composite?

Example 11. When you use Euclid's method to show that there are unboundedly many primes:

$$2 + 1 = 3$$
$$(2 \times 3) + 1 = 7$$
$$(2 \times 3 \times 5) + 1 = 31$$
$$(2 \times 3 \times 5 \times 7) + 1 = 211$$
$$(2 \times 3 \times 5 \times 7 \times 11) + 1 = 2311$$

you don't always get primes:

$$\begin{aligned}
(2 \times 3 \times 5 \times 7 \times 11 \times 13) + 1 &= 30011 = 59 \times 509 \\
(2 \times 3 \times 5 \times 7 \times 11 \times 13 \times 17) + 1 &= 510511 = 19 \times 97 \times 277 \\
(2 \times 3 \times 5 \times 7 \times 11 \times 13 \times 17 \times 19) + 1 &= 9699691 = 347 \times 27953
\end{aligned}$$

but if you go to the *next* prime, its difference from the product is always a prime:

$$\begin{gathered}
5 - 2 = 3 \\
11 - (2 \times 3) = 5 \\
37 - (2 \times 3 \times 5) = 7 \\
223 - (2 \times 3 \times 5 \times 7) = 13 \\
2333 - (2 \times 3 \times 5 \times 7 \times 11) = 23 \\
30047 - (2 \times 3 \times 5 \times 7 \times 11 \times 13) = 17 \\
510529 - (2 \times 3 \times 5 \times 7 \times 11 \times 13 \times 17) = 19 \\
9699713 - (2 \times 3 \times 5 \times 7 \times 11 \times 13 \times 17 \times 19) = 23
\end{gathered}$$

Example 12. From the sequence of primes, form the first differences, then the absolute values of the second, third, fourth, ... differences:

2 3 5 7 11 13 17 19 23 29 31 37 41 43 47 53 59 61 67
1 2 2 4 2 4 2 4 6 2 6 4 2 4 6 6 2 6
1 0 2 2 2 2 2 2 4 4 2 2 2 2 0 4 4 2
1 2 0 0 0 0 0 2 0 2 0 0 0 2 4 0 2
1 2 0 0 0 0 2 2 2 2 0 0 2 2 4 2 2
1 2 0 0 0 2 0 0 0 2 0 2 0 2 2 0
1 2 0 0 2 2 0 0 2 2 2 2 2 0 2 0
1 2 0 2 0 2 0 2 0 0 0 0 2 2 2
1 2 2 2 2 2 2 2 0 0 0 2 0 0 0
1 0 0 0 0 0 0 2 0 0 2 2 0 0
1 0 0 0 0 0 2 2 0 2 0 2 0 0
1 0 0 0 0 2 0 2 2 2 2 2 0

Is the first term in each sequence always 1?

Example 13. 2^n is never congruent to 1 (mod n) for $n > 1$. 2^n is congruent to 2 (mod n) whenever n is prime and occasionally when it isn't ($n = 341, 561, \ldots$). Is 2^n ever congruent to 3 (mod n) for $n > 1$?

Example 14. The good approximations to $5^{1/5}$, namely, the convergents to

$$1 + \frac{1}{2+}\frac{1}{1+}\frac{1}{1+}\frac{1}{1+}\frac{1}{2+}\cdots \quad \text{are} \quad \frac{1}{1}, \frac{3}{2}, \frac{4}{3}, \frac{7}{5}, \frac{11}{8}, \frac{29}{21}, \ldots$$

which have Fibonacci numbers for denominators and Lucas numbers for numerators.

Example 15.

$$
\begin{aligned}
(x+y)^3 &= x^3+y^3+3xy(x+y)(x^2+xy+y^2)^0\\
(x+y)^5 &= x^5+y^5+5xy(x+y)(x^2+xy+y^2)^1\\
(x+y)^7 &= x^7+y^7+7xy(x+y)(x^2+xy+y^2)^2
\end{aligned}
$$

Example 16. The sequences of **hex numbers** (so named to distinguish them from the **hexagonal** numbers, $n(2n-1)$) are depicted in Figure 2.

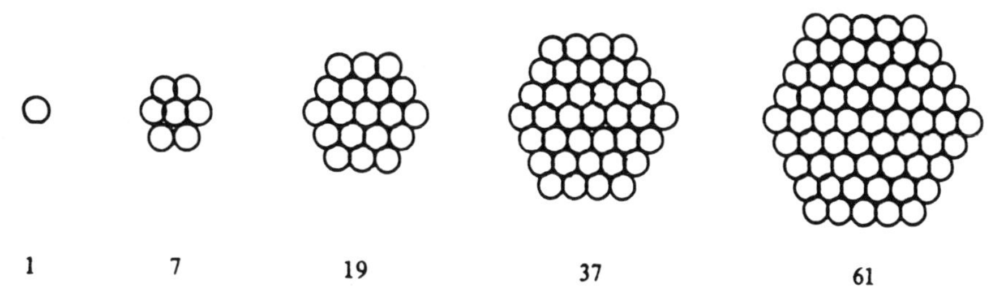

Figure 2: The hex numbers.

The partial sums of this sequence, 1, 8, 27, 64, 125, appear to be perfect cubes.

Example 17. Write down the positive integers, delete every second, and form the partial sums of those remaining:

1 ~~2~~ 3 ~~4~~ 5 ~~6~~ 7 ~~8~~ 9 ~~10~~ 11

1 4 9 16 25 36

Example 18. As before, but delete every third, then delete every second partial sum:

1 2 ~~3~~ 4 5 ~~6~~ 7 8 ~~9~~ 10 11 ~~12~~ 13 14 ~~15~~ 16

1 ~~3~~ 7 ~~12~~ 19 ~~27~~ 37 ~~48~~ 61 ~~75~~ 91

1 8 27 64 125 216

Example 19. Again, but delete every fourth, then every third partial sum, then every second of their partial sums:

1 2 3 ~~4~~ 5 6 7 ~~8~~ 9 10 11 ~~12~~ 13 14 15 ~~16~~ 17

1 3 ~~6~~ 11 17 ~~24~~ 33 43 ~~54~~ 67 81 ~~96~~ 113

1 ~~4~~ 15 ~~32~~ 65 ~~108~~ 175 ~~256~~ 369

1 16 81 256 625

Example 20. Again, but circle the first number of the sequence, delete the second after that, the third after that, and so on. Form the partial sums and repeat:

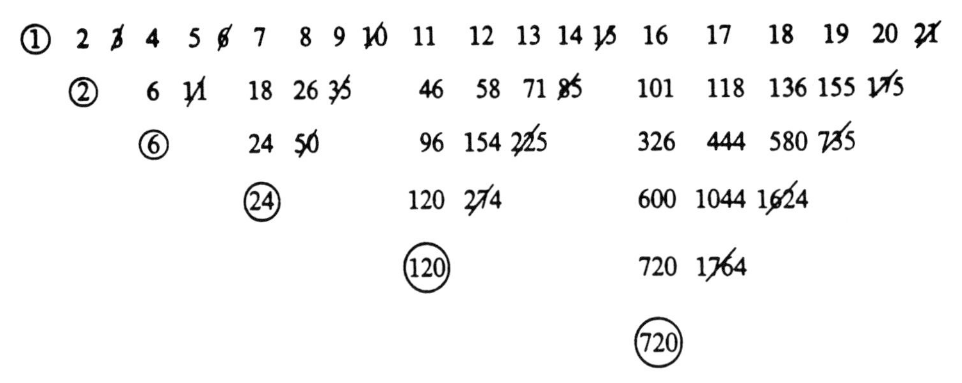

Example 21. Write down the odd numbers starting with 43. Circle 43, delete one number, circle 47, delete two numbers, circle 53, delete three numbers, circle 61, and so on. The circled numbers are prime (Figure 3).

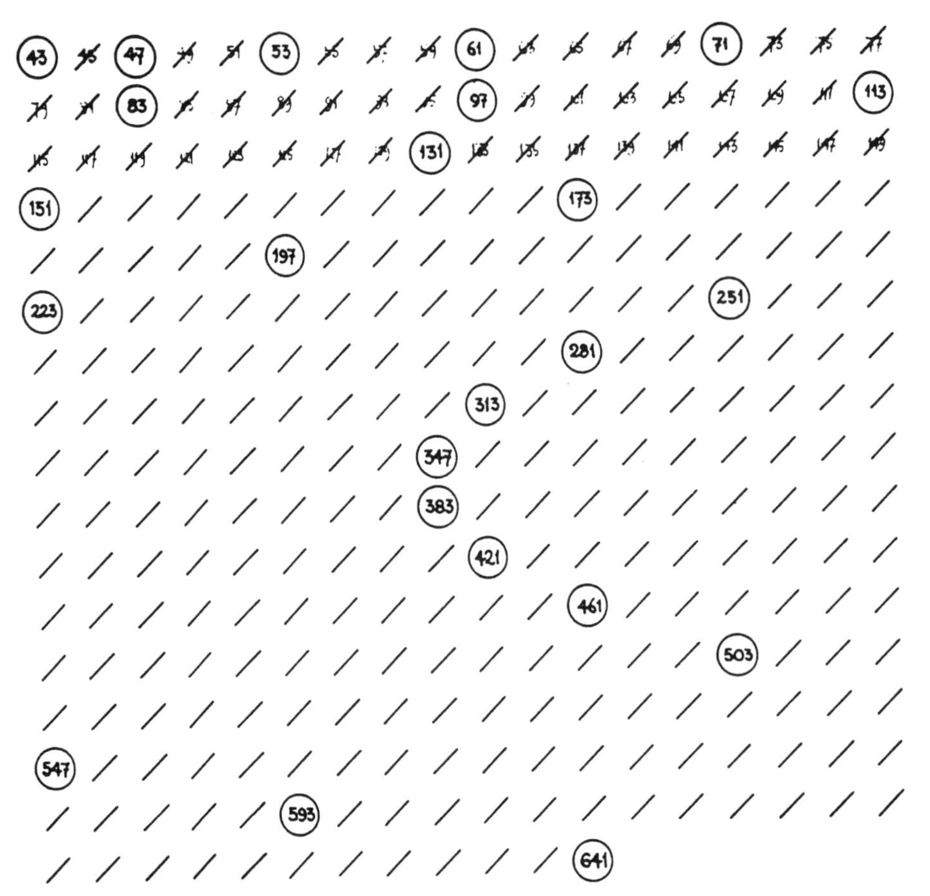

Figure 3: Parabolas of primes remain.

Example 22. In Table 1 the odd prime values of $n^4 + 1$ and of $17 \times 2^n - 1$ are printed in **bold**. They occur simultaneously for $n = 2, 4, 6, 16, 20$.

n	$n^4 + 1$	$17 \times 2^n - 1$
0	1	$16 = 2^4$
1	2	$33 = 3 \times 11$
2	**17**	**67**
3	$82 = 2 \times 41$	$135 = 3^3 \times$
4	**257**	**271**
5	$626 = 2 \times$	$543 = 3 \times$
6	**1297**	**1087**
7	$2402 = 2 \times$	$2175 = 3 \times$
8	$4097 = 17 \times$	$4351 = 19 \times$
9	$6562 = 2 \times$	$8703 = 3^2 \times$
10	$10001 = 73 \times$	$17407 = 13^2 \times$
11	$14642 = 2 \times$	$34815 = 3 \times$
12	$20737 = 89 \times$	$69631 = 179 \times$
13	$28562 = 2 \times$	$139263 = 3 \times$
14	$38417 = 41 \times$	$278527 = 223 \times$
15	$50626 = 2 \times$	$557055 = 3^2 \times$
16	**65537**	**1114111**
17	$83522 = 2 \times$	$2228223 = 3 \times$
18	$104977 = 113 \times$	$4456447 = 59 \times$
19	$130322 = 2 \times$	$8912895 = 3 \times$
20	**160001**	**17825791**
21	$194482 = 2 \times$	$35651583 = 3^4 \times$
22	$234257 = 73 \times$	$71303167 = 13 \times$
23	$279842 = 2 \times$	$142606335 = 3 \times$

Table 1

n	$21 \times 2^n - 1$	$7 \times 4^n + 1$
0	$20 = 2^2 \times 5$	$8 = 2^3$
1	**41**	**29**
2	**83**	**113**
3	**167**	**449**
4	$335 = 5 \times$	$1793 = 11 \times$
5	$671 = 11 \times$	$7169 = 67 \times$
6	$1343 = 17 \times$	$28673 = 53 \times$
7	**2687**	**114689**
8	$5375 = 5^3 \times$	$458753 = 79 \times$
9	$10751 = 13 \times$	$1835009 = 11 \times$
10	**21503**	**7340033**
11	$43007 = 29 \times$	$29360129 = 37 \times$
12	$86015 = 5 \times$	$117440513 = 3907 \times$
13	**172031**	**469762049**
14	$344063 = 17 \times$	$1879048193 = 11 \times$
15	$688127 = 11^4 \times$	$7516192769 = 29^2 \times$
16	$1376255 = 5 \times$	$30064771073 = 113 \times$
17	$2752511 = 19 \times$	$120259084289 = 379 \times$

Table 2

Example 23. In Table 2 the prime values of $21 \times 2^n - 1$ and of $7 \times 4^n + 1$ are printed in **bold**. They occur simultaneously for $n = 1, 2, 3, 7, 10, 13$.

Example 24. Consider the sequence

$$x_0 = 1, \qquad x_{n+1} = (1 + x_0^2 + x_1^2 + \cdots + x_n^2)/(n+1) \qquad (n \geq 0).$$

n	0	1	2	3	4	5	6	7	8	9	$\cdots$
x_n	1	2	3	5	10	28	154	3520	1551880	267593772160	$\cdots$

Is x_n always an integer?

Example 25. The same, but with cubes in place of squares:

$$y_0 = 1, \qquad y_{n+1} = (1 + y_0^3 + y_1^3 + \cdots + y_n^3)/(n+1) \qquad (n \geq 0).$$

Same question.

n	0	1	2	3	4	5	$\cdots$
y_n	1	2	5	45	22815	2375152056927	$\cdots$

Example 26. Also for fourth powers, $z_{n+1} = (1 + z_0^4 + z_1^4 + \cdots + z_n^4)/(n+1)$

n	0	1	2	3	4	$\cdots$
z_n	1	2	9	2193	5782218987645	$\cdots$

And for fifth powers, and so on.

Example 27. The irreducible factors of $x^n - 1$ are **cyclotomic polynomials**, i.e., $x^n - 1 = \prod_{d|n} \Phi_d(x)$, so that $\Phi_1(x) = x - 1$, $\Phi_2(x) = x + 1$, $\Phi_3(x) = x^2 + x + 1$, $\Phi_4(x) = x^2 + 1$. The cyclotomic polynomial of order n, $\Phi_n(x)$, has degree $\phi(n)$, Euler's totient function. It is easy to write down $\Phi_n(x)$ if n is prime, twice a prime, or a power of a prime, and for many other cases. Are the coefficients always ± 1 or 0?

Example 28. If two people play Beans-Don't Talk, the typical position is a whole number n, and there are just two options, from n to $(3n \pm 1)/2^*$, where 2^* means the highest power of 2 that divides the numerator. The winner is the player who moves to 1. For example, 7 is a $\mathcal{P}$-**position**, a previous-player-winning position, because the opponent must go to

$$(3 \times 7 + 1)/2 = 11 \quad \text{or} \quad (3 \times 7 - 1)/2^2 = 5$$

and 11 and 5 are $\mathcal{N}$-**positions**, next-player-winning positions, since they have the options $(3 \times 11 - 1)/2^5 = 1$ and $(3 \times 5 + 1)/2^4 = 1$.

If τ is the probability that a number is an $\mathcal{N}$-position, and there are no $\mathcal{O}$-positions (from which neither player can force a win), then the probability that a number is a $\mathcal{P}$-position is $1 - \tau$. This happens just if both options are $\mathcal{N}$-positions, so $1 - \tau = \tau^2$, and τ is the golden ratio $(\sqrt{5} - 1)/2 \approx 0 \cdot 618$.

So it is no surprise that 5 out of the first 8 numbers are $\mathcal{N}$- positions, 8 out of the first 13, 13 of the first 21, 21 of the first 34, and 34 of the first 55, since the ratio of consecutive Fibonacci numbers tends to the golden ratio.

Example 29. Does each of the two diophantine equations

$$2x^2(x^2 - 1) = 3(y^2 - 1) \quad \text{and} \quad x(x - 1)/1 = 2^n - 1$$

have just the five positive solutions $x = 1, 2, 3, 6$, and 91?

Example 30. Consider the sequence $a_1 = 1$, $a_{n+1} = \left\lfloor \sqrt{2a_n(a_n + 1)} \right\rfloor$ $(n \geq 1)$

n	1	2	3	4	5	6	7	8	9	10	11	12	13	14	15	16	17	18	19	20	21
a_n	1	2	3	4	6	9	13	19	27	38	54	77	109	154	218	309	437	618	874	1236	1748
		1		2		4		8		16		32		64		128		256		512	

Are alternate differences, $a_{2k+1} - a_{2k}$, the powers of two, 2^k?

Example 31. In the same sequence, are the even ranked members, a_{2k+2}, given by $2a_{2k} + \epsilon_k$, where ϵ_k is the kth digit in the binary expansion of $\sqrt{2} = 1.01101010000010\ldots$?

Example 32. Is this the same sequence as $a_1 = 1$, $a_2 = 2$, $a_3 = 3$, $a_{n+1} = a_n + a_{n-2}$ $(n \geq 3)$?

Example 33. The nth derivative of x^x, evaluated at $x = 1$, is an integer. Is it always a multiple of n? Values for $n = 1, 2, 3, \ldots$ are

$$1 \times 1,\ 2 \times 1,\ 3 \times 1,\ 4 \times 2,\ 5 \times 2,\ 6 \times 9,\ 7 \times (-6),\ 8 \times 118,\ 9 \times (-568),$$
$$10 \times 4716,\ 11 \times (-38160),\ 12 \times 358126,\ 113 \times (-3662088),\ 14 \times 41073096,$$
$$15 \times (-500013528),\ 16 \times 6573808200,\ 17 \times (-92840971200),$$
$$18 \times 1402148010528, \ldots.$$

Example 34. In how many ways, c_n, can you arrange n pennies in rows where every penny in a row above the first must touch two adjacent pennies in the row below?

n	0	1	2	3	4	5	6	7	8	9	10	11	12	13	14	15	16
c_n	1	1	1	2	3	5	9	15	26	45	78	135	234	406	704	1222	2120

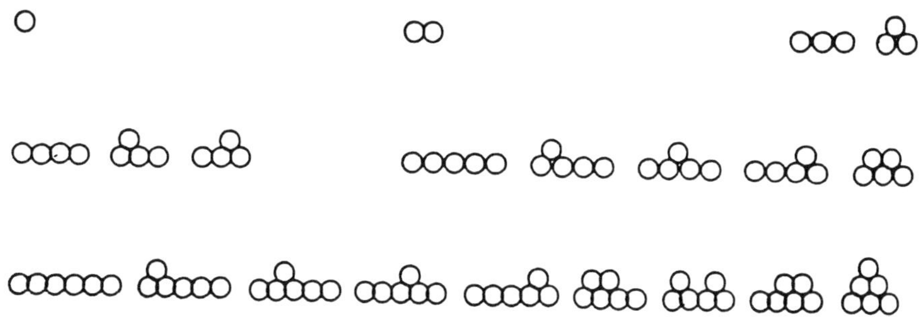

Figure 4: Propp's penny partitions.

To throw more light on such sequences, partition theorists often express their generating function

$$\sum_{n=0}^{\infty} c_n x^n = 1 + x + x^2 + 2x^3 + 3x^4 + 5x^5 + 9x^6 + 15x^7 + \cdots$$

as an infinite product,

$$\prod_{n=1}^{\infty}(1 - x^n)^{-a(n)}$$

In this case, $a(n)$ are consecutive Fibonacci numbers:

n	1	2	3	4	5	6	7	8	9	10	$\cdots$
$a(n)$	1	0	1	1	2	3	5	8	13	21	$\cdots$

Example 35. If p_k is the kth prime, $p_1 = 2, p_2 = 3, \ldots$, does

$$\prod_{k=1}^{\infty}(1 - x^{p_k})^{-1} = 1 + \sum_{n=1}^{\infty} \frac{x^{p_1+p_2+\cdots+p_k}}{(1-x)(1-x^2)\cdots(1-x^k)}?$$

Answers

1. No less a person than Fermat was fooled by the Strong Law! Euler gave the factorization $2^{32} + 1 = 641 \times 6700417$. All other known examples of Fermat numbers are composite; Jeff Young & Duncan Buell [32] have recently shown that $2^{2^{20}} + 1$ is composite.

2. There are very few **Mersenne primes**, $2^p - 1$. No one can prove that there are infinitely many; $2^{11} - 1 = 23 \times 89$ is not one. See A3 in [12] and sequence 1080 in [28].

3. In the "prime number race," $4k-1$ and $4k+1$ alternately take the lead infinitely often. This was proved by Littlewood [18]. For many papers on this subject see *N–12* of *Reviews in Number Theory,* for example Chen [4].

4. A theorem of Legendre (see [6], for example) states that if D_+ and D_- are the numbers of divisors of n of shapes $4k+1$ and $4k-1$, then the number of representations of n as the sum of two squares is $4(D_+ - D_-)$. So $D_+ \geq D_-$ for every number!

5. Before we reveal all, here is a circle (Figure 5) with ten points to further confuse you. It has 256 regions.

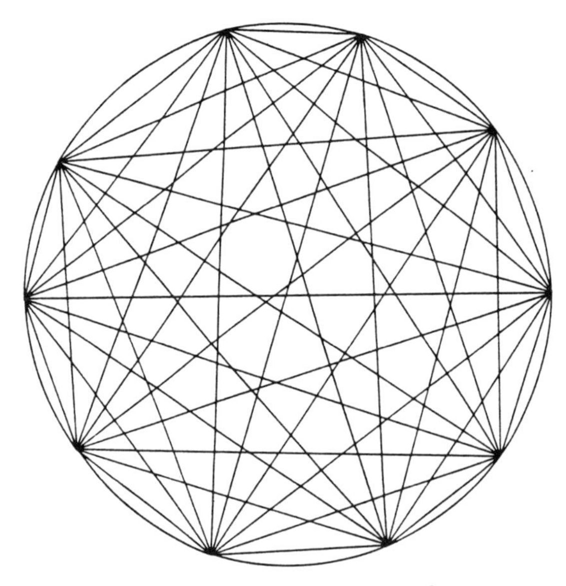

Figure 5: Circle partitioned in 2^8 regions.

If the circle has n points, there are $\binom{n}{4}$ intersections of chords inside the circle, since each set of four points gives just one such intersection. The number of vertices in the figure is $V = n + \binom{n}{4}$. To find the number of edges, count their ends. There are $n+1$ at each of the n points and four at each of the $\binom{n}{4}$ intersections, so $2E = n(n+1) + 4\binom{n}{4}$. By Euler's formula, the number of regions inside the circle is

$$\begin{aligned} E + 1 - V &= 2\binom{n}{4} + \frac{1}{2}n(n+1) + 1 - \binom{n}{4} - n \\ &= \binom{n}{4} + \frac{1}{2}n(n-1) + 1 \\ &= \binom{n-1}{4} + \binom{n-1}{3} + \binom{n-1}{2} + \binom{n-1}{1} + \binom{n-1}{0}. \end{aligned}$$

A direct proof, by labelling the regions with at most four of the numbers $1, 2, \ldots, n-1$, will appear in [5]. The answer is just five of the n terms in the binomial expansion of $(1+1)^{n-1}$. For $n < 6$, this is all the terms, and the number is a power of 2. For $n = 6$, only 1 is missing. For $n = 10$ just half the terms are missing, and the number of regions is $\frac{1}{2} \cdot 2^9 = 256$.

# of points =	1	2	3	4	5	6	7	8	9	10	11	12	13	14
# of regions =	**1**	**2**	**4**	**8**	**16**	31	57	99	163	**256**	386	562	794	1093

Some other famous numbers, e.g., 163 and 1093, also occur in this sequence, number 427 in [28].

6. No member of this sequence is divisible by 2, 3, 5, 7, 11, 13, or 37, as may be seen immediately from well known divisibility tests. On the other hand, 17, 19, 23, 29, 31, ... divide 33...331 just if the number of threes is respectively $16k + 8$, $18k + 11$, $22k + 20$, $28k + 19$, $15k + 1, \ldots$, while 41, 43, 53, 67, 71, 73, 79, ... divide no members of the sequence. I don't think that there is a simple description of which primes do, and which primes don't, divide. The next member 33333331, is also prime, but $333333331 = 17 \times 19607843$.

7. We've again given ourselves a good start, since $\sum_{k=1}^{n}(-1)^{n-k}k!$ is not divisible by any prime $\leq n$. However,

$$9! - 8! + 7! - 6! + 5! - 4! + 3! - 2! + 1! = 326981 = 79 \times 4139.$$

8. This example as well as example **5**, was first shown to me by Leo Moser, a quarter of a century ago. Row n is the list of denominators of the **Farey series** of order n, i.e., the set of rational fractions r, $0 \leq r \leq 1$, whose denominators do not exceed n. In getting row n from row $n - 1$, just $\phi(n)$ numbers are inserted, where $\phi(n)$ is Euler's totient function, the number of numbers not exceeding n which are prime to n. It is fortuitous that $1 + \sum_{k=1}^{n}\phi(k)$ is prime for $1 \leq n \leq 9$. As $\phi(10) = 4$, the number of numbers in row 10 is $29 + 4 = 33$, and is not prime.

9. The expression $7013 \times 2^n + 1$ is composite for $0 \leq n \leq 24160$ [15]. Duncan Buell & Jeff Young have sieved out 325 further candidates $n < 10^5$ which might yield a prime. None is known, though it's likely that there is one.

10. The number $78557 \times 2^n + 1$ is always divisible by at least one of 3, 5, 7, 13, 19, 37, 73 [26, 27]. For this and the previous example, see also B21 in [12].

11. R. F. Fortune conjectured that these differences are always prime: see [9], [10] and A2 in [12]. The next few are 37, 61, 67, 71, 47, 107, 59, 61, 109, 89, 103, 79. There's a high probability that the conjecture is true, because the difference can't be divisible by any of the first k primes, so the smallest composite candidate for $P = \prod p_k$ is p_{k+1}^2, which is approximately $(k \ln k)^2$ in size. The product of the first k primes is about e^k: to find a counterexample we need a gap in the primes near N of size at least $(\ln N \ln\ln N)^2$. Such gaps are believed not to exist, but it's beyond our present means to prove this.

12. This is N. L. Gilbreath's conjecture, which has been verified for $k < 63419$ [16]. Hallard Croft has suggested that it has nothing to do with primes as such, but will be true for any sequence consisting of 2 and odd numbers, which doesn't increase too fast, or have too large gaps: A10 in [12]. In an 87-08-03 letter Andrew Odlyzko reported that he had verified the conjecture for $k < 10^{10}$.

13. D. H. & Emma Lehmer discovered that $2^n \equiv 3 \pmod{n}$ for $n = 4700063497$, but for no smaller $n > 1$.

14. The kth Lucas number and the $(k+1)$th Fibonacci number are

$$\left(\frac{1+\sqrt{5}}{2}\right)^k + \left(\frac{1-\sqrt{5}}{2}\right)^k \quad \text{and} \quad \frac{1}{\sqrt{5}}\left\{\left(\frac{1+\sqrt{5}}{2}\right)^{k+1} - \left(\frac{1-\sqrt{5}}{2}\right)^{k+1}\right\}.$$

Their ratio, as k gets large, approaches $(5-\sqrt{5})/2 \approx 1\cdot381966011$, whereas $5^{1/5} \approx 1\cdot379729661$. The next few convergents to $5^{1/5}$,

$$\frac{40}{29}, \frac{109}{79}, \frac{912}{661}, \frac{1021}{740}, \frac{26437}{19161}, \frac{27458}{19901},$$

do not involve Fibonacci or Lucas numbers. Compare sequences 256 & 260 and 924 & 925 in [28]. This example goes back to 1866 [25].

15. This is quite fortuitous [30]. Put $x = y = 1$, giving $2^{2n+1} - 2 = (2n+1) \times 2 \times 3^{n-1}$. It's true that

$$2^2 - 1 = 3 \times 3^0, \qquad 2^4 - 1 = 5 \times 3^1, \qquad 2^6 - 1 = 7 \times 3^2$$

but it's clear that the pattern can't continue.

16. The $(n+1)$th hex number, $1 + 6 + 12 + \cdots + 6n = 3n^2 + 3n + 1$, when added to n^3, gives $(n+1)^3$, so the pattern is genuine. It is instructive to regard the nth hex number as comprising the three faces at one corner of a cubic stack of n^3 unit cubes (Figure 6).

17, 18, 19, and **20** are examples of **Moessner's process**, which does indeed produce the squares, cubes, fourth powers and factorials. Moessner's paper [20] is followed by a proof by Perron. Subsequent generalizations are due to Paasche [22]: see [19] for a more recent exposition.

21. A thinly disguised arrangement of Euler's formula, $n^2 + n + 41$, which gives primes for $-40 \le n \le 39$. For $n = 40$, $n^2 + n + 41 = 41^2$. See A1 and Figure 1 in [12]. For remarkable connexions with quadratic fields, continued fractions, modular functions, and class numbers, see [29].

22. The initial pattern is explained by the facts that if n is odd, $n^4 + 1$ is even, and $17 \times 2^n - 1$ is a multiple of 3. Thereafter it's largely coincidence until $n = 24$, for which $n^4 + 1 = 331777$ is prime, while $17 \times 2^n - 1 = 285212671 = 149 \times 1914179$. See [17], [24] and sequences 386 and 387 in [28].

23. This is also a coincidence, until we reach $n = 18$, for which $21 \times 2^n - 1 = 5505023$ is prime, while

$$7 \times 4^n + 1 = 481036337153 = 166609 \times 2887217.$$

See [31]. [23] and sequences 314 and 315 in [28].

24. A sequence introduced by Fritz Göbel. A more convenient recursion for calculation is $(n+1)x_{n+1} = x_n(x_n + n)$, $(n \ge 1)$. If you work modulo 43, you'll find that for

$n =$	0	1	2	3	4	5	6	7	8	9	10	11	12	13	14	15	16	17	18	19	20	21
$x_n \equiv$	1	2	3	5	10	28	25	37	10	20	15	38	19	42	36	34	2	35	39	31	13	2

$n =$	22	23	24	25	26	27	28	29	30	31	32	33	34	35	36	37	38	39	40	41	42
$x_n \equiv$	6	26	28	29	4	14	42	5	20	17	4	20	16	29	42	13	42	20	8	23	33

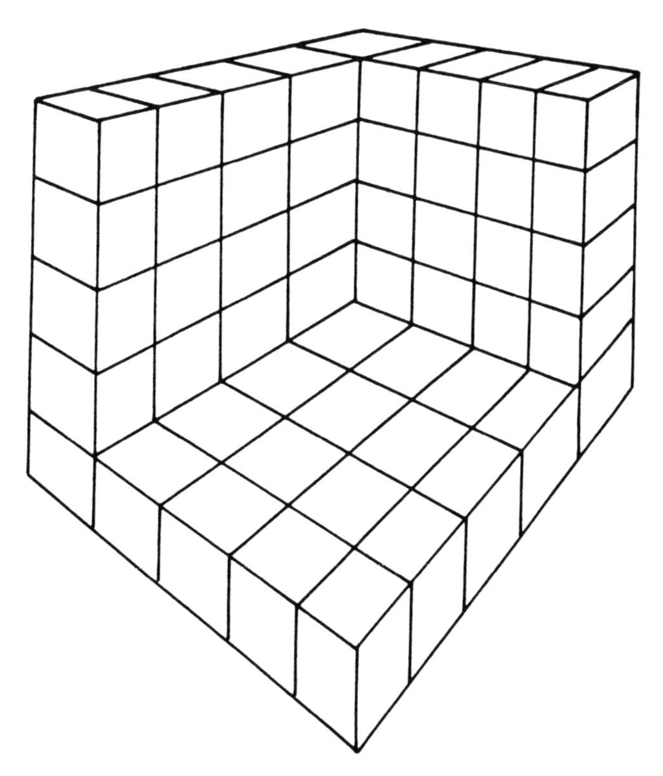

Figure 6: The fifth hex number

and $x_{42}(x_{42}+42) \equiv -10(-10+42) = -320$, which is not divisible by 43, so x_{43} is not an integer, although x_n is an integer for $0 \le n \le 42$.

25. Similar calculations, mod 89, using the relation $(n+1)y_{n+1} = y_n(y_n^2+n)$, show that y_{89} is not an integer. For this, and the previous example, see E15 in [12].

26. Since this question was asked, Henry Ibstedt has made extensive calculations, and found the first noninteger term, x_n, in the sequence involving kth powers, to be

k	2	3	4	5	6	7	8	9	10	11
n	43	89	97	214	19	239	37	79	83	239

He also found corresponding results with different initial values. The longest to hold out ($n = 610$) are the cubes ($k = 3$, Example **25**) with $x_0 = 1$, $x_1 = 11$.

27. The first cyclotomic polynomial to display a coefficient other that ± 1 and 0 is

$$\begin{aligned}\Phi_{105}(x) = {} & x^{48}+x^{47}+x^{46}-x^{43}-x^{42}-2x^{41}-x^{40}-x^{39}+x^{36}+x^{35}+x^{34}\\ & +x^{33}+x^{32}+x^{31}-x^{28}-x^{26}-x^{24}-x^{22}-x^{20}+x^{17}+x^{16}+x^{15}\\ & +x^{14}+x^{13}+x^{12}-x^{9}-x^{8}-2x^{7}-x^{6}-x^{5}+x^{2}+x+1\end{aligned}$$

Coefficients can be unboundedly large, but require n to contain a large number of distinct odd prime factors; see [8]. More recently, Montgomery & Vaughan [33] have shown that if $\Phi_n = \sum a(m,n)x^m$ and $L(n) = \ln \max_n |a(m,n)|$ then, for m large, $\frac{m^{1/2}}{(\ln 2m)^{1/4}} \ll L(n) \ll \frac{m^{1/2}}{(\ln m)^{1/4}}$.

28. This game was misremembered by John Conway from John Isbell's game of Beanstalk [13]. The Fibonacci pattern is not maintained: only 52 of the first 89 numbers, 81 of the first 144, 126 of the first 233, and 201 of the first 377, are $\mathcal{N}$-positions. The probability argument is fallacious: the probabilities of the status of the two options are *not* independent.

29. True, but why the coincidence?

30 and **31.** The patterns of powers of 2 and of binary digits of $\sqrt{2}$ both continue; see [11], [14], and sequence 206 in [28].

32. A different sequence, number 207 in [28], which agrees for $n < 9$, but then continues 28, 41, 60, 88, 129, 189, 277, 406, 595, 872, 1278, ...

33. If $y = x^x$ and $y_n(1)$ denotes the value of $d^n y/dx^n$ at $x = 1$, then

$$y_{n+1}(1) = y_n(1) + \binom{n}{1} y_{n-1}(1) - \binom{n}{2} y_{n-2}(1) + 2!\binom{n}{3} y_{n-3}(1) - 3!\binom{n}{4} y_{n-4}(1) + - + \cdots + (-1)^n (n-1)!.$$

This was not known to be a multiple of $n + 1$ when it was submitted to the Unsolved Problems section of the *American Mathematical Monthly* by Richard Patterson & Gaurar Suri. But in an 87-05-28 letter, Herb Wilf gives a proof, using the generating function for Stirling numbers of the first kind. His proof in fact shows that $n(n-1)$ divides $y_n(1)$ just if $n-1$ divides $(n-2)!$, which it does for $n \geq 7$, provided that $n-1$ is not prime.

34. This sequence was investigated by Jim Propp. Except that $a(12) = 55$, the pattern of Fibonacci numbers does not continue:

$n =$	11	12	13	14	15	16	17	18
$a(n) =$	35	55	93	149	248	403	670	1082

Since this was written, Wilf [21] has linked the generating function with Ramanujan's continued fraction, and he observes that the numbers of Propper partitions with k coins in the lowest row are yet another manifestation of the Catalan numbers, 1, 2, 5, 14, 42, ... [7]. These partitions are a variant of some considered by Auluck [1]. Auluck's partitions have the pennies contiguous in *every* row, not just the lowest. Their numbers 1, 1, 2, 3, 5, 8, ... are another good example of the Strong Law.

35. The expansion of the product as a power series, is

$$1 + x^2 + x^3 + x^4 + 2x^5 + 2x^6 + 3x^7 + 3x^8 + 4x^9 + 5x^{10} + 6x^{11} + 7x^{12} + 9x^{13} + 10x^{14}$$
$$+12x^{15} + 14x^{16} + 17x^{17} + 19x^{18} + 23x^{19} + 26x^{20}$$
$$+30x^{21} + 35x^{22} + 40x^{23} + 46x^{24} + 52x^{25} + 60x^{26} + 67x^{27} + 77x^{28} + 87x^{29} + \cdots$$

The sum is the same, until ...

$$+31x^{21} + 35x^{22} + 41x^{23} + 46x^{24} + 54x^{25} + 60x^{26} + 69x^{27} + 78x^{28} + 89x^{29} + \cdots.$$

This was entry 29 in Chapter 5 of Ramanujan's second notebook [2], [3]: but he had crossed it out!

Let me know if I've missed your favorite example! For 45 more examples of the Strong Law, see "The Second Strong Law of Small Numbers," *Math. Mag.*, **63** (1990) 3–20.

References

1. F. C. Auluck, On some new types of partitions associated with generalized Ferrers graphs, *Proc. Cambridge Philos Soc.*, 47(1951) 679–686; MR 13, 536.

2. Bruce C. Berndt, *Ramanujan's Notebooks,* Part 1, Springer-Verlag, 1985, p. 130.

3. Bruce C. Berndt & B. M. Wilson, Chapter 5 of Ramanujan's second notebook, in M. I. Knopp (ed.) *Analytic Number Theory,* Lecture Notes in Math. 899, Springer, 1981, pp. 49–78; MR 83i:10011.

4. W. W. L. Chen, On the error term of the prime number theorem and the difference between the number of primes in the residue classes modulo 4, *J. London Math. Soc.,* (2) 23(1981) 24–40; MR 82g:10058.

5. John Conway & Richard Guy, in preparation.

6. H. Davenport, *The Higher Arithmetic,* Hutchinson's University Library, 1952, p. 128.

7. Roger B. Eggleton & Richard K. Guy, Catalan strikes again! How likely is a function to be convex?, *Math. Mag.* 61(1988) 211–218.

8. P. Erdős & R. C. Vaughan, Bounds for the rth coefficients of cyclotomic polynomials, *J. London Math. Soc.* (2) 8(1974) 393–400; MR 40 #9835.

9. Martin Gardner, Mathematical games: patterns in primes are a clue to the strong law of small numbers, *Sci. Amer.,* 243 #6(Dec. 1980) 18, 20, 24, 26, 28.

10. Solomon W. Golomb, The evidence for Fortune's conjecture, *Math. Mag.* 54(1981) 209–210.

11. R. L. Graham & H. O. Pollak, Note on a linear recurrence related to $\sqrt{2}$, *Math. Mag.* 43(1970) 143–145; MR 42 #180.

12. Richard K. Guy, *Unsolved Problems in Number Theory,* Springer, 1981.

13. Richard K. Guy, John Isbell's game of Beanstalk and John Conway's game of Beans-Don't-Talk, *Math Mag.* 59(1986) 259–269.

14. F. K. Huang & S. Lin, An analysis of Ford & Johnson's sorting algorithm, *Proc. 3rd Annual Princeton Conf. on Info. Systems and Sci.*

15. Wilfrid Keller, Factors of Fermat numbers and large primes of the form $k \cdot 2^n + 1$, *Math. Comp.* 41(1983) 661–673.

16. R. B. Killgrove & K. E. Ralston, On a conjecture concerning the primes, *Math. Tables Aids Comp.* 13(1959) 121–122; MR 21 #4943.

17. M. Lal, Primes of the form $n^4 + 1$, *Math. Comp.* 21(1967) 245–247.

18. J. E. Littlewood, Sur le distribution des nombres premier, *C. R. Hebd. Séanc. Acad. Sci., Paris* 158(1914) 1868–1872.

19. Calvin T. Long, Strike it out—add it up, *Math. Mag.* 66(1982) 273–277. See also the *Amer. Math. Monthly* 73(1966) 846–851.

20. Alfred Moessner, Eine Bemerkung über die Potenzen der natürlichen Zahlen, *S.-B. Math.-Nat. Kl. Bayer. Akad. Wiss.* 1951, 29(1952); MR 14-353b.

21. Andrew M. Odlyzko & Herbert S. Wilf, n coins in a fountain, *Amer. Math. Monthly* 95(1988).

22. Ivan Paasche, Eine Verallgemeinerung des Moessnerschen Satzes, *Compositio Math.* 12(1956) 263–270; MR 17, 836g.

23. Hans Riesel, Lucasian criteria for the primality of $N = h \cdot 2^n - 1$, *Math Comp.* 23(1969) 869–875.

24. Raphael M. Robinson, A report on primes of the form $k \cdot 2^n + 1$ and on factors of Fermat numbers, *Proc. Amer. Math. Soc.* 9(1958) 673–681.

25. P. Seeling, Verwandlung der irrationalen Grösse $\sqrt[n]{\ }$ in einen Kettenbruch, *Archiv. math. Phys.* 46(1866) 80–120 (esp. p. 116).

26. J. L. Selfridge, Solution to problem 4995, *Amer. Math. Monthly* 70(1963) 101.

27. W. Sierpiński, Sur un problème concernant les nombres $k \cdot 2^n + 1$, *Elem. Math.* 15(1960) 73–74; MR 22 #7983; corrigendum, *ibid.* 17(1962) 85.

28. N. J. A. Sloane, *A Handbook of Integer Sequences,* Academic Press, 1973.

29. Harold M. Stark, An explanation of some exotic continued fractions found by Brillhart, *Computers in Number Theory,* Atlas Sympos. No. 2, Oxford, 1969, pp. 21–35, Academic Press, London, 1971.

30. Peter Taylor & Doug Dillon, Problem 3, *Queen's Math. Communicator,* Dept. of Math. and Stat., Queen's University, Kingston, Ont., Oct. 1985, p. 16.

31. H. C. Williams & C. R. Zarnke, A report on prime numbers of the forms $M = (6a+1)2^{2m-1} - 1$ and $M' = (6a-1)2^{2m} - 1$, *Math. Comp.* 22(1968) 420–422.

32. Jeff Young & Duncan A. Buell, The twentieth Fermat number is composite, *Math. Comp.* 50(1988) 261–263.

33. H. L. Montgomery & R. C. Vaughan, The order of magnitude of the mth coefficients of cyclotomic polynomials, *Glasgow Math. J.* 27(1985) 143–159; MR 87e: 11026.

The University of Calgary,
Calgary, Alberta
Canada T2N 1N4

Match Sticks in the Plane

Heiko Harborth

0 Introduction

Match sticks are one of the cheapest and simplest objects for puzzles which can be both challenging and mathematical. In most books on recreational mathematics and sometimes in newspapers puzzles with match sticks are to be found. Here I will consider only connected sets of match sticks in the plane, for example a table, where common points in general are endpoints and no crossings are allowed.

An example, which can be found in one of Gardner's books from *Scientific American* [1], is as follows: What is the minimum number of match sticks to stabilize a unit square? Figure 1 implies that 23 match sticks are sufficient. Whether fewer than 23 are possible is unknown, to my knowledge.

Another example, which is my own, was influenced by an unsolved problem of P. Erdős. Erdős asked for the minimum number of points in the plane, such that each point has distance 1 to exactly r other points. To avoid crossings, I ask for the smallest number of endpoints of match sticks, such that exactly r match sticks come together at each endpoint. Figure 2 shows ten points, with 15 match sticks, as an example for $r = 3$. Well, dear reader, have you got your match sticks? Try to find an example with less than 15 match sticks where exactly three match sticks touch one another at all their endpoints ($r = 3$).

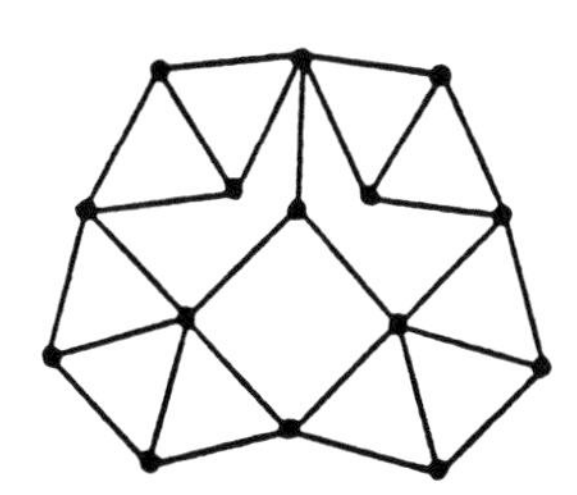

Figure 1: Stablizing a square

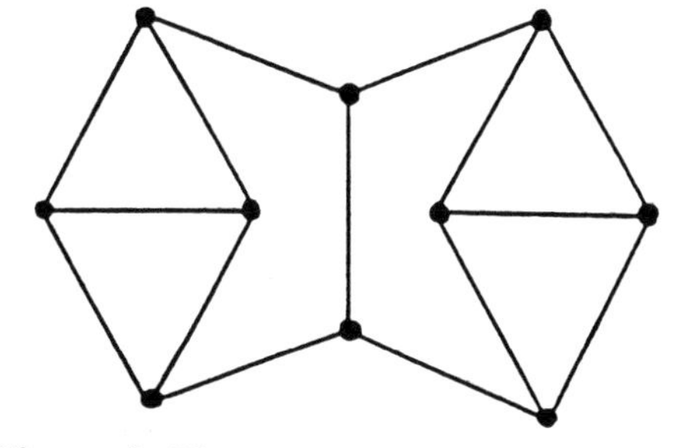

Figure 2: Three matches at each point

1 A General Task for Puzzlers

Produce all graphs which are possible with non-crossing match sticks on a table, that is, construct all planar graphs which have representations in the plane with straight edges of unit length. Figure 3 lists all 1, 2, 5, and 13 graphs for $p = 2, 3, 4$, and 5 points. Try to find all 50 possibilities for $p = 6$ points.

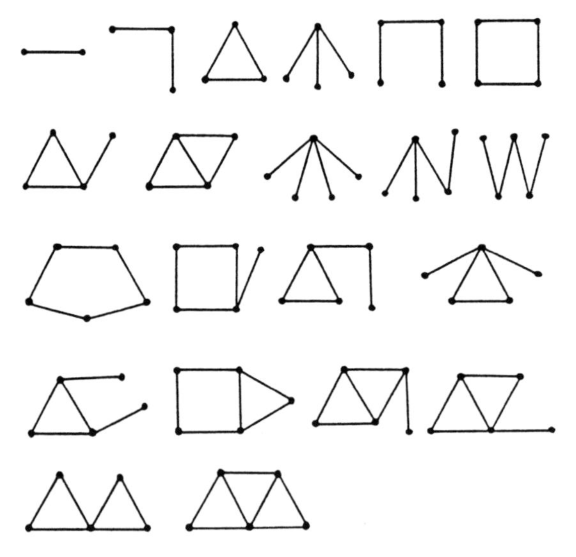

Figure 3: Planar graphs with p points.

If $q = 1, 2, 3, 4$, or 5 match sticks are given, then there are 1, 1, 3, 5, or 12 different graphs as in Figure 4. Do you find all 28 graphs with $q = 6$ match sticks?

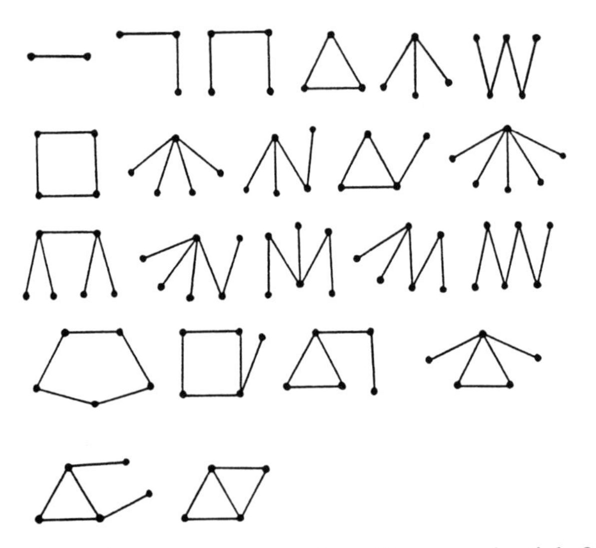

Figure 4: How many graphs with q match sticks?

In general these numbers of plane match stick representations of graphs with p points or q unit edges cannot be known since already the number of nonisomorphic trees with p points or q edges is still unknown. Nevertheless, you can puzzle as far as you have got patience, time and match sticks.

2 Extremal classes of planar match stick graphs

Now consider extremal classes of planar match stick graphs. Those with a minimum number $m(q)$ of points for q given match sticks fulfil the inequality

$$m(q) \leq 1 + \left\lceil \frac{1}{3}\left(q - 1 + \sqrt{4q+1}\right)\right\rceil ,$$

where $\lceil x \rceil$ denotes the smallest integer larger than or equal to x. For graphs with a maximum number $M(p)$ of match sticks, if p points are given, holds the inequality

$$M(p) \geq 3p - \left\lceil \sqrt{12p-3} \right\rceil .$$

Graphs with q match sticks can have a maximum number

$$n(q) \geq q - \left\lceil \frac{1}{3}\left(q - 1 + \sqrt{4q+1}\right)\right\rceil$$

of closed regions in the plane. Equality should be provable in all three cases. These equalities are proved (see **[4]**) if it is assumed that all pairs of points have distance at least 1, since then circles of radius $\frac{1}{2}$ with the p points as their centres form a packing of p congruent circles (coins), and the maximum number $M(p)$ of sticks is equivalent to the maximum number of touching points of the p circles. This number, however, was determined in **[4]**.

It is another nice puzzle to place p equal coins on a table such that the number of touching points is a maximum. Which shapes are possible? For $p = 9$ all three possibilities are shown in Figure 5. It is interesting to note that for all $p = 3t^2 + 3t + 1$, $t = 1, 2, \ldots$, there is a unique extremal case which has the shape of a large regular hexagon.

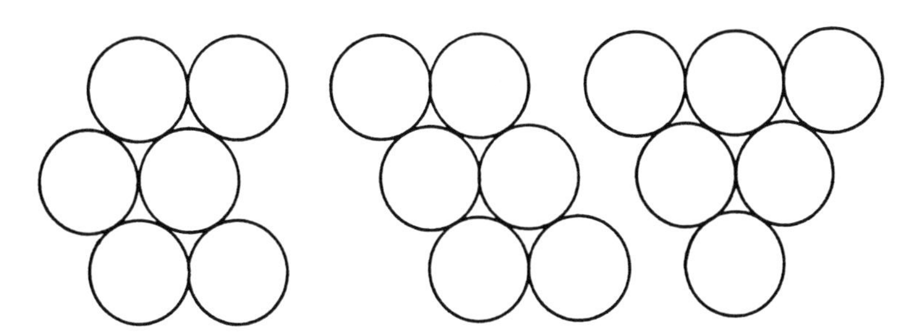

Figure 5: Maximum numbers of touching points.

3 Polyominoes

Well known planar graphs with straight edges of unit length are polyominoes or animals with n cells, that are simply-connected sets of n cells as parts of the three regular tessellations of the plane. Although the number of different polyominoes with n cells is unknown in general, it is challenging to find all polyominoes for small n by placing match sticks on a table.

Another puzzle task is to find all polyominoes with q given match sticks. Do they exist at all? They do in case of

triangles	for all $q \geq 3$,	unless	$q = 4, 6, 8, 10,$
squares	for all $q \geq 4$,	unless	$q = 5, 6, 8, 9, 11, 14,$
hexagons	for all $q \geq 6$,	unless	$q = 7, 8, 9, 10, 12, 13, 14, 17, 18, 22.$

A simple proof, for example in the case of squares, can proceed as follows: Figure 6 depicts polyominoes for $q = 4 + 3i$, $q = 12 + 3i$, and $q = 17 + 3i$, $i \geq 0$, so that square polyominoes for all $q \geq 15$ do exist. Polyominoes with $n \geq 5$ squares always need at least 15 sticks (see the formula below). Then the remaining polyominoes with $n \leq 4$ cells are easily checked.

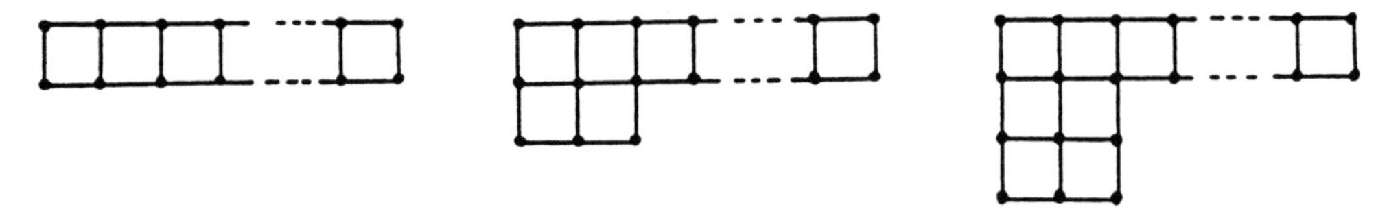

Figure 6: Polyominoes with q match sticks.

What are the maximum and minimum numbers of match sticks to produce a polyomino with n cells? The following sharp inequalities were found in [3]:

triangles	$n + \left\lceil \frac{1}{2}\left(n + \sqrt{6n}\right) \right\rceil$	$\leq q \leq 2n + 1$
squares	$2n + \left\lceil 2\sqrt{n} \right\rceil$	$\leq q \leq 3n + 1$
hexagons	$3n - \left\lceil \sqrt{12n - 3} \right\rceil$	$\leq q \leq 5n + 1$

The maximum is attained for polyominoes like trees. If the polyominoes grow up cell by cell like spirals then the minimum occurs. It is again a nice puzzle to find all different polyominoes with n cells and minimum q for small n. In Figure 7 all six shapes for $n = 10$ are drawn. A general formula is unknown.

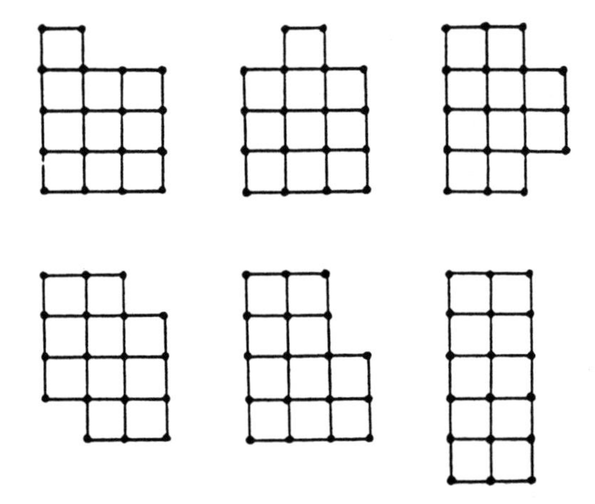

Figure 7: Polyominoes with n cells. Ten cells using 27 match sticks.

4 Regular Graphs

Another interesting kind of planar graph is a regular graph. By Euler's polyhedron formula, r-regular planar graphs exist only for $r \leq 5$. Are these graphs realizable by match sticks? What is the smallest number $p(r)$ of points? The corresponding number of match sticks then is $\frac{1}{2}rp(r)$.

For $r = 1$ one match stick and for $r = 2$ a triangle imply $p(1) = 2$ and $p(2) = 3$. It was your task, dear reader, to find for $r = 3$ the example of Figure 8. Did you get it? An unpublished result of A. Blokhuis proves the non-existence of $p(5)$. The value of $p(4)$ is unknown so far. The example of Figure 9 proves that $p(4) \leq 52$. Can you find an example for $r = 4$ with fewer than 104 match sticks?

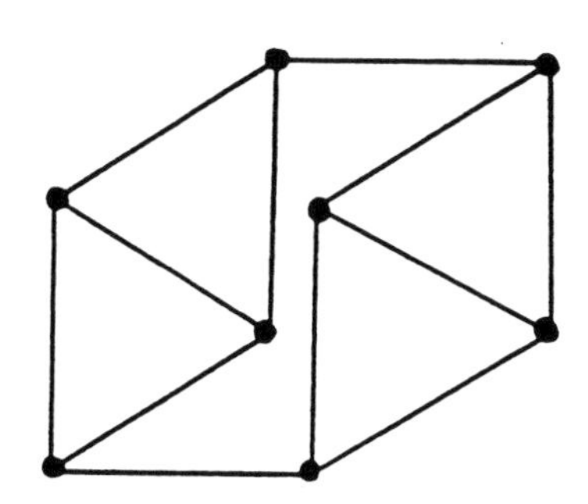

Figure 8: $p(3) = 8$

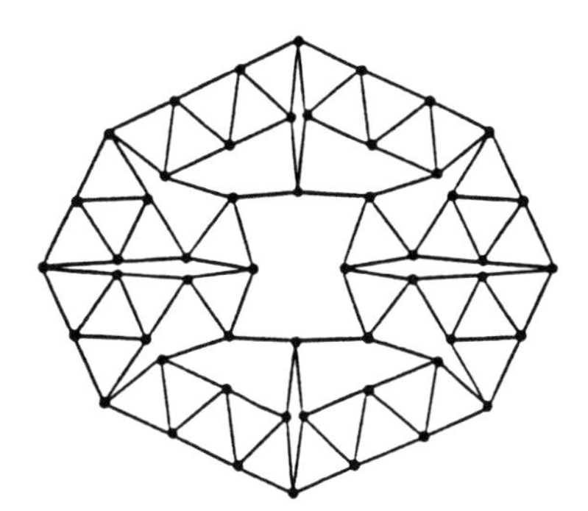

Figure 9: $p(4) \leq 52$

5 A Variation

As a variation of the preceding section I now allow at most d match sticks to be placed as a straight line segment of length d for edges of r-regular graphs. The endpoints within these straight lines are not counted as points of the graph.

For $r = 3$ and $d = 2$ the minimum graph has $p = 6$ points and $q = 15$ sticks (Figure 10). For $d = 2$ two other 3-regular examples are depicted in Figures 11 and 12.

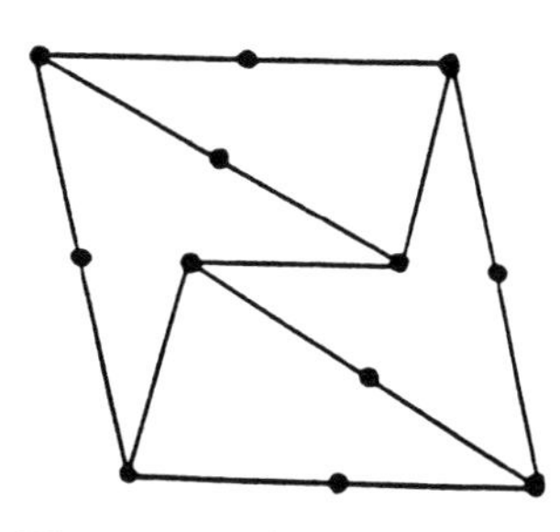

Figure 10: $(p, q) = (6, 15)$

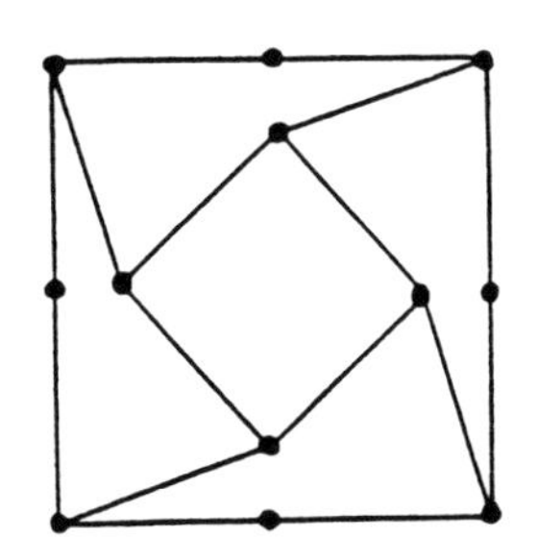

Figure 11: $(p, q) = (8, 16)$

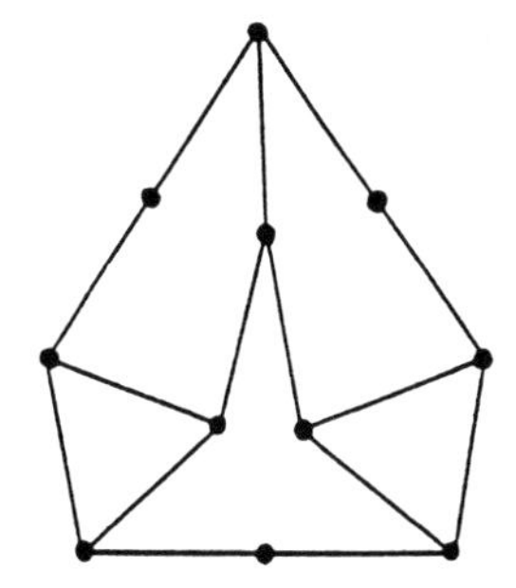

Figure 12: $(p, q) = (8, 15)$

The smallest 3-regular graph is the tetrahedron. Here you have to use a minimum of $d = 17$ match sticks in a straight line (see [5] and Figure 13, whose edges are 17, 17, 16, 10, 10, and 9 match sticks long).

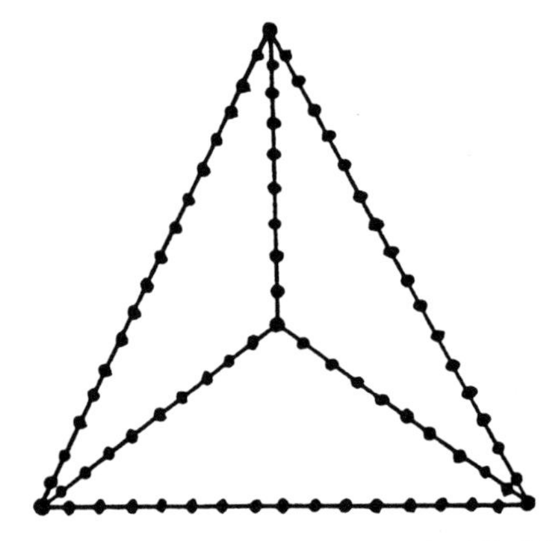

Figure 13: Tetrahedron with edges 17, 17, 16, 10, 10, 9.

For $r = 4$ three examples are (1) $d = 2, p = 21, q = 63$ (Figure 14), (2) $d = 3, p = 18, q = 60$ (Figure 15), and (3) $d = 4, p = 12, q = 70$ (Figure 16). The smallest 4-regular graph, the octahedron, needs a minimum of $d = 13$ match sticks in a straight line (see **[5]** and Figure 17). The smallest 5-regular graph is the icosahedron. It was proved by a computer that $d = 159$ match sticks in a straight line are necessary **[5]**.

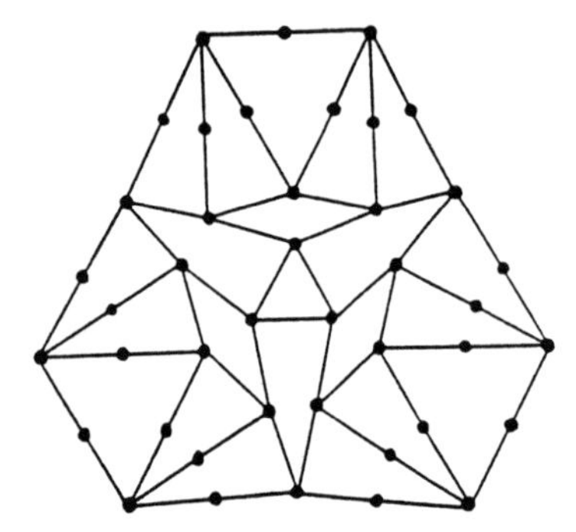

Figure 14: $(d, p, q, r) = (2, 21, 63, 4)$

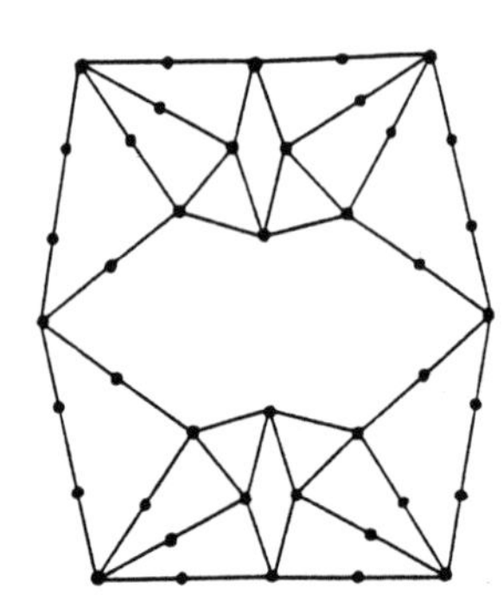

Figure 15: $(d, p, q, r) = (3, 18, 60, 4)$

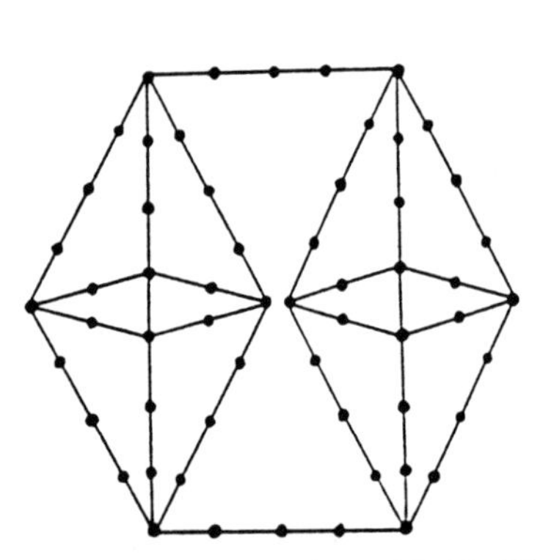

Figure 16: $(d, p, q, r) = (4, 12, 70, 4)$

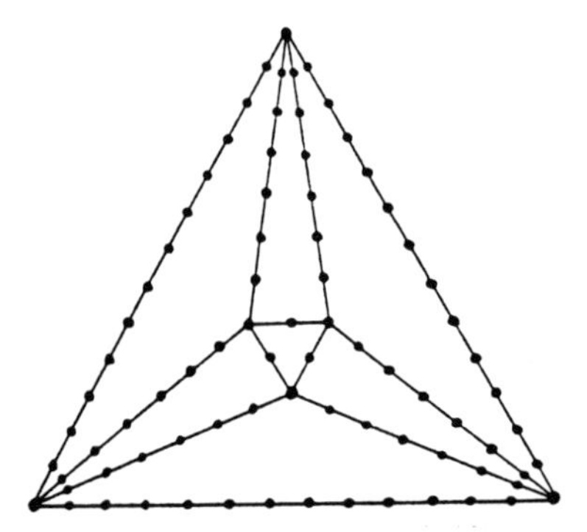

Figure 17: $(d, p, q, r) = (13, 6, 87, 4)$

It is only a conjecture that in general every planar graph can be represented in the plane with edges of integer length, that is, with edges as match sticks in straight line.

Other variations for r-regular planar graphs could be to puzzle for the smallest rigid examples, or to ask for examples without triangles. For $r = 3$ Figure 18 has no triangles, $p = 30$ points and $q = 45$ match sticks. For $r = 4$, triangles must occur by Euler's formula.

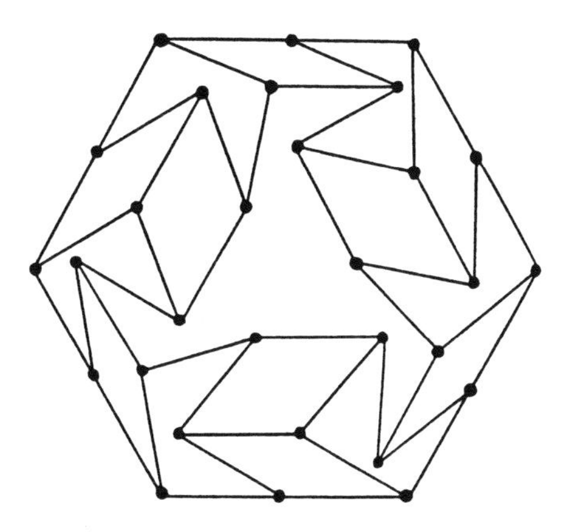

Figure 18: $(p, q, r) = (30, 45, 3)$

6 Conclusion

I conclude with a special case of 4-regular graphs, where exactly two triangles of match stick side length meet at each point of the graph. Do graphs of this kind exist at all? Instead of match sticks you can puzzle also with congruent nonoverlapping equilateral triangles cut from paper. The smallest example I have found so far uses 42 triangles, that is 63 points or 126 match sticks (Figure 19). In another context this figure also occurs in **[2]**.

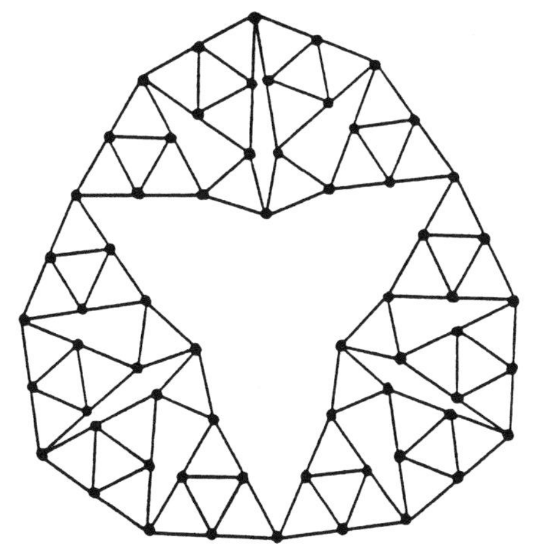

Figure 19: $(p, q, r) = (63, 126, 4)$

However, there occur triangles in Figure 19 which are not puzzle triangles. To avoid these, I ask now for examples where three puzzle triangles never enclose a triangle. The smallest example known to me then needs 3800 triangles, that means 11400 match sticks **[6]**. However, in all known cases you have to allow that several points of the graph touch another edge at an inner point. If this is forbidden, too, then I conjecture without any idea for a proof that examples do not exist.

At the end I hope that the reader was stimulated both to puzzle and to reflect on some of the indicated underlying solved and unsolved mathematical problems.

References

1. Martin Gardner, *Sixth Book of Mathematical Games from Scientific American,* W.H. Freeman and Co., San Francisco, 1971.

2. Branko Grünbaum & G.C.Shephard, Les charpentes de plaques rigides [Rigid plate frameworks], *Structural Topology*, **14**(1988) 1–8; MR **89f**:51084.

3. Frank Harary & Heiko Harborth, Extremal animals, *J. Combin. Inform. System Sci.*, **1**(1976) 1–8; MR **56** #15471.

4. H. Harborth, Solution of unsolved problem 664A proposed by O.Reutter, *Elem. Math.*, **29**(1974) 14–15.

5. Heiko Harborth, Arnfried Kemnitz, Meinhard Möller & Andreas Süßenbach, Ganzzahlige planare Darstellungen der platonischen Körper, *Elem. Math.*, **42**(1987) 118–122; MR **88k**:51043.

6. H. Harborth, Planar four-regular graphs with vertex-to-vertex unit triangles, *Discrete Math.* (to appear).

Bienroder Weg 47
D-3300 Braunschweig, Germany

Misunderstanding My Mazy Mazes May Make Me Miserable

Mogens Esrom Larsen

My first title of this story was an unintended maze, *the mazery of the harestones*. I thought *mazery* would mean a *collection of mazes* and *harestones* would point the attention in the direction of the misunderstanding of our hereditary antiquities, i.e., these grey lichen–covered stones once wrongly taken for Druidical remains, but now interpreted as the marks of a boundary line. To make up a maze accidentally like that only confirms my believe in the wisdom of Heraclitus as pointed out in the note on Rubik's cube (this volume).

During the last three years I have been smithing puzzles for the popular science magazine *Illustreret Videnskab*, published monthly in Denmark, Sweden, Norway, and Finland. The Danish version of the magazine is filed in the Strens Collection.

Of course, the puzzles are mostly classical matters from H. E. Dudeney, Sam Loyd etc., but I try to sneak in a little mathematics here and there. E.g., in the problem of the jeep crossing the desert, I added the question, "how big a desert can we cross?" Hopefully some reader will prove the divergence of the harmonic series.

To tease computer-freaks I like to ask questions with very large solutions. A source to such problems is the Pell equation. Some of these problems I have discussed in [2], but another is the following:

"The pride of the republic of Inner Urdistan was the army. Each year every one of the 60 regiments sends 16 soldiers to the parade. They marched in 60 squares. Then the general M. Urder joined the forces and all of them formed one big square together.

After the revolution the new leaders founded a new regiment making a total of 61. But the general M. Urder wants to form the parade all the same. So he asked the 61 regiments each to send a square number of soldiers such that they could all together including himself form a new big square."

"How many soldiers did each regiment send?"

In this case we must solve the equation

$$x^2 - 61 \cdot y^2 = 1$$

The smallest solution is

$$511456 22669840400.$$

Among the numbers up to 100, the biggest solution is required by 61.

Counting is good, but formulas are better. One thing is to draw a heptagon with all diagonals and count its 287 triangles. (See Figure 1.) Another is to find the number of triangles in an n-gon in general position with vertices on a circle, see [3], the solution is

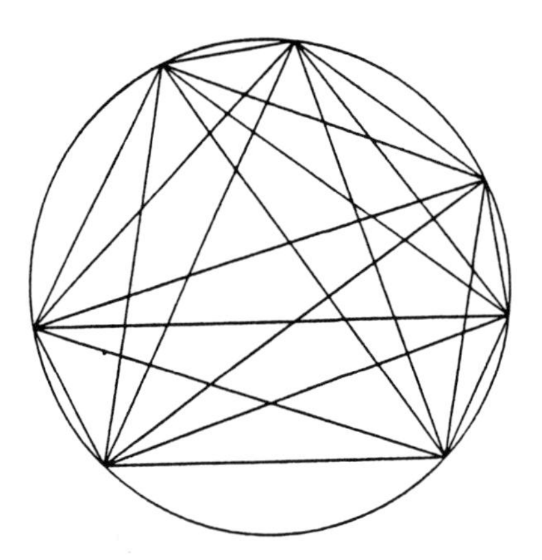

Figure 1: How many triangles?

$$\binom{n}{6} + 5\binom{n}{5} + 4\binom{n}{4} + \binom{n}{3}$$

to be found in [1].

As stressed by Richard K. Guy in his answer to [1] this formula is evident looking at Figure 2 carefully.

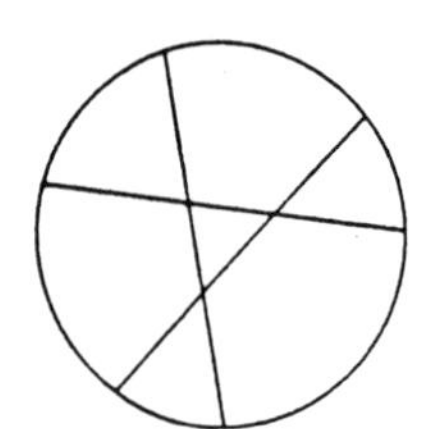
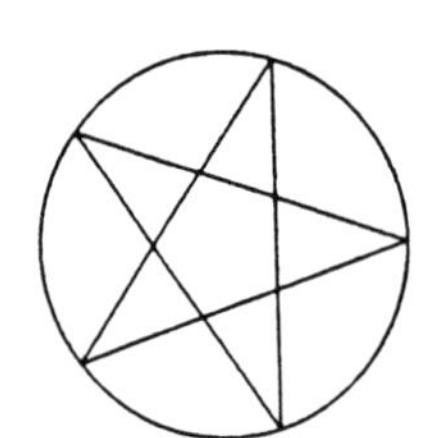
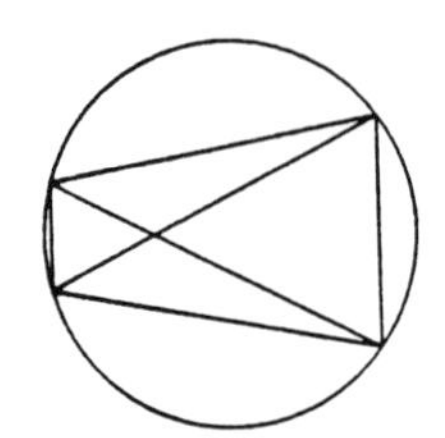
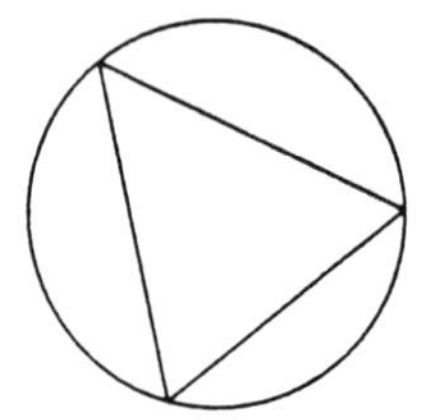

Figure 2: How to see the answer.

It is even more fun to count the number of triangles in a regular triangular lattice, see Figure 3, an old reference is [5], but see [4] on the general formula:

$$\left\lfloor \frac{n(n+2)(2n+1)}{8} \right\rfloor$$

Theorems are still better than formulas. Take a paper with the ordinary square lattice. Take any 5 lattice points and prove that there must be two of those 5 points, such that the interval from the one to the other goes through a lattice point. (The pigeonhole principle).

As computation can give the security of a result, and this is second to the understanding obtained from the logical deduction of a theorem, the latter is second to the *insight*, the experience of seeing right through the problem as can happen in geometry. Draw two intersecting circles and ask for that line through one of their cutting-points which is longest: in Figure 4 we ask for the choice of line through A, such that the segment BC is longest.

The solution is surprisingly simple. Draw the triangle $\triangle BCD$, where D is the other cutting point of the circles. See Figure 5.

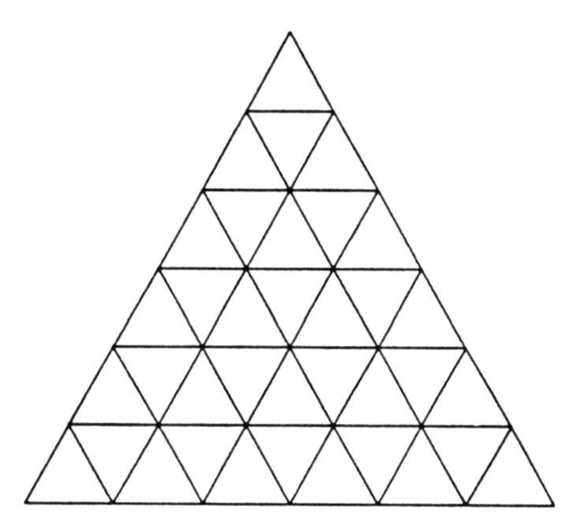

Figure 3: How many triangles?

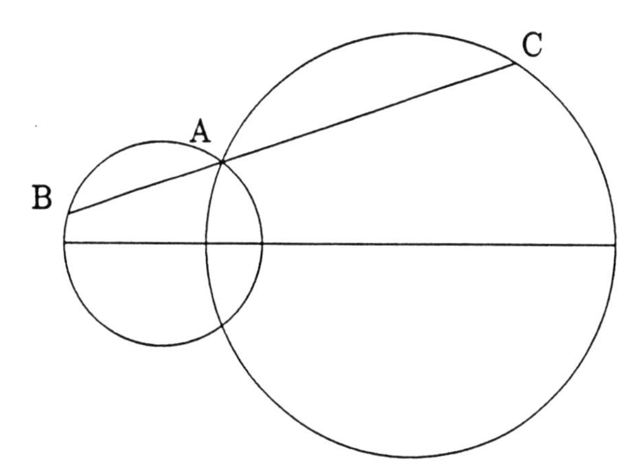

Figure 4: Which line through A is longest?

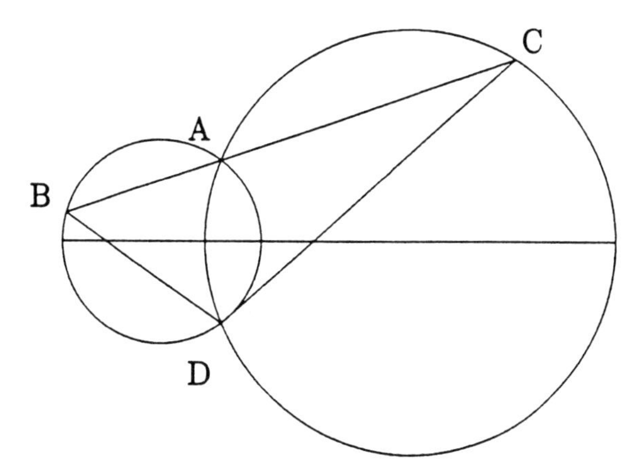

Figure 5: Draw the triangle.

Then it is obvious that the angles at B and C are independent of the particular choice of line through A. All the different triangles are similar. This means that we obtain a maximal distance BC if we chose a maximal distance DC (or BD). And this is easy, we have to chose one (and then both) as the diameter from D. See Figure 6.

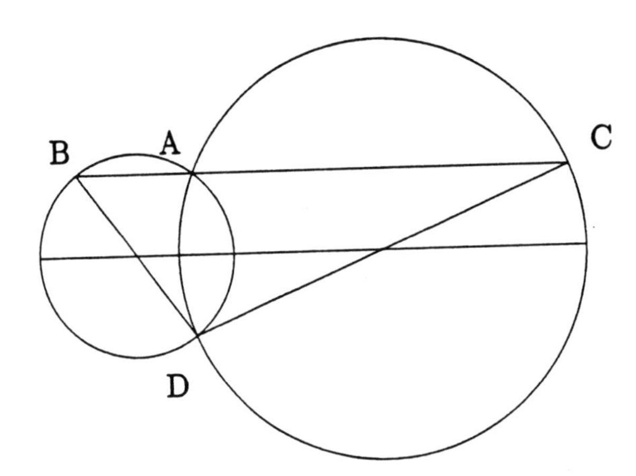

Figure 6: The solution found.

Then BC becomes parallel to the centerline.

Even more fun than geometry will be topology. Everybody knows the impossibility of joining three utilities with three houses in the plane without crossings. It is to say, the complete graph between two sets of three elements is not a planar one. But if we change the conditions to a non-planar surface, e.g., a torus, a Möbius strip, or the projective plane, then it might be solvable. And it is seen in Figure 7 below.

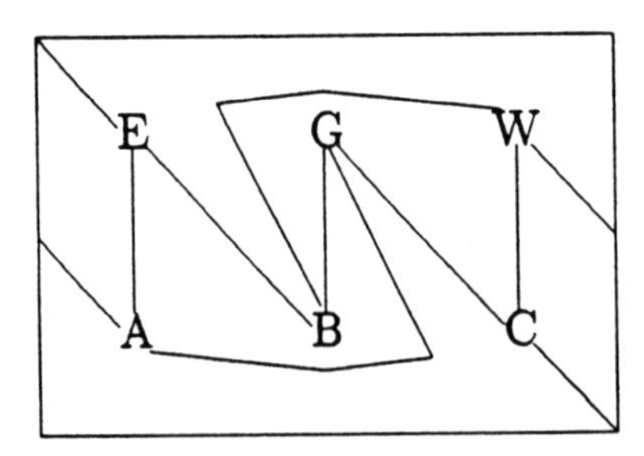

Figure 7: Electricity, gas and water off the plane.

This solution works equally well no matter which one of the interpretations of the rectangle you prefer.

This shows some of the types of entertainments that I use in my efforts to spread the joy of mathematics to the Fenno–Scandinavian population.

REFERENCES

1. R. J. Cormier, R. B. Eggleton, Counting by correspondence, *Math. Mag.* **49** (1976) 181–186.

2. Mogens Esrom Larsen, Pell's equation: a tool for the puzzle-smith, *Math. Gazette* **71** (1987) 261–265.

3. Mogens Esrom Larsen, Problem 72A, *Math. Gazette* **72** (1988) 53.

4. Mogens Esrom Larsen, The eternal triangle—a history of a counting problem, *College Math. J.* **20** (1989) 370–384.

5. Sam Loyd, *Sam Loyd's Cyclopedia of 5,000 Puzzles, Tricks and Conundrums with Answers,* Corwin Books, New York, 1914, pp. 284, 378.

Københavns universitets matematiske institut
Universitetsparken 5
DK-2100 København Ø
Denmark

Henry Ernest Dudeney
Britain's Greatest Puzzlist

Angela Newing

Henry Ernest Dudeney was born at Mayfield, East Sussex, in the southeastern part of England on 10th April 1857. He was the third of ten children of Gilbert Dudeney, a schoolmaster in a local private academy, and came from a long line of Dudeneys—a Sussex family whose traditional occupation had been farming and shepherding on the Sussex Downs (a local range of chalk hills).

This branch of the family had turned from farming to the teaching profession largely because of the influence exerted by Gilbert Dudeney's uncle, John Dudeney (1782–1852). John had been a shepherd and farm labourer like his father and grandfather from the age of eight: but he was an extremely intelligent young man who taught himself mathematics, science and several languages while looking after the sheep on the hills. At the age of 21 he gave up shepherding and joined a Lewes printer, Mr. Baxter, to work more closely with the books he had come to love. A year later, he founded an evening school for boys in a Lewes cellar, and subsequently became a schoolmaster instead of a printer when he took over the pupils and premises of a retiring schoolmaster, Mr. Inskip.

John Dudeney built up a marvelous reputation as a good teacher and his school grew rapidly. In his youth he had imparted some of what he knew to his youngest brother Henry—known as Harry (1790–1843)—and Harry became an assistant master in John's school. Harry's first wife died after only a very few years of marriage. In 1820 he married for a second time. His new wife, Edith, came from Mayfield, some 15 miles northeast of Lewes near to the Sussex–Kent border. She was a Wesleyan Methodist, a religion which had interested both John and Harry.

Following the marriage, Harry left Lewes to start a school of his own for the children of Methodists in Mayfield. Gilbert (born 1825) was the second child of this second marriage, and he followed in his father's footsteps to become assistant master of the Mayfield School.

It is not known how Gilbert met the lady who became his wife at Christmas 1853. Lucy Ann Rich was the daughter of a match maker (the sort you strike to make a fire) from Bridgwater in Somerset, more than 200 miles from Mayfield. The marriage took place at Bridgwater and the couple then returned to set up their first home in Mayfield. Just over a year later, their first child, Thomas James, was born; then, early in 1856, a daughter, Edith Mary. Henry Ernest arrived in April 1857, and another son, George Seaton, in 1859.

One imagines that, with four children to feed and clothe, the salary of an assistant master was rather stretched. One can also imagine that Lucy's mother, far away in Somerset, was disappointed at being unable to be close to her daughter and grandchildren, especially since she was now a widow. One can conjecture that she suggested that Gilbert should bring Lucy and the family to

Somerset where the population was growing with the popularity of seaside resorts such as Weston-super-Mare.

One of the new seaside towns was Burnham-on-Sea, between Bridgwater and Weston-super-Mare. Early in 1860, the young Dudeneys left Sussex and Gilbert founded a school of his own in College Street, Burnham. (The road may even have been named after the school.) The family grew. Isabel Lucy was born in 1860. Annie and Lucy Ellen (Isabel having died as a baby) in 1861, and three more daughters, Kate Dixon, Emily Gertrude and Alice Mary in 1862, 1863 and 1865. The 1861 census, taken only 18 months after their arrival in Burnham, recorded that Gilbert's school already had two assistant teachers, a nurse, a servant and 18 pupils. These 18 may have been residents and the numbers might have been increased by the addition of day boys.

During childhood, Edith, George, Annie and Lucy died, leaving only five children, Thomas, Henry Ernest, Kate, Emily and Alice to grow to adulthood. Both names, Henry and Ernest, are being used deliberately, as there is plenty of evidence to suggest that he did indeed use both names. "Henry" was by far the most common name given to Dudeney male children, so his family called him Ernest to distinguish him from the others. He himself preferred "Henry" and always referred to himself as Henry whenever he had the chance to do so.

Henry Ernest was, in many ways, very much like his great-uncle John. He was usually lost in thought and showed great interest in problems of all kinds. He was undoubtedly a precocious child, recalling in later life how, while still in his pram, he tried to figure out how it moved when his nurse pushed the handles, but when she stopped to talk to a friend in the park, he was unable to get it to move by pushing the handles himself.

He was often taken with his brothers and sisters to visit his grandmother at Bridgwater. She had a "Solitaire" board set out with glass marbles which she kept on a side table. As soon as he was tall enough to reach it, Henry Ernest began to play Solitaire, and this instilled in him a lifelong fascination with the game. Years later, he published several puzzles based on Solitaire, and in 1908 produced a minimal 19-move solution which he felt could not be improved upon. By then his work was so popular that much interest in the game was engendered, and in 1912, Ernest Bergholt managed to find a better, 18-move, solution.

Thomas and Henry Ernest both learnt to play the piano as small boys and both of them retained a great interest in music throughout their lives. They learned the violin and the organ as well and were able to play the organ at the local chapel for Sunday services when required.

Henry Ernest learned to play chess at a young age too and this became a hobby which lasted a lifetime. He also read widely on the history of chess and became interested in chess-problems. Another hobby was conjuring. He was introduced to this when, at the age of 6 or 7, he went to watch an "illusionist" perform in the local village hall and was able to reason how a vanishing trick was done. This encouraged him to practise various tricks of his own which he performed before the critical audience of his brothers and sisters.

At the age of 9 he began composing puzzles to amuse the rest of the family. One of his relatives saw some of his puzzles and encouraged Henry Ernest to submit them for possible publication in a boys' paper. When he sent them in he was delighted to find that they were accepted and he was paid 5 shillings (25p) for each one that was published.

At this time he was receiving only the standard education provided which was fairly elementary, but he developed a particular interest in mathematics and its history and spent much of his spare time reading about it and learning more advanced mathematics. He had knowledge of Eulerian paths as is illustrated by one of his early puzzles which is shown in Figure 1.

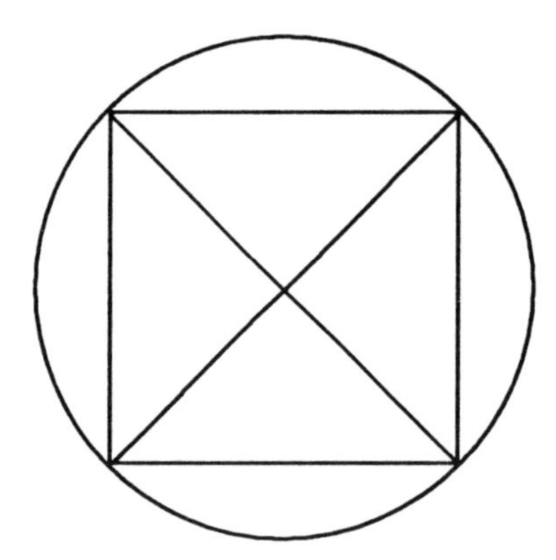

Figure 1: Trace the figure without lifting the pencil from the paper.

The solver is required to trace the figure without lifting his pencil from the paper. This is obviously impossible without some trickery because the diagram has four odd vertices. The trick is to fold the paper so that the two vertical lines lie on top of each other: they can then be drawn together with one stroke of the pencil. The paper is then unfolded without moving the pencil from one of the vertices, and the remainder of the figure can be traced.

His study of geometry led Henry Ernest to an interest in paperfolding tricks and in the technique of geometric dissection. He also learned about Möbius strips and how they were being used by conjurers. At the age of 11, he put on a one-man show in aid of a local charity. This included a piano solo, a few pieces on the violin, some conjuring tricks and a ventriloquial sketch.

When he was 13 he started work as a clerk in the Civil Service. The chief gain from this, apart from a small salary, was that he was moved fairly soon to an office in London. This gave him access in his spare time to the museums and libraries of the capital, and to the bookshops which he had been unable to visit while living in Somerset.

As well as persevering with his mathematical studies, he joined a chess club where he was rapidly accepted as a gifted newcomer. His brother Thomas was also working in London and the two of them began to go to concerts and to the opera where he was introduced to the music of Wagner who became his favorite composer.

He did not enjoy his job which he found very tiresome. However, Henry Ernest soon turned his attention to short story writing, and in so doing opened the door to the world of journalism. Towards the end of the nineteenth century people had begun to have more leisure time and more money, and there was a boom in the publication and sale of weekly and monthly magazines as well as newspapers. Publishers were on the lookout for new writers to help them to fill their columns, and Henry Ernest's entry into the world of fiction writing was thus extremely fortunately timed. Some of his early stories were published in the magazine "Tit-Bits", one of the many periodicals produced by George Newnes. This brought Henry Ernest into contact with other contemporary writers such as Arthur Conan Doyle (who was almost exactly the same age as Henry Ernest), and this group of authors met together regularly for discussion and socializing. Before long they formed themselves into a formal group which they called "The Society of Authors". Later they acquired premises in Whitehall Place, London.

At about this time, Henry Ernest began to submit puzzles for publication in the various journals with which he was involved. He used the pseudonym "Sphinx" (and it is interesting to recall that this was the name chosen many years later for the first journal devoted entirely to recreational mathematics) and his puzzles became more popular than his stories had been. Puzzles in periodicals were uncommon at that time in England. Lewis Carroll had a few mathematical puzzles printed as a series in "The Monthly Packet" from 1880, but that was a magazine for young people. Carroll had also published some word puzzles.

In the early 1880's Henry Ernest was introduced by a mutual friend to Alice Whiffin, a young lady 7 years younger then he, who was an aspiring author of short stories. They were immediately drawn to each other. She was pretty and talkative and keen to succeed. He was tall, broad-shouldered and handsome and they had much in common. They were married in October 1884 at St Andrew's Church, Holborn, London—a church where Henry Ernest often played the organ.

The young couple rented a house in Great James Street, Bedford Row, close to the printing houses. Henry Ernest was, by this time, devoting all his time to being a journalist and had given up his work as a clerk. Alice wrote a few short stories for publication but was also much occupied in making a home for her husband. Their first child was born in May 1887 but, unfortunately, lived for only four months. At that time infant mortality was high but, nevertheless, Alice was distraught at the loss. To help to take her mind off the dead baby, she took a job as assistant secretary to Sir Wemyss Reid, head of the Cassells publishing firm. She much enjoyed the literary atmosphere there and before long began writing in earnest. Three of her short stories were published within a short space of time in various Cassells journals and she was launched in the world of fiction writing.

Alice's connexions at Cassells gave Henry Ernest another outlet for his work. He had begun providing chess-problems as well as mathematical and logical ones to several periodicals, and also the occasional word puzzle. (Although he frowned on these as trivialities, he was still able to justify the invention of several new types of word puzzle.)

The Dudeney's second child, Margery Janet, was born in 1890. They decided that the time had come to move away from London—but not too far away, so that Henry Ernest could easily reach his publishing houses, the Author's Club and the chess club, and so that he could continue his regular visits to the opera. Alice too did not wish to be far from her contacts among the publishers. They rented a house near the Surrey–Sussex border in a hamlet north of Billingshurst and soon found that they enjoyed life in the country.

By the mid 1890's, they were better off financially than ever before. Alice had begun writing full length novels which were very popular with readers, and Henry Ernest was well established as a puzzlist. They bought a three-acre plot of land on the outskirts of Horsell, near Guildford in Surrey, and, with the help of Henry Ernest's brother-in-law, Maurice Pocock (husband of his sister Kate) who was living in Chertsey, they planned a splendid house which was built for them. They moved to this grand house, called Littlewick Meadow, in 1897 and were able to employ several servants to help them to run it. Both of them had an interest in antique furniture and, by attending sales in the area they were able to furnish their house with many fine examples of Jacobean and later antiques.

Henry Ernest's interest in chess drew him to try to solve any chess problem which came his way. In 1894, a chess-problem from America was published by the New York Chess Association and circulated world-wide. The puzzle had been composed by Sam Loyd, and a large prize was offered to anyone who could solve it in 50 moves. Sam Loyd, 16 years older than Henry Ernest, had a big reputation in the States both as a composer of chess puzzles and of problems of the sort which Henry Ernest was publishing in periodicals. On this occasion, Sam was confident that no-one would better his own 53-move solution, and was therefore surprised to receive a 50-move solution from Henry Ernest within the specified time limit. This may have been the first contact between the two men, although they knew of each other's existence before this.

Sam was very artistic as well as being America's foremost chess problemist. He had edited a chess-problem column from the age of 16 and had subsequently produced many popular American advertising gimmicks—such as Barnum's Trick Donkeys, and his famous "Get Off The Earth" Puzzle. From there it had been an easy step to mathematical and logical puzzles. However Sam was not a brilliant mathematician and built his reputation on puzzle ideas produced by other people as well as problems for which he had no answer until the solutions came in from his readers. He would offer a huge prize for the best solution to arrive by a particular date. Americans would send in their answers in their thousands and Sam's secretaries would select the best-looking ones for him to see. Money was made from the remainder by selling the names and addresses at so much per thousand to advertising agencies.

In 1893, Sam decided to extend his net to the British Press. By some quirk of fate he chose to write to "Tit-Bits" in which Henry Ernest had a fairly regular column. (Indeed, perhaps that is why he chose it.) Sam offered "Tit-Bits" a puzzle which he said was so difficult that only one other person in the world besides himself knew the answer. The other person was described as a "Professor Fiske" of Columbia College. Sam was so confident, that the arrangement struck with the magazine was that he should receive a £100 fee provided that no correct solutions were submitted, whereas the £100 would go to the sender if, by chance, anyone should produce the answer. £100 was a very considerable sum at that time.

The editor was impressed and, having heard of Sam's reputation he printed the puzzle with suitable comments about the unlikelihood of anyone's finding the answer. The puzzle was as follows:

> *Find how to arrange the figures .4.5.6.7.8.9.0. in an arithmetical sum which adds up the nearest to 82.*

The solution required only a knowledge of recurring decimals.

$$80.\dot{5} + .\dot{9}\dot{7} + .\dot{4}\dot{6} = 82 \quad \text{exactly.}$$

[A solution involving only six of the dots is $80.\dot{5} + .9\dot{7} + .4\dot{6} = 82$—Eds.]

In the event, a large number of correct answers arrived and "Tit-Bits" gave the £100 to the first one opened and felt bound to reward some of the others as well by giving £5 each to the next ten competitors.

Three weeks later some scathing comments—not signed "Sphinx" but probably written by Henry Ernest—appeared. These included reference to recurring decimals having been taught in schools in England "*from time out of mind*" and suggested that "*as soon as Mr. Loyd receives this paper, he will doubtless send a reply to the criticism we have ventured to make . . .* ".

Sam's response was to offer £10 of his own money as a goodwill gesture for another puzzle on the same theme. The controversy rumbled on till the end of the year when Sam submitted another puzzle for a modest prize of £20. Two years later "Tit-Bits" obviously felt that the situation had calmed down. Henry Ernest and Sam collaborated in a regular column in the journal with Sam providing puzzles and Henry Ernest, using the name Sphinx, writing commentary, providing the answers, and awarding the prizes. This exchange continued for a year with Sam using puzzles which he had earlier published in America: some 'old chestnuts' and some which were new.

At this time, Henry Ernest sent whole batches of his own puzzles to Sam, presumably for Sam's own interest, and because Sam was sending puzzles to England. He was to regret this for many years to come. Although Henry Ernest regularly used puzzles and puzzle ideas given to him by others, he always credited them to their authors and presumably expected that, if Sam wanted to use an idea of his, he would do the same. This was not in Sam's nature, and Henry Ernest became increasingly more infuriated as he saw his puzzles passed off in the States as Sam's own. There are dozens of illustrations of this in Loyd's "Cyclopedia" and elsewhere.

Henry Ernest's daughter, Margery, was at a very impressionable age. She recalled her father raging and seething with anger to such an extent that she was very frightened and, thereafter, equated Sam Loyd with the devil. Alice, too, noticed a great change in her husband. He often became moody and irritable and she did not like these changes.

Alice was, by now, a famous person in her own right. She had produced some 15 or 16 novels which were well reviewed and widely read. The house at Littlewick regularly filled, at weekends, with guests who felt privileged to be invited. A frequent visitor was David Paul Hardy, illustrator of some of Alice's books—who also did the drawings for Henry Ernest's first bound puzzle collection, "The Canterbury Puzzles". He was married, and his wife often came to Littlewick too, but he and Alice had much in common and were greatly attracted to each other. Her life with Henry Ernest was far from exciting. They would both spend the mornings writing in their separate studies and then meet over lunch. She would want to discuss her latest novel or something else which he thought trivial, while he had some weighty mathematical idea on his mind. The same thing would happen at dinner, and consequently they often had acrimonious arguments. He used to spend one day a week in London where he found the atmosphere at the Authors Club and his chess club more congenial than at home, so he would often stay overnight.

In these circumstances Alice felt drawn to David Hardy, and it appeared that he had the same feelings for her. Events came to a head when Alice left home and moved to Angmering in Sussex to be closer to David's home at Storrington. Left on his own, Henry Ernest was devastated.He had been very ill six months before with bronchitis, pleurisy and a suspected stomach ulcer (he had been a heavy smoker for some time) so was low in spirits. He still loved Alice in his own way and genuinely did not realize that he had been alienating her. He felt a complete failure.

Margery, now aged 21, was teaching art in London during the week and, with her mother's move, she spent her weekends in Angmering. It was decided to sell the lovely house at Littlewick. Although he was very fond of it, it did not seem worthwhile for Henry Ernest to live there on his own. He went back to London and took a flat near to his old haunts in Holborn. Margery acted as the go-between taking messages between her parents.

With so much more time to himself, Henry Ernest threw himself into his work and produced more and more new and ingenious puzzles and developed various mathematical ideas which were investigated by 'main stream' mathematicians. He spent much time at his chess club and wrote on chess for "The Strand", a magazine in which he had had a regular puzzle column for some time. During his early married years, Alice had introduced him to the Anglo-Catholic church of St Alban's Holborn where he had been 'converted' from being a nominal non-conformist to a high Anglican. He renewed his contact with St Alban's on his return to London and was pleased to find that a lot of the people he formerly knew were still connected with the church. He became the assistant organist once again, and his long talks with the curate, Father Stanton, helped him to put the breakdown of his marriage in perspective.

Alice's relationship with David did not develop as she had expected. She thought that he would leave his wife for her but, although he became a regular visitor and took her around the area to show her his favorite haunts, he remained with his wife. Alice took up her fiction writing as if nothing much had happened, and her next few novels were written around the part of Sussex she was living in and the characters of her new neighbors. She was ostracized by Henry Ernest's sisters but, interestingly, his nephew John Pocock, son of their Littlewick architect, helped her to renovate the historic house she had leased in Angmering.

In June 1912, Margery became engaged to Christopher Fulleylove, a young man from Lewes. Both her parents liked him but a minor crisis developed a year later when the young couple announced their intention of emigrating to Canada. Christopher had an uncle who had recently returned from Canada with glowing stories of life on that side of the Atlantic, and both Christopher and Margery felt that they wanted to go and live there. She was their only surviving child, and neither Alice nor Henry Ernest wanted to lose Margery to the other side of the world. A family conference was called at Angmering and Henry Ernest stayed there for the weekend for the conference to take place. Presumably, it was their joint disapproval of the proposal which drew them together: one can imagine them trying, jointly, to dissuade the young couple from going. They did not succeed, however, and, in fact, the atmosphere must have been very unhappy, because Christopher and Margery decided to emigrate to Canada and get married when they got there rather than marrying in England.

The surprising effect of all this was that Alice and Henry Ernest were reconciled to each other. Presumably the outbreak of the 1914–18 war had some effect as well, as did their mutual feeling of loss with Margery's departure. It was decided to give up the house at Angmering, by now restored to its earlier glory, and move to the county town of Lewes.

They bought another historic house in the upper High Street in Lewes and, before long had settled in almost as if no separation had ever taken place. They resumed their visits to furniture and antique sales and the house was soon filled with expensive antique furniture just as Littlewick had been. Alice's next novel was a strong advertisement for marriage and Henry Ernest's best known puzzle collection, "Amusements in Mathematics" was soon published. He joined the Lewes and Brighton Chess Clubs and maintained his weekly visits to London. He joined the congregation at the nearby Anglo-Catholic church of St Michael where he again became the assistant organist.

In 1921 the Dudeneys moved again, but this time only a short distance to a delightful house within the castle precincts at Lewes, called Castle Precincts House. (The house is today divided into two dwellings and is called Brack Mound House.) He was now almost 65 and less active than in his earlier years, but he was still supplying regular puzzles to about a dozen periodicals including "The Strand", where his column ran for a total of 30 years, "The Daily News", "The Weekly Dispatch" and many more. He worked on a collection of his word puzzles which was published in 1925, and his third mathematical problem book, "Modern Puzzles", which appeared a year later.

In 1922, Margery paid a visit from Canada with her two children. The family were obviously reconciled to each other and Alice and Henry Ernest were delighted with their two grandsons. The visit lasted 3 months and brought much happiness.

Henry Ernest still played croquet and bowls but, each winter he suffered from bronchitis and often pleurisy and Alice became increasingly worried about his health. His ability at chess was not impaired and his output of puzzles did not diminish. In the winter of 1929–30 he became seriously ill. He did not recover with the coming of spring and died on 24th April 1930. He had been sending off his usual regular puzzles to the papers until the previous week, and the "Daily News" obituary referred to a substantial collection of as yet unpublished "Breakfast Table Problems" which they proposed to continue until they ran out.

After a funeral service of Anglo-Catholic splendour at St Michael's church, he was buried in Lewes cemetery.

After getting over her grief, Alice spent a period of two months in America with Margery and her family, who had moved to New York. She now had four grandchildren. She then returned to Lewes and set about collecting her late husband's unpublished puzzles. She persuaded one of Henry Ernest's friends, James Travers, a London headmaster, to help her to edit this collection, and in 1931 the first posthumous book, "Puzzles and Curious Problems", was published. Travers, described by Alice as a charming and talkative Irishman, subsequently collaborated in a second posthumous book, "A Puzzle Mine".

Alice continued to live at Castle Precincts House, and also to buy up various nearby properties as they became vacant. One cottage on the lane behind the house became her studio, and others she let to friends. She continued to write novels for several years and died on 21st November 1945.

The Rectory,
Brimpsfield,
Gloucester GL4 8LD,
England.

From Recreational to Foundational Mathematics

Victor Pambuccian

Problems from the recreational mathematics folklore that ask for the possibility of fair cuts of either pancakes or ham-sandwiches have proofs that do not enable us actually to perform the cuts even for pancakes that have fairly simple shapes, such as triangles.

To the intuitionist, the proofs do not prove that the cuts are possible, because they are based on the intermediate value property of continuous functions, the truth of which is derivable only by means of *tertium non datur*. There are, moreover, examples of functions for which the "intermediate value" is not"constructible" either from an intuitionistic (see [**6**]) or from a formalist (see [**1**]) point of view.

To the formalist, the proofs are valid in a very "rich" Euclidean geometry, but not in its "poorer" constructive variants.

To be more specific about the nature of provability and the distinction just made, we must call on some notions from mathematical logic, (see [**2**], for example) including specifying the formal language for the geometry in question as well as the class of structures which may act as models. The first order language L_{BD} has two non-logical predicates, a ternary predicate $Bxyz$, whose interpretation is that y lies between x and z; and a quaternary relation $Dwxyz$ for which the intended interpretation is that the distance between w and x equals that between y and z. Given an ordered field F, and a positive integer n, one can associate a structure $C_n(F)$ for L_{BD} in the obvious way. The interpretation of B is the set of triples (a, b, c) of n-tuples from F^n for which b lies between a and c (i.e., there is t with $0 < t < 1$ and $b - a = t(c - a)$).

The interpretation of D is the set of quadruples (a, b, c, d) of n-tuples from F^n for which $\|a - b\| = \|c - d\|$, where $\|u\|$ is the square of the normal Euclidean distance, $\|u\| = \sum_{i=1}^{n} u_i^2$.

We use the square because the ordered field may not have square roots of arbitrary positive numbers. Of course, this sacrifices the triangle inequality. The class of ordered fields is denoted by F. An ordered field is said to be **Pythagorean** if, for each x and y, there is a z such that $x^2+y^2 = z^2$. The class of Pythagorean ordered fields is denoted by P. An ordered field is **Euclidean** if every positive number has a square root, and the class of these fields is denoted by E. The real closed fields, denoted by RC, are those Euclidean fields in which each odd degree polynomial has a root. Trivially $F \supseteq P \supseteq E \supseteq RC$.

Based on these classes of fields we may consider families of theories. If K is a class of ordered fields then $TK(n)$ is the theory which is the set of sentences (in the language L_{BD}) which are true in $C_n(F)$ for each F in K. In other words every model of $TP(n)$, $(TE(n), TRC(n))$ is isomorphic to an n-dimensional Cartesian space over a Pythagorean ordered (respectively Euclidean ordered, real closed) field. Axiomatizations for $TP(n)$, $TE(n)$ and $TRC(n)$ can be found in **[12]**, and also for $n = 2$ in [7] and **[8]**. The difference between $TP(2)$ and $TE(2)$ is that the sentence "the circle, drawn with radius greater than the distance from its centre to a given line, intersects the line", belongs to $TE(2)$ but not to $TP(2)$. For this reason $TP(2)$ (or $TP(3)$) are closest to what Euclid postulates at the beginning of Book I of his "Elements" and $TE(2)$ ($TE(3)$) corresponds to the theory in which he actually "works".

One of the basic questions that is asked about a theory is whether it is decidable (i.e., whether there is an algorithmic (or mechanical) procedure for deciding, given a sentence, whether or not it is a theorem). Tarski proved in **[12]** that $TRC(n)$ is complete and decidable. (Grigor'ev gave a faster algorithm in **[5]**). By contrast, $TE(n)$ and $TP(n)$ are known to be undecidable **[15]**. Therefore, finding a proof (or refutation) of a sentence in each of these two theories remains essentially a creative act, as in number theory, while the same problem for $TRC(n)$ is essentially mechanical and not creative (see also **[4]**). Today elementary geometry problems are widely considered to belong to high-school mathematics and unworthy of serious research. However, the importance of effective construction in Greek geometry (see **[14]**) suggests that one should approach the "elementary" geometry problems from an axiomatization of either $TP(3)$ or $TE(3)$. This involves essentially creative research.

The recreational problems we were referring to earlier have positive solutions in $TRC(3)$, but no solution in $TP(3)$ or $TE(3)$ is known. The problems are:

(1) The pancake problems ([**3**], Theorems 11.1 and 11.2, pp. 65 & 68),

(2) The ham-sandwich problem ([**3**],Theorem 33.1, p. 120; see also **[10]**, **[11]**),

(3) The Emch-Schnirelman Theorem **[9]**.

Stated as elementary geometry problems they read:

(1a) Given two disjoint convex polygons in the plane, there is a straight line that cuts both polygons in half by area;

(1b) A convex polygon can be divided by two orthogonal cuts (straight lines) into four equal area parts;

(2) Given three disjoint convex polyhedra in space, there is a single plane which divides each exactly in half by volume;

(3) For every convex polygon, there are four points on its boundary that are the vertices of a square.

If we specify the number of vertices of the polygons and polyhedra, then the above "theorems" become L_{BD}-sentences. Even for the simplest cases—two triangles in (1a), a triangle in (1b), three tetrahedra in (2) and a quadrilateral in (3)—the corresponding sentences are not known to be true in either $TP(3)$ or $TE(3)$.

I conjecture that they are false in general. For example, to be specific in (1b), for some triangles it is provably possible to perform the cut in $TE(3)$—but probably not in $TP(3)$—and for the others it is not.

It is remarkable how few unsolved problems we found, compared to the innumerable unsolved number-theoretic problems. In order to find $TE(3)$-unsolved problems, one apparently has to look at geometry problems the solutions of which rely on continuity assumptions.

REFERENCES

1. Günter Baigger, Die Nichtkonstruktivität des Brouwerschen Fixpunktsatzes, *Arch. für math. Logik Grundlag.*, **25**(1985), 183–188; MR **88h**:03082.

2. C.C. Chang & H.J. Keisler, *Model Theory,* North-Holland, 1973; MR **53** #12927.

3. W.G. Chinn & N.E. Steenrod, First Concepts of Topology: The Geometry of Mappings of Segments, Curves, Circles, and Disks, New Math. Library, **18**, Random House, New York, 1966; MR **34** #6705.

4. Y.S. Chou, Mechanical Geometry Theorem Proving, D. Reidel, Dordrecht, 1988.

5. D.Yu. Grigor'ev, Complexity of deciding Tarski algebra, *J. Symbolic Comput.*, **5**(1988) 65–108; MR **90b**:03054.

6. A. Heyting, *Intuitionism: An Introduction,* North-Holland, Amsterdam, 1956; MR **17**, 698j.

7. Victor Pambuccian, An axiom system for plane Euclidean geometry, *Bull. Polish Acad. Sci. Math.*, **35**(1987), 333–335; MR **89h**:51024.

8. Rudolf Schnabel & Victor Pambuccian, Die metrisch-euklidische Geometrie als Ausgangspunkt für die geordnet-euklidische Geometrie, *Exposition. Math.*, **3**(1985), 285–288; MR **88a**:51022.

9. L.G. Schnirelman, On certain geometrical properties of closed curves (Russian), *Uspekhi Mat. Nauk*, **10**(1944), 34–44; MR **7**, 35b.

10. H. Steinhaus, Sur la division des ensembles de l'espace par les plans et des ensembles plans par les cercles, *Fund. Math.*, **33**(1945), 245–263; MR **8**, 164d.

11. A.H. Stone & J.W. Tukey, Generalized "sandwich" theorems, *Duke Math. J.*, **9**(1942), 356–359; MR **4**, 75c.

12. A. Tarski, *A Decision Method for Elementary Algebra and Geometry,* 2nd edition, University of California Press, Berkeley & Los Angeles CA, 1951; MR **13**, 423a.

13. A. Tarski, What is elementary geometry?, in The Axiomatic Method, North-Holland, Amsterdam, 1959, pp. 16–29; MR **21** #4919.

14. H.G. Zeuthen, Die geometrische Konstruktion als 'Existenzbeweis' in der antiken Geometrie, *Math. Ann.*, **47**(1896), 222–228; Jbuch **27**, 35.

15. M. Ziegler, Einige unentscheidbare Körpertheorien, *Enseign. Math.*, **28**(1982) 269–280; MR **83m**:03040b.

Department of Mathematics,
The University of Michigan,
Ann Arbor MI 48109-1003

Alphamagic Squares*

Adventures with turtle shell and yew between the mountains of mathematics and the lowlands of logology.

Lee C. F. Sallows

The history of *magic squares* is a venerable one, reaching back into the legendary past of ancient China. So it is that the simplest, oldest, and most famous square of all, the so-called *Lo shu (shu* meaning *writing, document),* is said to have first been revealed on the shell of a sacred turtle which appeared to the mythical Emperor Yü from the waters of the Lo river in the 23rd century B.C. (This is discussed in Camman's article cited at the end of this paper.)

The celebrated turtle's shell must have looked like Figure 1. In fact, modern sinology identifies these signs as a pseudo-archaic invention of the tenth century A.D., although indirect references to the essential structure date from as early as the fourth century B.C. Translating them into Arabic numerals yields the square shown as Figure 2.

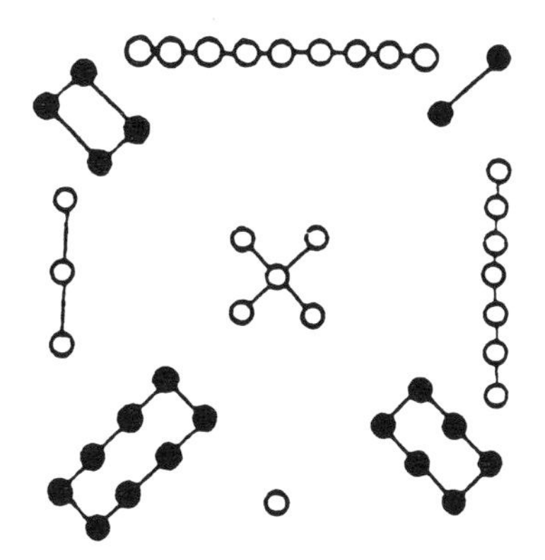

Figure 1: The *Lo shu*, revealed on the shell of a sacred turtle, according to Chinese Legend.

4	9	2
3	5	7
8	1	6

Figure 2: Translation of Figure 1

As many readers will not need to be told, the magical property distinguishing such squares is that the sum of the numbers in every row, column, and diagonal is the same; in the *Lo shu* this magic constant is 15. The literature on these recondite curiosities is amazingly prodigious and deeply ramified, not to say widely and haphazardly dispersed. It is clear that the spell cast by the elegant symmetries reflected in these interlocking number patterns has held countless devotees in thrall, eminent mathematician and lowliest layman alike. Hardly a turtle shell has been left

*Reprinted with permission from *Abacus,* Vol. 4, No. 1 (Fall 1986) 28–45. ©1986 Springer- Verlag.

The opening page from *The Origin of Tree Worship,* a work shrouded in mystery. Privately published in England, 1887, a copy was placed in the British Museum but disappeared soon after. Now rediscovered after 98 years, its pages reveal—among other riddles—the unprecedented *Li shu*, a mathematico-linguistic formula of demonstrably magical power.

[Courtesy of the British Library, London.]

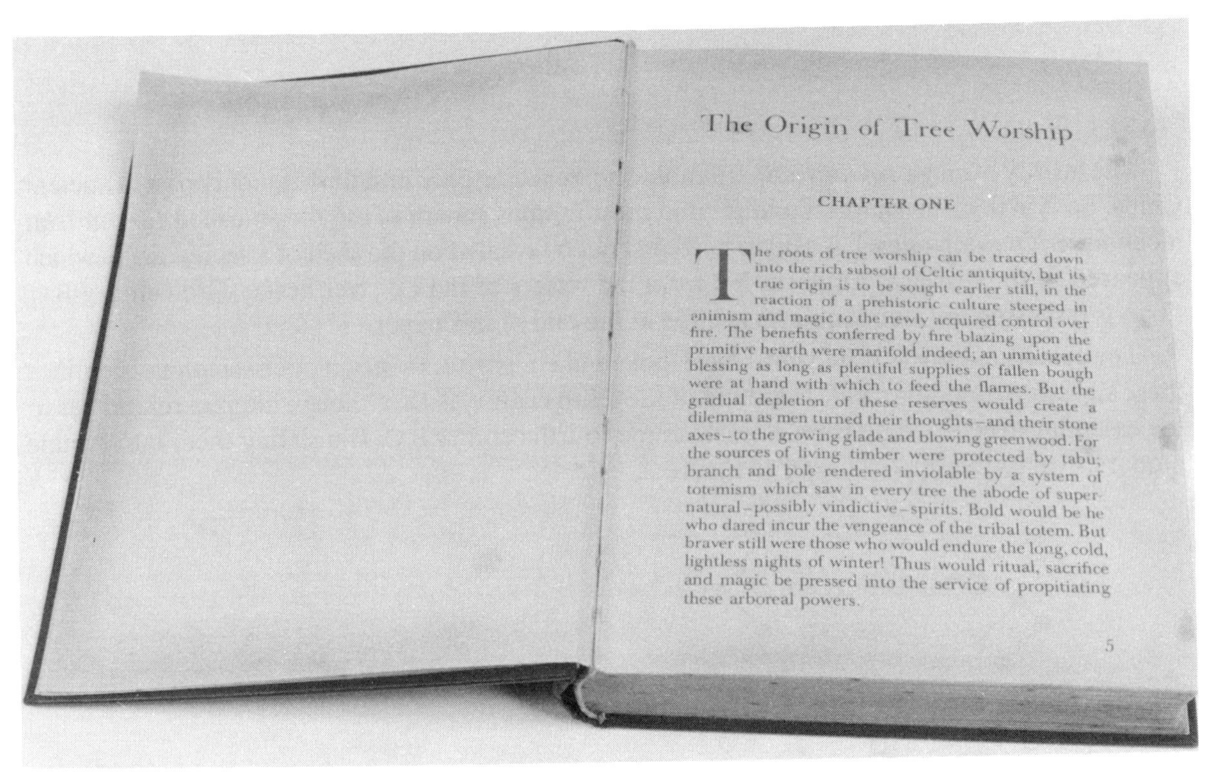

The Origin of Tree Worship

CHAPTER ONE

The roots of tree worship can be traced down into the rich subsoil of Celtic antiquity, but its true origin is to be sought earlier still, in the reaction of a prehistoric culture steeped in animism and magic to the newly acquired control over fire. The benefits conferred by fire blazing upon the primitive hearth were manifold indeed; an unmitigated blessing as long as plentiful supplies of fallen bough were at hand with which to feed the flames. But the gradual depletion of these reserves would create a dilemma as men turned their thoughts–and their stone axes–to the growing glade and blowing greenwood. For the sources of living timber were protected by tabu; branch and bole rendered inviolable by a system of totemism which saw in every tree the abode of supernatural–possibly vindictive–spirits. Bold would be he who dared incur the vengeance of the tribal totem. But braver still were those who would endure the long, cold, lightless nights of winter! Thus would ritual, sacrifice and magic be pressed into the service of propitiating these arboreal powers.

5

unturned in exploring variations on the central theme, so that articles and even books abound devoted to special categories of squares, as well as magic triangles, rectangles, circles, stars, antimagic squares, prime-number squares, multiplicitively magic squares, magic cubes, N-dimensional arrays, and so on. Not least in adding spice to the subject is the variety of simply stated yet peculiarly intractable mathematical problems they give rise to. Nobody knows, for instance, how many distinct consecutive-number specimens there are for any square larger than 5×5. Barely credible, but true!

A new development of unexpected relevance to this topic is the recovery during 1985 of a unique book, bringing to light an extraordinary parallel between an episode in the reign of King Mī, a historically dubious late-fifth(?)-century tribal chieftain of North Britain, and the Chinese legend of the *Lo shu.* Apparently misplaced in or about 1888, this book, *The Origin of Tree Worship,* a privately printed nineteenth-century work of scholarship devoted to a study of Druidical practices and the spread of the yew cult among Celtic and Germanic peoples in pre-Christian Europe, recently surfaced again during a reorganization of bookshelves at the British Library (formerly the British Museum) in London.

Mysteriously abandoned after the preliminary publication in a sparse edition of just six sample copies, the rediscovered volume is in all probability the only surviving exemplar (see photo), and its reappearance after nearly one hundred years has caused a considerable ripple in philological circles. The reason for this lies in a wealth of unmistakable internal evidence showing that the author must have been borrowing from medieval manuscript material previously believed lost in the fire that destroyed so much of the famous Cottonian collection of priceless, early English documents, while it was housed at Little Deans Yard, Westminster, in 1731. As such, *The Origin of Tree Worship* is presently the subject of minute scrutiny by experts and quite apart from the urgent questions thrown up by the provenance of its cited material, is already shedding light in several areas of paleological research. Readers interested in further details (including a review of conflicting evidence as to the real identity of its author) may care to consult the *British Library Department of Occidental Manuscripts Internal Report No. 2704/729,* as well as the forthcoming article by J. Allardyce and M. Sandeford, scheduled to appear in the *Journal of English and Germanic Philology.*

Returning to our present purpose: among other previously unrecorded Celtic myths alluded to in *The Origin of Tree Worship* is an account of a pilgrimage made by King Mī to a sacred grove in Ēohdalir, Valley of the Yews, where, following pious observance of symbolical pagan rites, a runic charm or magical formula is revealed to him, scored on the bole of the hallowed *Li,* eldest of yews. Runes, it will be recalled, are thin angular characters suited to incision on wood, stone, metal, and so forth; their employment by primitive (chiefly Scandinavian) tribes was seldom for practical purposes of communication, but almost always bore magico-ritualistic significance. An excellent survey of the subject is *Runes: An Introduction* by R. W. V. Elliott (Manchester University Press, 1980).

As an amateur runologist fortunate enough to have been granted a privileged view of this exciting find (a facsimile edition is presently in preparation), I was naturally drawn to deciphering the runic charm reproduced in the book along with the narrative of King Mī (Figure 3). It is a mark of the great advances made in paleography over the intervening years that a problem that seems to have baffled solution in 1887 (the date of publication) offers little difficulty to the modern investigator. In Figure 4, modern usage replaces Old English orthography.

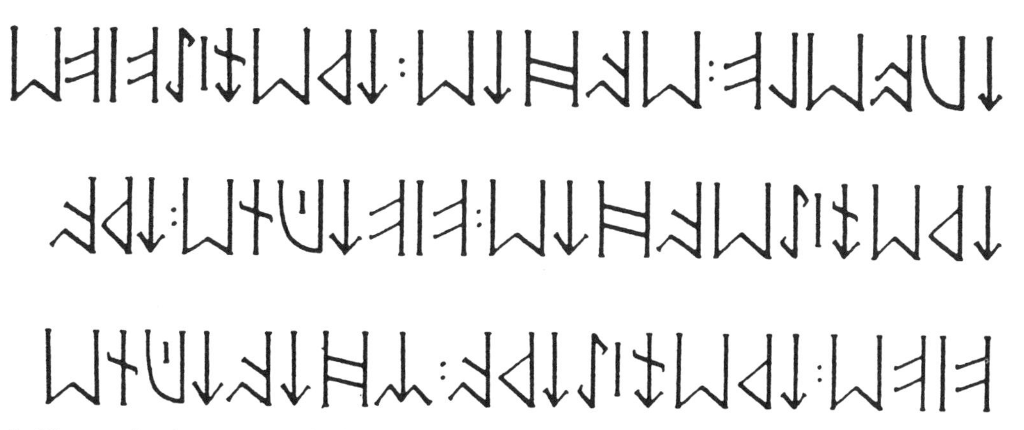

Figure 3: The runic charm revealed to King Mī, scored on the bole of *Li,* eldest of the sacred yew trees.

Five	:	Twenty-two	:	Eighteen
Twenty-eight	:	Fifteen	:	Two
Twelve	:	Eight	:	Twenty-five

Figure 4: Figure 3 in Modern English

5	22	18
28	15	2
12	8	25

Figure 5:
. . . and in Numerals

At first I was much puzzled by the pattern of cardinal number-names thus disclosed, and it was only on writing them out in more perspicuous form that understanding eventually dawned (see Figure 5). As the reader can easily verify, the sum of the three elements occurring in every row, column, and diagonal is the same: 45. What we have here, in other words, is a familiar 3×3 *magic square.*

Fascinating as this parallel with the *Lo shu* legend is, however, it remains worth noting that although distinct, the nine numbers appearing in the runic square fail to form a consecutive series, as in their Chinese counterpart. Nevertheless, the *Li shu* (as I suppose it can hardly otherwise be called) bears closer examination. Seeking for something to warrant a supernatural manifestation on a sacred yew tree, and having already been prompted through registering a small coincidence while transliterating the runes, I soon discovered that the *number of runes*—and, thus, by chance, the number of *modern English letters*—making up the three words used in every row, column, and diagonal is also identical: there are twenty-one! (The coincidence between modern and archaic word lengths will seem of less moment to readers familiar with the normal course of etymological development from old English forms: two = *twă,* five = *fifè,* eight = *eahte,* twelve = *tuoelf,* and so on).

4	9	8
11	7	3
6	5	10

Figure 6:
Letter Counts
from Figure 4

Moreover (and here I began to appreciate the potency of this singular thaumaturgical device), writing out the rune or letter totals associated with each number-name not only results in a second magic square, the numbers now emerging do indeed comprise an unbroken consecutive series (Figure 6). Furthermore, since no English cardinal number-name, old or new, is shorter than three letters—the smallest number occurring here—this square even embodies the lowest consecutive sequence imaginable on purely lexical grounds.

Astonished by this unlooked-for revelation of the secrets inherent in the *Li shu,* I quickly turned back again to *The Origin of Tree Worship* in hope of finding further details. Alas, nothing of interest is to be found there, save a bare record of the legend quoted as evidence of yew practices in Northumbria at that period, together with a conjecture that the formula had probably been credited with healing powers and would have been worn on talismans to ward off evil. Nor has any external enquiry succeeded in eliciting further amplification. The *Li shu,* it appears, exists as a unique, isolated prototype, and any subsequent developments it may have given rise to have long since been lost to us, buried in the dust of history.

Obscure as its origins remain, clearly the rediscovery of this fantastic formula immediately provokes a host of tantalizing questions and contingencies quite independent of the historical, mythological, philological, and indeed, criminological issues raised in connection with *The Origin of Tree Worship* itself. In fact, as the following will show, the *Li shu* furnishes a point of departure into an exciting new genre, a hitherto undreamed-of field, perhaps best described as a kind of recreational department of Computational Linguistics. I refer to the exploration of *alphamagic squares.*

3×3 Alphamagic Squares

Alphamagic is the word I use to describe any magic array (whether square, rectangular, triangular, N-dimensional, etc.) that remains magic when all of its entries are replaced by numbers representing the word length, in letters, of their conventional written names (thus, *one* becomes 3). Plainly, a square that is alphamagic in one language need not be so in another (and nonalphabetic languages are irrelevant in this context). It will be convenient to refer to the letter-count of a number-word as the *logorithm*—or *log*, for short—of the original number (*logos* = word, *arithmos* = number). Logorithm should not be confused with a similar word coined by a Scotchman called Napier in 1614. Where unstipulated, "natural" logs or $log_{\text{e(nglish)}}$ will be assumed; hence $log\ 15 + log\ 3 = log_{\text{french}}69$ since $7 + 5 = 12$, the number of letters in *soixante-neuf.*

By *magic* I shall mean any arrangement producing a constant sum along its various orthogonals and diagonals, regardless of whether the elements involved are distinct or not. Naturally, a square showing repeated entries is less interesting than one in which all are different. The *order* of a square refers to its size—the number of cells on a side. The *Li shu* is thus an English alphamagic square of order 3, having as additional properties that its logorithms are distinct, consecutive, and *minimal* (that is, they comprise a set of the smallest possible non-alike logorithms existent in English). Order-1 and order-2 squares need not detain us, as a moment's consideration will show. With these few conventions established, we are ready to pursue the main theme.

$a+b$	$a-b-c$	$a+c$
$a-b+c$	a	$a+b-c$
$a-c$	$a+b+c$	$a-b$

Figure 7: General Formula

As a tentative entry into the unfamiliar terrain, it is natural to wonder if there are any 3×3 alphamagic squares other than the one produced above. Useful in this connection is the general formula for order 3 shown in Figure 7 (and due to Édouard Lucas), since both numbers and logorithms in an alphamagic square must satisfy the relations it exemplifies.

Note that the three elements on each straight-line bisector through the center form a set of four 3-term arithmetic series (that is, they show a constant difference between adjacent terms: $[a - b + c] - a = a - [a + b - c]$, for instance). Then one obvious initial step is to search for arithmetic triples whose logorithms share the same property.

Figure 8 lists the cardinal numbers from 0 to 35 together with their English logorithms. Taking for illustration a center number C of 15, consider in turn arithmetic triples formed by C and its equidistant neighbors $C - 1$ and $C + 1$, $C - 2$ and $C + 2$, and so on. Note down those cases in which $log(C - N)$, $log\ C$ (= 7), and $log(C + N)$ also form arithmetic triples. When $N = C$ we can go no further, since $C - N = 0$. By now we shall have a list of pairs of associated arithmetic triples (Figure 9).

If there are any 3×3 alphamagic squares with a center number of 15 (and we already know there is one), at least *four* of these five cardinal-number triples must appear in it: one along each straight-line bisector, including diagonal bisectors.

A Table of Natural Logs

numbers

0 1 2 3 4 5 6 7 8 9 10 11 12 13 14 15 16 17 18 19 20 21 22 23 24 25 26 27 28 29 30 31 32 33 34 35 . . .

4 3 3 5 4 4 3 5 5 4 3 6 6 8 8 7 7 9 8 8 6 9 9 11 10 10 9 11 11 10 6 9 9 11 10 10 . . .

logorithms

Figure 8: The natural numbers 0 through 35 together with their logorithms. Connecting lines indicate triples in which both numbers and logorithms form regular arithmetic series.

Numbers				English Logorithms		
12	15	18	:	6	7	8
11	15	19	:	6	7	8
8	15	22	:	5	7	9
5	15	25	:	4	7	10
2	15	28	:	3	7	11

Figure 9:

Numbers:

12		11
	15	
19		18

English Logorithms:

6		6
	7	
8		8

Figure 10:

Selecting now the first two triple-pairs on the list for closer scrutiny, write them into the diagonals of corresponding matrices (Figure 10). The choice of *diagonals* here is not critical; alternative linear cell-groups might be used. We argue that since the latter will have to be occupied by two of the listed cardinal-number triples, testing each pair in turn in these positions will comprise an exhaustive check of all possibilities. Note that changing the order in which a given pair is written into the diagonals merely creates rotations or reflections of the same configuration.

Referring back to the general formula, we find that the magic constant of any square is always 3 times its center number. Therefore, if the left-hand matrix is to be magic, the middle cell in its top row will have to contain $(3 \times 15) - (12 \times 11) = 22$. Similarly, if the square is to be *alphamagic,* the corresponding cell in the right-hand matrix will hold $(3 \times 7) - (6 + 6) = 9$. Now, does *log* 22 = 9? Yes, it does. So far so good. Consider next the middle cell, right-hand column. Does *log* (45 − 29) = 21 − 14? Again, *yes.* Fine; next take the bottom row. Does *log* (45 − 37) = 21 − 16? *Yes!* This is too good to last; cross your fingers and try the last vacant cell. Does *log* (45 − 31) = 21 − 14? *No!* Yuck . . . .

So far, however, we have considered only the first pair of triples, and there remain nine other such combinations to be tried. A few minutes with pencil and paper will reward interested readers with a second alphamagic square, less elegant than the *Li shu* but still having 15 as its center number. Try it; one doing is worth a hundred seeings (old Northumbrian proverb). But what about all the other possible center numbers? To canvass all cases systemically, we need to begin with $C = 4$ (a lower number would be pointless, at least 4 distinct triples being demanded in any square), considering in turn $C = 5$, $C = 6, \ldots$, for as long as we wish to pursue the problem. Clearly, if ever a task was made for a computer, this is it.

The algorithm sketched above represents just one possible method, here incorporated into the simple Basic program labeled ALPHA.BAS; a Pascal form, AlphaMagicSquaresOfOrder3 was later prepared by my colleague Victor Eijkhout (see pages 316–319). Once the program was running, I was able to amuse myself over several weeks by exploring the alphamagic realm of order 3. It is a pursuit I can recommend to others. As one proceeds, the impression slowly grows of having ventured into a space offering almost unlimited recreational potential.

Besides the two examples already signalled, are there many other 3 × 3 English alphamagic squares? The answer is yes—an infinity of them. To see why, consider what happens if each of the *Li Shu* entries is prefixed with the words *one hundred.* The addition of a uniform constant to both numbers (100) and logorithms (10) means that the resulting matrix (Figure 11) will again be alphamagic.

5	22	18
28	15	2
12	8	25

fundamental

105	122	118
128	115	102
112	108	125

second harmonic

Figure 11: Li Shu

Such a square forms an example of what I call the *second harmonic* of the fundamental (first harmonic) square. Using *two hundred* instead of *one hundred* would result in the third harmonic, and so on. Subharmonics ("*zero point* ...")are conceivable too, if a little far-fetched. The harmonic phenomenon thus gives rise to an endless progression of alphamagic squares, none of them claiming our serious further interest (save perhaps in specialized contexts) when once their fundamentals have been identified. What about the latter?

Figure 12 presents (in numerical form) the first ten English alphamagic squares of order 3; rotations and reflections of the same square are counted identical. Alphamagics using repeated numbers I deem trivial; repetitions in their logorithm squares (shown alongside) are not. The ten are put in sequence firstly by magic constant, which for order 3 is equivalent to ranking by center number, and secondly by the lowest number occurring: 2 in the first square, 5 in the second, and so on. Extendable to higher orders, this system attaches a unique index number to every square, thus providing a convenient method of reference. Where the lowest numbers of different squares coincide, ranking will depend on the second lowest, and so on. As with ordinary magic squares, standard practice is to reproduce examples so that the smallest corner number appears in the top left-hand position, with the smaller of its two immediate neighbors oriented to the top row (middle cell). Where different squares employ identical numbers, as may occur with higher orders, this latter convention will determine rank.

Looking over the list, certain characteristic features emerge. As intuition might have led one to surmise, No. 1, the primordial Anglo-Saxon square, being the smallest and simplest (as well as oldest) exemplar in the language, is indeed none other than the *Li shu,* the Arkenstone among alphamagic gems, unmatched in revealing consecutive, minimal logorithms. Aside from its harmonics (Nos. 17, 26, 126, ...), we have to ascend to the 91st square (magic constants = 885;60) before finding another consecutive specimen (Figure 13). There is only one other such funda-

Alphamagic Squares Nos. 1–10

Index Numbers	Alphamagic Squares			→			Logorithm Squares		
No. 1 (the *Li shu*)	5	22	18	five	twenty-two	eighteen	4	9	8
	28	15	2	twenty-eight	fifteen	two	11	7	3
	12	8	25	twelve	eight	twenty-five	6	5	10
No. 2	8	19	18	eight	nineteen	eighteen	5	8	8
	25	15	5	twenty-five	fifteen	five	10	7	4
	12	11	22	twelve	eleven	twenty-two	6	6	9
No. 3	15	72	48	fifteen	seventy-two	forty-eight	7	10	10
	78	45	12	seventy-eight	forty-five	twelve	12	9	6
	42	18	75	forty-two	eighteen	seventy-five	8	8	11
No. 4	18	69	48	eighteen	sixty-nine	forty-eight	8	9	10
	75	45	15	seventy-five	forty-five	fifteen	11	9	7
	42	31	72	forty-two	thirty-one	seventy-two	8	9	10
No. 5	21	66	48	twenty-one	sixty-six	forty-eight	9	8	10
	72	45	18	seventy-two	forty-five	eighteen	10	9	8
	42	24	69	forty-two	twenty-four	sixty-nine	8	10	9
No. 6	4	101	57	four	one hundred one	fifty-seven	4	13	10
	107	54	1	one hundred seven	fifty-four	one	15	9	3
	51	7	104	fifty-one	seven	one hundred four	8	5	14
No. 7	44	61	57	forty-four	sixty-one	fifty-seven	9	8	10
	67	54	41	sixty-seven	fifty-four	forty-one	10	9	8
	51	47	64	fifty-one	forty-seven	sixty-four	8	10	9
No. 8	5	102	58	five	one hundred two	fifty-eight	4	13	10
	108	55	2	one hundred eight	fifty-five	two	15	9	3
	52	8	105	fifty-two	eight	one hundred five	8	5	14
No. 9	45	62	58	forty-five	sixty-two	fifty-eight	9	8	10
	68	55	42	sixty-eight	fifty-five	forty-two	10	9	8
	52	48	65	fifty-two	forty-eight	sixty-five	8	10	9
No. 10	46	78	101	forty-six	seventy-eight	one hundred one	8	12	13
	130	75	20	one hundred thirty	seventy-five	twenty	16	11	6
	49	72	104	forty-nine	seventy-two	one hundred four	9	10	14

Figure 12: The first ten English alphamagic squares of order 3, together with their logorithm squares

mental square among the 217 alphamagics constructible from the English number-names up to *five hundred,* No. 120 (magic constants = 897;60) (Figure 14).

As we see, increasing word length entails that neither of these, nor in fact any beyond, are minimal. In our language, therefore, this is the exclusive property of the *Li shu.* The spirit of the yew tree knew well its errand to King Mī.

215	372	298
378	295	212
292	218	375

17	22	21
24	20	16
19	18	23

Figure 13: Alphamagic Square No. 91 and its Logorithms

249	320	328
378	299	220
270	278	349

19	18	23
24	20	16
17	22	21

Figure 14: Alphamagic Square No. 120 and its Logorithms

A	B	C
B	C	A
C	A	B

Figure 15: Latin Square

$A+a$	$B+b$	$C+c$
$B+c$	$C+a$	$A+b$
$C+b$	$A+c$	$B+a$

Figure 16: Greco-Latin Square

Glancing next at Square No. 7 (Figure 12), one detects the essential structure underlying the formation of 3×3 alphamagic squares: the well-known mathematical structure known as the *greco-latin* or *Eulerian* square. By a *latin* square of order N we mean one having N^2 entries of N different elements, none of them occurring twice in any row or column (Figure 15). A *greco-latin* square is one formed by superimposing two suitable latins such that each cell becomes occupied by a *distinct* entry. The term *greco-latin* derives from the once-common practice of using Greek and Roman letters to distinguish their two components; squares of this kind were first investigated in the 1770s by the great mathematician Leonhard Euler. It is easy to prove that order 3 admits of just one possibility—the square shown as Figure 16. (In combining a pair of latins it is not essential to *add* their separate elements, as is done here, but merely to append the contents of corresponding cells.)

Comparing this with Square No. 7 (among others, see also Nos. 1, 3, 6, 8, 9), the identity of form is immediately apparent (you will note the correspondence: $A \leftrightarrow 4$, $B \leftrightarrow 6$, $C \leftrightarrow 5$, $a \leftrightarrow 4$, $b \leftrightarrow 1$, $c \leftrightarrow 7$). Rows and columns (but not diagonals) in numerical representations of these squares are therefore composed of different permutations of the same set of *digits.* I leave it to readers to show that if $1 = (b + c)/2$ and $C = (A + B)/2$ (the conditions necessary for magic diagonals), the resulting matrix is isomorphic with Lucas's formula. We shall have more to discuss about greco-latins later.

Staying with Square No. 7 for a moment, observe that the distribution of 1s, 4s, and 7s in the units' position of every entry has a curious consequence. Due to the chance that $log\ 1 = log\ 2$, $log\ 4 = log\ 5$, and $log\ 7 = log\ 8$, adding 1 to every number in the matrix results in a second alphamagic square: No. 9. Squares Nos. 6 and 8 form a similar related dyad. There are sixteen of these pairs—some adjacent, some more widely separated—among the first 100 squares.

The alphamagic properties of Square No. 7 are not yet entirely exhausted. Although trivial, the magic (latin) square formed by its logorithms (which I shall designate by *log* [No. 7]) is worth a closer look. Writing out *log* [No. 7] in full, we have Figure 17 as a result. Viewed thus, a natural question arises: could *log* [No. 7] by any chance be alphamagic, albeit trivial, too? The answer, of course, is yes, the magic constant of *log*[*log*[No. 7]] being 12 (see Figure 18).

nine	eight	ten
ten	nine	eight
eight	ten	nine

Figure 17: Log[No. 7]

At this point it is difficult not to wonder whether this second latin (magic) square is in turn alphamagic itself. Alas, repetition of the same process yields only a semi-magic derivative. Leaving apart superficial cases where the initial logorithm square is made up of nine identical numbers (a far from uncommon occurrence), I have been unable to find any such instance among the first few hundred English squares. No. 7 shares its distinction with Nos. 5, 9, and 36.

4	5	3
3	4	5
5	3	4

Figure 18: Log[Log[No. 7]]

There is an interesting computer project here that ambitious readers may like to follow up. Ideally, of course, we seek a square giving rise to an unbroken chain of alphamagic derivatives, culminating, as any chain eventually must do, in a closed loop. The shortest and most elegant such alphamagic loop would be a self-reproducing square—Figure 19. I leave more complicated loops to the contemplation of interested parties. Lest the ground to be explored here seem unduly narrow, bear in mind that we are under no compunction to remain in the same language at each stage in the derivation process. What, for instance, might be the longest chain of *multilingual* alphamagic links constructible? In any case, the search for ever more potent magic "spells" of this and other kinds soon encourages a glance beyond the confines of English.

4	5	3
3	4	5
5	3	4

Figure 19: Self-Reproducing Square

Exotic Squares

The exact number of alphabetic languages used throughout the world has perhaps never been estimated. Clearly there are many. Besides those like our own employing Roman letters, there remain others using the Greek, Hebrew, and Cyrillic alphabets. The work of collecting and collating alphamagic squares in the various tongues and dialects opens a wide (if decidedly recondite) area of research. One has only to think of the enormous literature on ordinary magic squares, with its endless ramifications, almost all of which become reapplicable to alphamagic squares, to catch a glimpse of the undeveloped possibilities. My own peregrinations in the field having been superficial, I shall present here only a few examples of order 3.

Investigating 3×3 alphamagic squares in different languages calls for no alterations to the program already described, save in loading appropriate logorithm data into memory. Having had some experience in this line of late, I can report that ascertaining the correct spelling of foreign cardinals is often trickier than one suspects. Books supposedly supplying this information should be treated circumspectly (in French, is 101 *cent un* or *cent et un*?). Typing in word lengths without introducing errors is another task requiring perseverance and concentration; a subprogram for calculating letter-counts from the words themselves is advisable. Without care in this preparatory phase, interpretation of the printout is troubled with doubts.

Taking French as an initial object of study, I was intrigued to discover only a single alphamagic square using number-names in the range up to *deux cents* (200). It seemed that Gallic orthography combined with a vigesimal (twenty-based) system of counting to produce similar effects on the alphamagic plane. Thinking what a rare collector's item this must represent if it turned out to be the sole existing French alphamagic square of order 3, I quickly extended the search up to *trois cents,* only to be glutted with a sudden deluge of 255 new specimens! Square No. 14 (magic constants = 336;27) was the first (of 3) to show consecutive logorithms (Figure 20). [In parentheses, the numerical representation of each number-word is followed by the *log* or letter-count.]

Quinze (15;6)	Deux cent six (206;11)	Cent quinze (115;10)
Deux cent douze (212;13)	Cent douze (112;9)	douze (12;5)
Cent neuf (109;8)	dix huit (18;7)	deux cent neuf (209;12)

Figure 20: French Square No. 14

A curiosity worth remarking is the prevalence of prime numbers among French alphamagics, a by-product of frequent *un, trois,* and *sept* terminations. Even so, a square composed uniquely of primes, in this or in any other language, has yet to be identified. The urge to uncover specialist items of this kind will probably prove a stimulus to logophiles for some time to come. Serious aficionados will hardly rest until the Tower of Babel has been ransacked from roof to basement.

Following the French experience, I was better prepared for a foray into German. After entering the new logorithms and typing RUN, within seconds the printer chirped into life and began spitting out alphamagic squares in a steady rhythmical tattoo evocative of massed hordes on the march. The reason for this regularity was soon apparent: every one of the 221 squares resulting from number-names under *hundert* (100) employs nine double-digit numbers; with few exceptions, the adjacently printed logorithms of every one of these nine were the same: 14.

Many readers, I imagine, will be surprised to learn of hundreds of alphamagic squares extant in three different languages. How is this prodigality made possible? The answer lies, simply enough, in the (inevitable) regularity of our naming systems for cardinals higher than *twenty,* the designations beyond this point being exact verbal counterparts of their decimal-point representations (*twenty-one*= 20 + 1, *twenty-two*= 20 + 2, and so on). Thus, the combinative properties of numbers are often paralleled in their logorithms, with the result that many an unexceptional magic square (of which there are myriads, contrary to expectation), is automatically rendered alphamagic. In German—an extreme case, where the words for 1, 2, 3, 4, 5, 8, and 9 all have four letters, and those for 20, 30, 40, 50, 60, 70, 80, 90, and 100 all have seven—this factor issues in a rash of uniform logorithm squares, few of them revealing any redeeming feature of interest. A typical example is No. 72, shown in Figure 21 (magic constants = 165;42).

Günfundvierzig (45;14)	Zweiundsechzig (62;14)	Achtundfünfzig (58;14)
Achtundsechzig (68;14)	Fünfundfünfzig (55;14)	Zweiundvierzig (42;14)
Zweiundfünfzig (52;14)	Achtundvierzig (48;14)	Fünfundsechzig (65;14)

Figure 21: German Square No. 72

The trouble with squares generated by this parallel effect is their structural transparency, which robs them of logological charm. As logophiles we prize cunning arrangements exploiting unsuspected linguistic fortuity. In almost any language, therefore, the vast majority of squares will fail to command admiration. In general, of course, as in Gardner's marvelous anagram prefacing this article, alphamagic elegance resides in small numbers.

Wearying of pedestrian languages, I turned next to some of the less familiar tongues. Keeping research within manageable bounds, surveys were limited to cardinals in the range up to 100. Figure 22, a recherché anthology if ever there was one, records the numbers of squares discovered in each case. Totals are generally modest, which is not to say they would remain so if the census were extended further. Raising the ceiling to 200, for instance, second harmonics will account for a doubling in figures, at the very least.

A study of squares in foreign languages can hardly proceed by far before an obvious contingency springs to thought. Has anyone noticed, I wonder, that the German square No. 72 given above is a perfect *translation* of English square No. 9? As a matter of fact, both the German translation of $\log_e$ [No. 9] and the English translation of $\log_g$ [No. 72] are themselves alphamagic, like their originals; but here we are straying into a less central, even frivolous hinterland. Once glimpsed, of course, the notion of such a (primary) correspondence soon urges systematic comparison among squares, alphamagic translations forming yet another branch to explore in the logological labyrinth. Figure 22 includes a résumé of the interlingual connections so far established.

Two of the languages listed show no alphamagic squares at all in the range investigated. Extending examination of the first of these discovers six Danish squares using numbers below *tohundrede* (200). Likewise, in the second case, four squares are brought to light, No. 4 (magic constants

Alphamagic Squares Around the World

	Number of alphamagic squares	Translations										Total number of translations
Danish	0											0
Dutch	6	4										1
English	7		9									1
Esperanto	6											0
Finnish	13			6								1
French	1											0
Gaelic	1											0
German	221	77	72	12	14							4
Icelandic	3	2										1
Indonesian	1											0
Italian	1											0
Latin	0											0
Maltese	3											0
Norwegian	12		16	12	2							3
Portuguese	2											0
Samoan	9				2	4	5					3
Spanish	14							9				1
Swahili	11								4	6	8	3
Swedish	5								1	2	3	3
Turkish	17		25		3				24	34	41	5
Welsh	26	7		12		9	11	15	14	24	46	8

Figure 22: What is the total number of alphamagic squares with cardinals not higher than 100? This chart shows the answer for squares in different languages (left). In the column marked "translations," a circled figure is the index number of a given square, with lines linking numerically identical squares (mutual translations). Thus, Dutch Square No. 4 is a translation of German Square No. 77.

SEPTEM ET NONAGINTA (XCVII;XVII)	CENTUM SEPTUAGINTA (CLXXV;XXIV)	CENTUM TRIGINTA (CXXXIX;XIX)
CENTUM SEPTUAGINTA NOVEM (CLXXIX;XXII)	CENTUM TRIGINTA SEPTEM (CXXXVII;XX)	QUINQUE ET NONAGINTA (XCV;XVIII)
CENTUM TRIGINTA QUINQUE (CXXXV;XXI)	NOVEM ET NONAGINTA (XCIX;XVI)	CENTUM SEPTUAGINTA SEPTEM (CLXXVII;XXIII)

Figure 23: Latin Square No. 4

Ottantasette (87;12)	*Cento Sessantacinque* (165;19)	*cento ventinove* (129;12)
Cento sessantanove (169;17)	*Cento ventisette* (127;15)	*Ottantacinque* (85;13)
Cento venticinque (125;16)	*Ottantanove* (89;11)	*Cento sessantasette* (167;18)

Figure 24: Italian Square No. 3

= 411;60) being a rare consecutive-logorithm curio using odd numbers only (Figure 23). Here the influence of underlying latin squares is unmistakable. Likewise, early Roman influence is perhaps responsible for the consecutive logorithms to be found in Figure 24, a *modern* Italian ("I, a latin") square—No. 3 (magic constants = 381;45).

Note the constant difference between corresponding entries at both numerical and logorithm levels in this geographically related pair.

Oddly, of all the languages so far examined, there is one which stands out as peculiarly rich in alphamagic *translations.* English is poor, yielding only the example previously cited. French, together with others, has none. Norwegian and Samoan show three, as do Swedish and Swahili, an alliterative duo remarkable in that Nos. 1, 2, and 3 in the former translate into Nos. 4, 6, and 8 in the latter (as shown in Figure 25). German yields no less than four, which is not surprising in view of its total of 221 squares. And Turkish delights in five, three of them correlating with squares in the most prolific source of all: it is the language of the west Britons, the language of the Bards, Welsh.

Alphamagic Translations between Swedish and Swahili

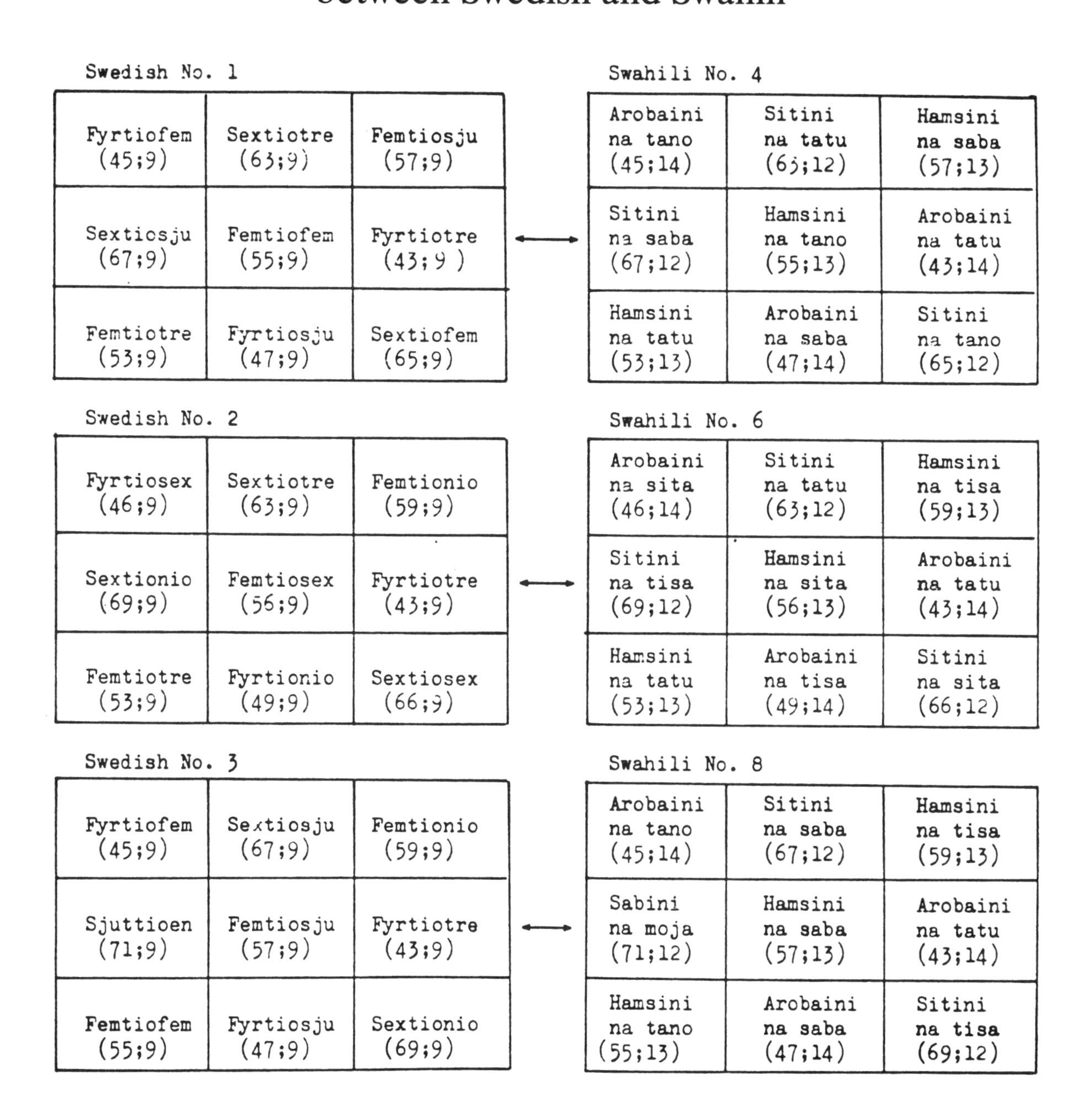

Swedish No. 1

Fyrtiofem (45;9)	Sextiotre (63;9)	Femtiosju (57;9)
Sextiosju (67;9)	Femtiofem (55;9)	Fyrtiotre (43;9)
Femtiotre (53;9)	Fyrtiosju (47;9)	Sextiofem (65;9)

↔

Swahili No. 4

Arobaini na tano (45;14)	Sitini na tatu (63;12)	Hamsini na saba (57;13)
Sitini na saba (67;12)	Hamsini na tano (55;13)	Arobaini na tatu (43;14)
Hamsini na tatu (53;13)	Arobaini na saba (47;14)	Sitini na tano (65;12)

Swedish No. 2

Fyrtiosex (46;9)	Sextiotre (63;9)	Femtionio (59;9)
Sextionio (69;9)	Femtiosex (56;9)	Fyrtiotre (43;9)
Femtiotre (53;9)	Fyrtionio (49;9)	Sextiosex (66;9)

↔

Swahili No. 6

Arobaini na sita (46;14)	Sitini na tatu (63;12)	Hamsini na tisa (59;13)
Sitini na tisa (69;12)	Hamsini na sita (56;13)	Arobaini na tatu (43;14)
Hamsini na tatu (53;13)	Arobaini na tisa (49;14)	Sitini na sita (66;12)

Swedish No. 3

Fyrtiofem (45;9)	Sextiosju (67;9)	Femtionio (59;9)
Sjuttioen (71;9)	Femtiosju (57;9)	Fyrtiotre (43;9)
Femtiofem (55;9)	Fyrtiosju (47;9)	Sextionio (69;9)

↔

Swahili No. 8

Arobaini na tano (45;14)	Sitini na saba (67;12)	Hamsini na tisa (59;13)
Sabini na moja (71;12)	Hamsini na saba (57;13)	Arobaini na tatu (43;14)
Hamsini na tano (55;13)	Arobaini na saba (47;14)	Sitini na tisa (69;12)

Figure 25: Swedish alphamagics Nos. 1, 2, and 3 translate into Swahili alphamagics Nos. 4, 6, and 8. Word-game players will note that in Swahili, *sits* (6) is an anagram of *tisa* (9).

Together with its sister tongues Breton and Cornish, Welsh belongs to the Celtic family of languages, which include Irish, Manx, and Gaelic. In former times the vigesimal system was current, but except in reading the clock, present-day Welsh has replaced this with decimal usage. Whether the originators of this reform had any premonition of its alphamagic consequences must remain conjectural, but the effects have been remarkable indeed. Old-fashioned (vigesimal) Welsh, which I have also examined up to *cant* (100), manifests no alphamagic squares whatever. Modern Welsh, on the other hand, rejoices in twenty-six squares in this range, no less than eight of them corresponding to translations of squares in either Turkish (3 cases), Samoan (2 cases), Spanish, Icelandic, or Norwegian. The latter instance furnishes a striking consecutive-logorithm cameo using even numbers only (magic constants = 216;33); see Figures 26 and 27.

Chwech deg dau (62;12)	*Wyth deg* (80;7)	*Saith det pedwar* (74;14)
Wyth deg pedwar (84;13)	*Saith deg dau* (72;11)	*Chwech deg* (60;9)
Saith deg (70;8)	*Chwech deg pedwar* (64;15)	*Wyth deg dau* (82;10)

Figure 26: Welsh Square No. 12

Sekstito (62;8)	*Atti* (80;4)	*Syttifire* (74;9)
Attifire (84;8)	*Syttito* (72;7)	*Seksti* (60;6)
Sytti (70;5)	*Sekstifire* (64;10)	*Attito* (82;6)

Figure 27: Norwegian Square No. 12

Amazingly, as many as *six* of these twenty-six Welsh squares show consecutive logorithms, a staggering total considering that of the remaining 333 squares spread over twenty languages in Figure 22, there is but a single instance of another consecutive-logorithm square: the *Li shu* (the French, Latin, and Italian examples given earlier lying outside our two- digit range). None of the Cambrian six are minimal, however, the shortest number-name in Welsh containing two letters, while the smallest series of logorithms occurring runs from 7 up to 15.

A detailed treatment of the unnumbered curiosities and secondary correlations to be found among alphamagic squares across the different languages is beyond the scope of a single article.

Leaving the field to enthusiasts who may like to pursue these researches—seeking, perhaps, what I failed to discover, a *triple*-language alphamagic translation—it is time to return to English and a look at the higher orders.

References

S. Camman, The magic square of three in old Chinese philosophy and religion, *History of Religions* 1 (Summer 1961) 37–80.

J. Allardyce and M. Sandeford, New evidence for the survival of codex 221(b) [MSS. Cotton Catullus B XIV], *Journal of English and Germanic Philology* 87 (in press).

British Library Department of Occidental Manuscripts: Internal Report No. 2704/1729 (second series), 1985.

ALPHA.BAS

```
10 ' ***************************************************************************************
20 ' *      Program:  ALPHA.BAS         (GWBASIC)                                            *
30 ' *      Purpose:  To generate and print out all 3 x 3 alphamagic                         *
40 ' *                squares formable using 9 distinct cardinals in                         *
50 ' *                the range 0 - NMAX. Index numbers and logorithm                        *
60 ' *                squares are printed alongside.                                         *
70 ' *      Author:   Lee C.F. Sallows                                                       *
80 ' *      Date:     Guy Fawkes Day (November 5th), 1985                                    *
90 ' ***************************************************************************************
100 ' ---------------------------------------------------------------------------------------
110 '        Array definitions:
120 '        A       Logorithms of 0 - NMAX (loaded as data)
130 '        B       Arithmetic triples showing arithmetic logorithms
140 '        C       Logorithm-triple counterparts to numbers in B
150 ' ---------------------------------------------------------------------------------------
160     NMAX=109        'In this example
170     DIM A(NMAX):   DIM B(100,3):   DIM C(100,3)
180 ' ---------------------------------------------------------------------------------------
190 '   Data is loaded into A; A(n) thus corresponding to log n.
200 '   Contingent triple CENtre numbers are considered in turn.
210 ' ---------------------------------------------------------------------------------------
220     FOR I=0 TO NMAX:   READ A(I):   NEXT I
230     FOR CEN = 4 TO NMAX-4
240 ' ---------------------------------------------------------------------------------------
250 '   Starting with the two highest-lowest values possible (BOUND)
260 '   and working inwards, A-elements equidistant about A(CEN) are
270 '   checked with A(CEN) to see if they form arithmetic triples. If so,
280 '   COUNT is incremented, the number-triple is stored in B and its
290 '   associated log-triple stored in C.
300 '   Provided at least 4 triple-pairs are found we proceed to the next
310 '   stage; otherwise reset COUNT and take the next CENtre number.
320 ' ---------------------------------------------------------------------------------------
330     IF CEN<(NMAX-CEN) THEN BOUND=CEN ELSE BOUND=(NMAX-CEN)
340     IF A (CEN)-A(CEN-BOUND) <>A(CEN+BOUND)-A(CEN) THEN GOTO 380
350     COUNT=COUNT+1
360     B(COUNT,1)=(CEN-BOUND): B(COUNT,2)=CEN: B(COUNT,3)=(CEN+BOUND)
370     C(COUNT,1)=A(CEN-BOUND): C(COUNT,2)=A(CEN): C(COUNT,3)=A(CEN+BOUND)
380     BOUND=BOUND-1: IF BOUND<>1 THEN GOTO 340
390     IF COUNT<4 THEN GOTO 960
400 ' ---------------------------------------------------------------------------------------
410 '   B now contains 4 or more arithmetic triples, C their associated
420 '   logorithms. Using I and J to address every possible pair of
430 '   B-triples in turn, we deal with them as though written into the
440 '   diagonals of a 3 x 3 test matrix, thus:
450 '                          -----------------------
460 '                          B(I,1)  . . .  B(J,3)
470 '                          -----------------------
480 '                            . . .  CEN  . . .
490 '                          -----------------------
500 '                          B(J,1)  . . .  B(I,3)
510 '                          -----------------------
520 '   Magic-fulfilling values are now calculated for the remaining empty
530 '   cells, checking one at a time to see if their logorithms also
540 '   satisfy magic conditions in the associated log-matrix.
550 ' ---------------------------------------------------------------------------------------
560     FOR I=1 TO COUNT-1
570     FOR J=I+1 TO COUNT
580     CL=3*CEN-B(I,1)-B(J,1)       'CL = centre left column number
590     IF CL>NMAX OR CL <0 THEN GOTO 950 'entries must be within limits
600     IF A(CL) <>3*A(CEN)-C(I,1)-C(J,1) THEN GOTO 950 'check left column log.
```

```
610       CT=3*CEN-B(I,1)-B(J,3)   'CT = centre top row number
620       IF CT>NMAX OR CT<0 THEN GOTO 950   'within limits?
630       IF A(CT) <>3*A(CEN)-C(I,1)-C(J,3) THEN GOTO 950 'check top row log.
640       CR=3*CEN-B(I,3)-B(J,3)   'CR = centre right column number
650       IF CR>NMAX OR CR<0 THEN GOTO 950   'within limits?
660       IF A(CR) <>3*A(CEN)-C(I,3)-C(J,3) THEN GOTO 950 'check right column log.
670       CB=3*CEN-B(I,3)-B(J,1)   'CB = centre bottom row number
680       IF CB>NMAX OR CB<0 THEN GOTO 950   'within limits?
690       IF A(CB) <>3*A(CEN)-C(I,3)-C(J,1) THEN GOTO 950 'check bottom row log.
700   ' -----------------------------------------------------------------
710   '   Any triple-pair surviving the above tests gives rise to an
720   '   alphamagic square. Duplicate entries may occur, however, but
730   '   in that case it can be shown that CT = B(J,3)
740   ' -----------------------------------------------------------------
750       IF CT=B(J,3) THEN GOTO 950
760   ' -----------------------------------------------------------------
770   '   An advantage of the particular algorithm here employed is that
780   '   solutions are discovered in order of their Index No. We print
790   '   this, together with the solution (in standard normal form)
800   '   and its logorithm square alongside. TL = Top Left, etc.
810   ' -----------------------------------------------------------------
820       INDX=INDX+1: PRINT "No."INDX
830       TL=B(I,1): TR=B(J,3): BL=B(J,1): BR=B(I,3) 'shorthand
840       PRINT USING "####";TL,CT,TR,:PRINT  "  ",  'matrix formatting
850       PRINT USING "####";A(TL),A(CT),A(TR)       'matrix formatting
860       PRINT USING "####";CL,CEN,CR,:PRINT "  ",  'matrix formatting
870       PRINT USING "####";A(CL),A(CEN),A(CR)      'matrix formatting
880       PRINT USING "####";BL,CB,BR,:PRINT  "  ",  'matrix formatting
890       PRINT USING "####";A(BL),A(CB),A(BR)       'matrix formatting
900       PRINT
910   ' -----------------------------------------------------------------
920   '   There may be still other squares with the same CENtre number.
930   '   If not, reset COUNT, take next case and search further.
940   ' -----------------------------------------------------------------
950       NEXT J: NEXT I
960       COUNT=0: NEXT CEN
970       PRINT "All possibilities up to" NMAX " examined"
980   ' -----------------------------------------------------------------
990   '   The first 110 English logorithms:
1000  ' -----------------------------------------------------------------
1010         DATA 4,3,3,5,4,4,3,5,5,4
1020         DATA 3,6,6,8,8,7,7,9,8,8
1030         DATA 6,9,9,11,10,10,9,11,11,10
1040         DATA 6,9,9,11,10,10,9,11,11,10
1050         DATA 5,8,8,10,9,9,8,10,10,9
1060         DATA 5,8,8,10,9,9,8,10,10,9
1070         DATA 5,8,8,10,9,9,8,10,10,9
1080         DATA 7,10,10,12,11,11,10,12,12,11
1090         DATA 6,9,9,11,10,10,9,11,11,10
1100         DATA 6,9,9,11,10,10,9,11,11,10
1120         DATA 10,13,13,15,14,14,13,15,15,14
1130         STOP
```

AlphaMagicSquaresOfOrder3

```
{-------------------------------------------------------------------------------}
{ ==================== AlphaMagic Squares Detection ==================== }
{-------------------------------------------------------------------------------}
{                              design  : Lee Sallows                            }
{                      implementation  : Victor Eijkhout                        }
{                                                                               }
{                          written in                                           }
{                             Turbo Pascal Version 3;                           }
{                             Borland International                             }
{-------------------------------------------------------------------------------}

Program AlphaMagicSquaresOfOrder3;

Const Range=109;
      Logorithm : Array[0..Range] Of Integer
      = ( 4,3,3,5,4,4,3,5,5,4,
          3,6,6,8,8,7,7,9,8,8,
          6,9,9,11,10,10,9,11,11,10,
          6,9,9,11,10,10,9,11,11,10,
          5,8,8,10,9,9,8,10,10,9,
          5,8,8,10,9,9,8,10,10,9,
          5,8,8,10,9,9,8,10,10,9,
          7,10,10,12,11,11,10,12,12,11,
          6,9,9,11,10,10,9,11,11,10,
          6,9,9,11,10,10,9,11,11,10,
          10,13,13,15,14,14,13,15,15,14);

Var Square : Array[-1..1 , -1..1] Of Integer;

Var center,counter : Integer;

Function Min( x,y : Integer ):Integer;
Begin If x<y Then Min:=x Else Min:=y End;

{-------------------------------------------------------------------------------}
{ ===================== output of completed square ===================== }
{-------------------------------------------------------------------------------}
Procedure ReportAlphaMagicSquare( c,d1,d2, t,l,r,b : Integer );

  Var i,j : Integer;

Begin
  Square[-1,0]:=l; Square[1,0]:=r; Square[0,-1]:=b; Square[0,1]:=t;
  Square[-1,-1]:=c-d2; Square[1,1]:=c+d2;
  Square[-1,1]:=c-d1; Square[1,-1]:=c+d1;
  Square[0,0]:=c;
  Writeln(' alphamagic square No. ',counter);
  For i:=-1 To 1
  Do Begin For j:=-1 To 1
          Do Write( Square[j,i]:6 );
          Write('       <->       ');
          For j:=-1 To 1
          Do Write( Logorithm[ Square[j,i] ]:6 );
          Writeln(' ')
     End;
  Writeln(' ')
End; {======= procedure ReportAlphaMagicSquare ============}
```

```
{-----------------------------------------------------------------------------}
{====================== test alphamagicality ==================== }
{-----------------------------------------------------------------------------}
Procedure MaybeAlphaMagicSquare( cen,dist1,dist2 : Integer );

   Var MagicConstant, AlphaMagicConstant,
       left,right,top,bottom : Integer;

   Function MagicTriple( x,y : Integer; Var mid : Integer): Boolean;
   Begin MagicTriple := False;
         mid := MagicConstant -x-y;
         If ( mid>0 ) And ( mid<=Range ) And ( mid <> y )
           { the third test eliminates trivial solutions }
         Then MagicTriple := ( AlphaMagicConstant =
                                      Logorithm[ x]
                                    + Logorithm[ mid ]
                                    + Logorithm[ y ]   )

   End;

Begin
  MagicConstant := 3*cen;
  AlphaMagicConstant := 3*Logorithm[ cen ];
  If MagicTriple( cen-dist1,cen-dist2, left )
  Then If MagicTriple( cen-dist1,cen+dist2, top )
       Then If MagicTriple( cen+dist1,cen-dist2, bottom )
            Then If MagicTriple( cen+dist1,cen+dist2, right )
                 Then Begin counter:=counter+1;
                            ReportAlphaMagicSquare( cen,dist1,dist2,
                                                 top,left,right,bottom )
                      End
End; {======= procedure MayBeAlphaMagicSquare =========}

{-----------------------------------------------------------------------------}
{====================== generate squares around ==================== }
{====================== a given center number ======================}
{-----------------------------------------------------------------------------}
Procedure GenerateSquaresAroundCenter( c : Integer );

   Var k,l : Integer;

   Function LogoArithmeticTriple( cen,dist : Integer ): Boolean;
   Begin LogoArithmeticTriple :=
            Logorithm[ cen ] - Logorithm[ cen-dist ]
                             =
            Logorithm[ cen+dist ] - Logorithm[ cen ]
   End;

Begin
  For k:=Min( c,Range-c) DownTo 1
  Do If LogoArithmeticTriple( c,k )
     Then For l:=Min( c,Range-c) DownTo k+1
          Do If LogoArithmeticTriple( c,l )
             Then MaybeAlphaMagicSquare( c,k,l )
End; {==== procedure GenerateSquaresAroundCenter ========}

{-----------------------------------------------------------------------------}
                          Begin {======Main Program========}
                            counter:=0;
                            For center:=4 To Range-4
                            Do GenerateSquaresAroundCenter( center )
                          End.  {======Main Program========}
{-----------------------------------------------------------------------------}
```

Alphamagic Squares,*
Part II

More adventures with abacus and alphabet, extending explorations into the untrodden realms of computational logology.

Lee C. F. Sallows

Higher Orders

Readers who may be regretting that most of the really worthwhile nuggets have already been culled from the alphamagical goldfield are in for a pleasant surprise. With the transition from order 3 to order 4, and higher, comes a concomitant jump in the perplexities confronting our advance, since hindsight reveals order 3 as a special, unusually tractable case. The problems involved having largely resisted solution thus far, this higher ground has been barely surveyed, let alone exhausted. As a result, it is no exaggeration to say that for programmers and pencil-owners alike, there remain rich pickings to be had, given ingenuity and the will to explore. Before turning to the difficulties imposed, however, it will be well to distinguish between logologist's gold and fool's gold.

In the previous issue we looked at greco-latin squares, noting obvious isomorphisms between the single instance of order 3 and certain 3×3 alphamagics. [a *latin* square of order N, it will be recalled, is defined as one comprising N^2 entries of N distinct elements, each occurring exactly once in every row and column. *Greco-latins* are formed when two suitable latin squares are appended so that all the resultant compound entries are unique. Only one square of 3×3 exists.] For all higher orders, however, assorted kinds of greco-latin squares exist—in particular, those using latin squares in which the N distinct elements also appear along both diagonals; see Figure 2.

Now an interesting if obvious property of these matrices is that their elements are always replaceable by appropriate numbers so as to produce a nontrivial magic square. In Figure 3, for example, $A = 20$, $B = 30$, $C = 40$, $D = 50$, $a = 6$, $b = 7$, $c = 8$, $d = 9$, and the magic constant is 170. A "*diagonal* greco-latin square," in other words, is a recipe for certain types of magic square. Less prominent perhaps, but equally true, is that it yields a recipe for certain types of alphamagic square too.

There is a neat trick that can be used for trapping friends into scornful expressions of baseless incredulity. You show someone the *Li shu* and explain its properties. While your subject is still

*Reprinted with permission from *Abacus,* Vol. 4, No. 2 (Winter 1987) 20–29. ©1987 Springer-Verlag.

Five	:	Twenty-two	:	Eighteen
Twenty-eight	:	Fifteen	:	Two
Twelve	:	Eight	:	Twenty-five

5	22	18
28	15	2
12	8	25

Figure 1: The Ancient Northumbrian *Li Shu* and the Ancient Chinese *Lo shu*

4	9	2
3	5	7
8	1	6

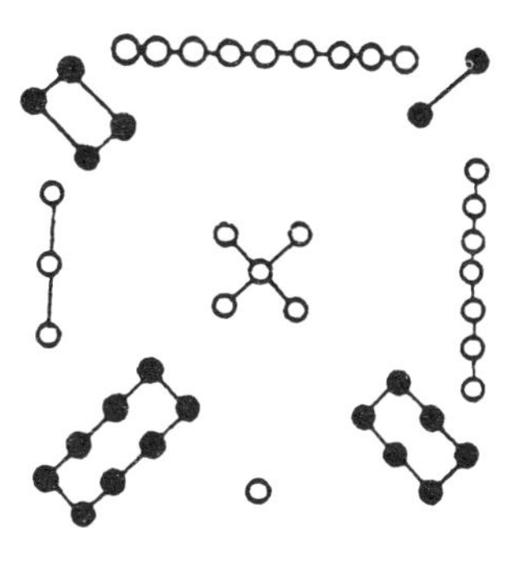

$A+a$	$B+b$	$C+c$	$D+d$
$C+d$	$D+c$	$A+b$	$B+a$
$D+b$	$C+a$	$B+d$	$A+c$
$B+c$	$A+d$	$D+a$	$C+b$

Figure 2: Diagonal 4×4 Greco-Latin Square

26	37	48	59
49	58	27	36
57	46	39	28
38	29	56	47

Figure 3: Greco-Latin-Based Magic Square

Twenty-six	*Thirty-seven*	*Forty-eight*	*Fifty-nine*
Forty-nine	*Fifty-eight*	*Twenty- seven*	*Thirty-six*
Fifty-seven	*Forty-six*	*Thirty-nine*	*Twenty-eight*
Thirty-eight	*Twenty-nine*	*Fifty-six*	*Forty-seven*

Figure 4: Literal Version of Figure 3

goggling under its impact, mention casually that this is only kid's stuff; you yourself have produced an alphamagic cube of order 8. "An $8 \times 8 \times 8$ alphamagic *cube* using entirely different number-words in every single position?" comes the unbelieving response. "Sure," you reply. "Not only that, the rows, columns, pillars and diagonals are all perfect *anagrams* of each other." Keeping a straight face at this point, be prepared to return any searching glances. Eventually your victim will be forced into a demurral. "But that's nothing," you retort, "my cube even retains all its alphamagic properties when translated into French . . . ." Puzzlement mixed with skepticism will now spill over into indignation at the leg-pulling. It is time suavely to produce your piece of paper showing the three superimposed order-8 latin cubes.

Order-8 greco-latin cubes, in fact, offer little difficulty in construction; for example, the interested reader can consult *Latin Squares and Their Applications* by Denes and Keedwell. Constructions of this size being cu(m)bersome, however, let us content ourselves with an order-4 square, a literal equivalent of the last square we examined. In Figure 4, as promised, orthogonals and diagonals share the same set of 39 letters. An anagrammatic, alphamagic square will also survive, following translation into French. Or Swedish. Or Transylvanian. Or . . . ,

Index Numbers	Alphamagic Squares		
No. 1 (the *Li Shu*)	5	22	18
	28	15	2
	12	8	25
No. 2	8	19	18
	25	15	5
	12	11	22
No. 3	15	72	48
	78	45	12
	42	18	75
No. 4	18	69	48
	75	45	15
	42	21	72
No. 5	21	66	48
	72	45	18
	42	24	69
No. 6	4	101	57
	107	54	1
	51	7	104
No. 7	44	61	57
	67	54	1
	51	47	64
No. 8	5	102	58
	108	55	2
	52	8	105
No. 9	45	62	58
	68	55	42
	52	48	65
No. 10	46	78	101
	130	75	20
	49	72	104

Figure 5: Alphamagic Squares Nos. 1–10

This is now what I mean by fool's gold: easily constructible greco-latin-based, higher-order alphamagic squares (or cubes, or whatever) exhibiting magical properties beyond the wildest fantasies of logomania. Seemingly marvelous trinkets that are nonetheless just so many gewgaws, so much logological junk. And the reason resembles that in the case of the 3×3 German squares: limpidity robs them of interest; even casual inspection soon exposes their cheap reliance on the pantographic coupling between cardinals above 19 and the numbers of letters in their names. Worse still, in reality they are nothing more than tediously redundant diagonal greco-latin squares whose single identifiers have been expanded into words. All the "magic" they possess is entirely due to this underlying pattern, which guarantees (multi-level) uniformity of row, column, and diagonal content, no matter *what* (composite) entities replace its variables. Substituting chemical compounds for the latter, for instance, the distribution of chemical *elements* resulting would inevitably be magic as well. Greco-latins are of serious mathematical interest in their own right, to be sure, but in the guise of alphamagic squares they are only worthless imitations of the precious metal sought. We (and our friends) will need to be wary of these ironic pyrites.

Understand that the defect of such squares lies not exactly in their greco-latin morphology per se, but in their failure to disguise, to cover the traces of that foundation. A simple card-trick that defies explanation will continue to excite wonder as long as its mechanism remains invisible. The ingredient of concealment, of presenting a remarkable effect without giving away how it is achieved, is a sine qua non in any manifestation of "magic": magic minus mystery means "mundane."

Consider the *Li Shu,* for instance, which, like any 3×3 magic square, is itself an instantiation of an order-3 greco-latin square—a characteristic that emerges quite clearly on checking the pattern of 1s, 2s, 5s, and 8s in its numerical representation (Figure 5, No. 1). But traces of that underlying structure are far from evident in the real or literal *Li shu,* a circumstance due to the usage of *twelve* instead of "ten-two," *fifteen* instead of "ten-five," and *eighteen* rather than "ten-eight"—usage, be it noted, that nevertheless preserves the word length of these more rational alternatives. Here then, in contrast to the transparency of the order-4 square above, we witness linguistic accident at work in the service of subterfuge, in helping to camouflage the tell-tale pattern that discloses its formative principle. Therein, in part, resides the power of the square, its claim to be Alphamagic—in the sense of ranking first.

Bear in mind, incidentally, that—unlike the case in higher orders—the nonexistence of any 3×3 *diagonal* greco-latin square means that even the most conspicuous of order-3 alphamagics (No. 6 is the first) is always something more than a mere substitution of number-words for identifiers. Cardinals occurring along diagonals can never be just an alternative ordering of those composing every row and column; $a = (b + c)/2$ and $C = (A + B)/2$, recall. [These are the extra conditions to be satisfied in using the order-3 greco-latin square as a template for constructing a 3×3 magic square.] Comparison of literal versions of squares in the figure is instructive here; note that in terms of concealment, No. 2 betters No. 1. With our minds now alerted to these lesser greco-latin alloys, we return to the search for logologist's gold.

Following the successful approach used in deriving order-3 alphamagics, a good plan now would seem to be examination of the general formula for magic squares of order 4. This we shall do, but first let us glance at an intriguing possibility that is bound to suggest itself to anybody familiar with traditional magic-square theory.

Normal Squares

Sometime before 1675 (the date of his death), a French ecclesiastic, Bernard Frénicle de Bessy, first established that there are 880 distinct normal magic squares of order 4, excluding rotations and reflections. By *normal* is intended squares using the natural consecutive series $1, 2, 3, \ldots, 16$. A complete listing of the 880 was first published in 1693; ever since, they have attracted close attention, forming the subject of endless deliberations. The list can be found, for instance, in Benson and Jacoby's *New Recreations With Magic Squares* (Dover, 1976). Now, could it be that one of these traditional gems might prove to be alphamagic too?

Note that this question, natural enough for order 4 (and higher) does not arise with order 3, for which there exists only one normal square—clearly nonalphamagic—the *Lo shu.* But how it is to be answered? At first sight the problem presents no insuperable difficulty since, in the last resort, a program could be written to generate and test every square in turn, a feasible if artless approach. However, the same question will reappear with order 5, for which Richard Shroeppel showed in 1973 that there are exactly 275,305,224 normal squares (again, excluding rotations and reflections), a figure for all practical purposes ruling out the brute-force method, on a personal computer at least. (For further details of Shroeppel's work, see Martin Gardner's excellent account in *Scientific American,* January 1976.) How then are we to determine whether one or more of these is alphamagic?

Surprisingly, an absurdly simple solution is to hand. Taking order 4 to begin with, notice that in any normal alphamagic square the words, *one, two, three, . . ., sixteen* would appear. The total number of letters involved is thus $3 + 3 + 5 + \cdots + 7 = 81$. Hence the magic constant in the logorithm square, the number of letters occurring in all four rows (and all four columns), must equal one-fourth of this total. [Remember, the *logorithm* of a number is the number of letters in its written name.] But 81 is not divisible by 4. Therefore, there are no normal alphamagic squares of order 4!

And what of the higher orders? What, in particular, is the lowest order N to fulfil the necessary (but not yet sufficient) condition

$$\left(\sum_{n=1}^{n=N^2} \log_e n\right) \bmod N = 0?$$

Alas, not one of those 275,305,224 squares of order 5 could be alphamagic. Nor, indeed, will any of the unknown but assuredly astronomical number of order-6 squares answer. The astonishing fact is that we have to go up to order 14 (logorithm square magic constant = 189) before encountering an undisqualified candidate! And there yet remains the little matter of trying to identify an actual 14×14 normal alphamagic square.

The chance of success in seeking for such a monster seems remote in the extreme. Nevertheless, the problem is there and conceivably, closer attention by intrepid programmers may discover means for delimiting the search so as to bring within the scope of practical computer investigation.

The dispiriting result thus arrived at applies only to english squares, of course. Perhaps other languages will admit of lower-order solutions, an unexamined possibility some readers may like to explore—a more encouraging prospect for research, certainly, than the problem proposed above. What language, one wonders, will turn out to provide the lowest-order normal alphamagic square? Alternatively, how about near-normal squares using consecutive numbers, or even just arithmetic series other than $1 - n^2$? Here are nice opportunities for chalking up some exotic "firsts" in computational logology. In any case, besides disposing of a seductive contingency, this digression has furnished a good example of how even seemingly sticky problems in the alphamagic sphere can unexpectedly yield to a cunning mixture of simple arithmetic and logo-logic. We shall now have need of all the cunning we can muster; it is time to turn to the ticklish problems indicated at the outset.

Formulae for Order 4

p	q	r	s
t	u	v	$p+q+r+s$ $-t-u-v$
w	$p+t+w$ $-s-v$	$q+r+2s$ $-t-u-w$	$u+v-w$
$q+r+s$ $-t-w$	$r+2s+v$ $-t-u-w$	$p+t+u+w$ $-r-s-v$	$t+w-s$

Figure 6: General Formula for Order 4

Reverting now to our original course, the obstacles to producing 4×4 alphamagic squares become clear on examining the general formula for magic squares of order-4 (see Figure 6). The method of construction here used, first described by J. Chernick in 1938 (*American Mathematical Monthly,* Volume 45, pages 172–75) and applicable to squares of any order, is simple and easy to follow. Starting at top left and filling in crosswise and downwards, cells in the top row are assigned independent variables p, q, r, and s; the magic constant C thus becoming $(p + q + r + s)$. Now t, u, and v follow in the next row, but if this is to total C, its final cell must contain $(C - t - u - v)$. Similarly, with w entered next, the lowest cell in the left-hand column becomes $(C - p - t - w)$, following which its immediate diagonal neighbor can then be calculated. And so on. At a later stage, a further variable x must be introduced, only to be replaced by a compound expression later. (Try it; one Northumbrian proverb is worth a hundred of the Chinese variety.)

Looking over the formula, a few features common to all 4 × 4 magic squares quickly emerge: the four corner cells, the four center cells, the four inside cells of the outer rows, and the four inside cells of the outer columns all total to the magic constant $(p + q + r + s)$. Beyond these simple observations, however, there is little to be added. With order 3 we were able to point to cell groups comprising arithmetic triples, relying on this characteristic to narrow the area searched, but no such restriction is imposed on numbers appearing in order-4 (or larger) squares. On the contrary, in looking at cells containing single independent variables, we find that as many as half the entries may be numbers selected entirely at random.

It is precisely this freedom, this absence of stricture in the choice of elements, that makes for the greatest difficulty in devising an alphamagic-divining algorithm. For without some further qualification regarding the properties of candidate entries (considered either individually or in relation to each other), the range of possible cases to be examined is simply boundless. And lacking criteria to limit the sweep of our search, is there any more reason to start looking in one direction than in another? Leaving aside number-crunching on a juggernaut scale, I for one have been unable to come up with a workable scheme for a computer program able to sift for solutions systemically, in any way analogous to the successful method evolved for order 3. Of course, others may yet succeed where I have failed.

Defeated, then, in attempting to comb methodically for larger squares in general, as well as in trying to derive an example using the restricted set of numbers 1–16, the problem of how to produce even a single nontrivial 4 × 4 alphamagic square of any kind soon formed the focus of attention. Readers may judge the eventual success of this mission from Figure 7, less a nugget retrieved from the ground than a product of patient alchemy. Appropriately, the Philosopher's Stone or essential catalyst necessary to this synthesis was revealed in a magical formula (see Figure 8).

Greco-Latin-Based 4 × 4 Alphamagic Square

Eighteen (8)	*Twelve* (6)	*Twenty- three* (11)	*Five* (4)
Three (5)	*Twenty-five* (10)	*Nineteen* (8)	*Eleven* (6)
Sixteen (7)	*Thirteen* (8)	*One* (3)	*Twenty-eight* (11)
Twenty-one (9)	*Eight* (5)	*Fifteen* (7)	*Fourteen* (8)

Figure 7: An English alphamagic square of order 4. Entries are transposable to form no fewer than 144 different alphamagics, every one of them exhibiting the 24 constellations here discoverable with magic constants 58;29.

Readers unfamiliar with magic-square material may be unaware that general formulae can appear in a variety of forms. Figure 8, for instance, is algebraically synonymous with Figure 6, the latter-named being a more redundant expression of exactly the same information. In fact, Figure 8, previously unpublished, is an example of what I term a *minimal matrix;* that is, one in which each

$A+a$	$B+b$	$C+c$	D
$C-x$	$D+c$	$A+b$	$B+a+x$
$D+b+x$	$C+a$	B	$A+c-x$
$B+c$	A	$D+a$	$C+b$

Figure 8: A Minimal Formula for Order 4

of the eight necessary independent variables appears no more than four times in the square, the least possible number. (For a similar formula for order 5, see D. E. Knuth and L. Sallows, Problem 1296 in the *Journal of Recreational Mathematics,* Vol 16, No. 2, 1983–84.) (*JRM,* by the way, should not be confused with a rival publication, the sadly maligned *Journal of Rejected Manuscripts.*) The derivation of minimal matrices, incidentally, is a small chapter of magic-square theory in itself; see Figure 9. But how can the minimal formula be of use in creating alphamagic squares?

For an answer to this, examine the placing of a, B, C, and D in Figure 8: a pattern comprising a diagonal latin square. The same is true of a, b, and c, although here the fourth expected identifier d is missing. Notice that no two cells are alike in content. Besides this, two positions are occupied by x and two by $-x$, an arrangement leaving the magic constant $(A+B+C+D+a+b+c)$ everywhere unaffected. Thus, carefully considered, the message contained in the matrix is that every 4×4 (alpha)magic square is decomposable into a diagonal greco-latin square (in which it will turn out that element $d = 0$, and hence need not appear)—distorted slightly, as it were, by a quantity corresponding to the simple zero-totalling pattern of x's. Then, recalling a comparable analysis of the *Li shu,* a *significant* or *interesting* order-4 alphamagic square could only be one in which this predominantly greco-latin substrate had been largely obscured. And properly assimilated, the effect of this insight is to suggest an entirely novel approach to the construction of such squares: calculated exploitation of linguistic accident with a view to *transforming* a trivial, easy-to-construct square.

It is here that pencil and paper can often supplement keyboard and screen, as the process of creation partly involves experimental tinkering and serendipity, a factor notoriously intransigent to algorithmic encapsulation. Detailed elucidation of the technique is therefore to some extent an exercise in rationalized reconstruction. The procedure can be illustrated, though, through a reduplication of Figure 7, itself retraceable to a primary diagonal latin (and thus alphamagic) square, shown as Figure 10.

Here, in effect, we have an instantiation of the general formula in which $A = 8, B = 1, C = 3, D = 5$, while $a = b = c = x = 0$. That the numbers 1, 3, 5, and 8 have not been selected without careful premeditation is seen from the following relations:

$$\log 1 + \log 10 = \log 11$$
$$\log 3 + \log 10 = \log 13$$
$$\log 5 + \log 10 = \log 15$$
$$\log 8 + \log 10 = \log 18$$

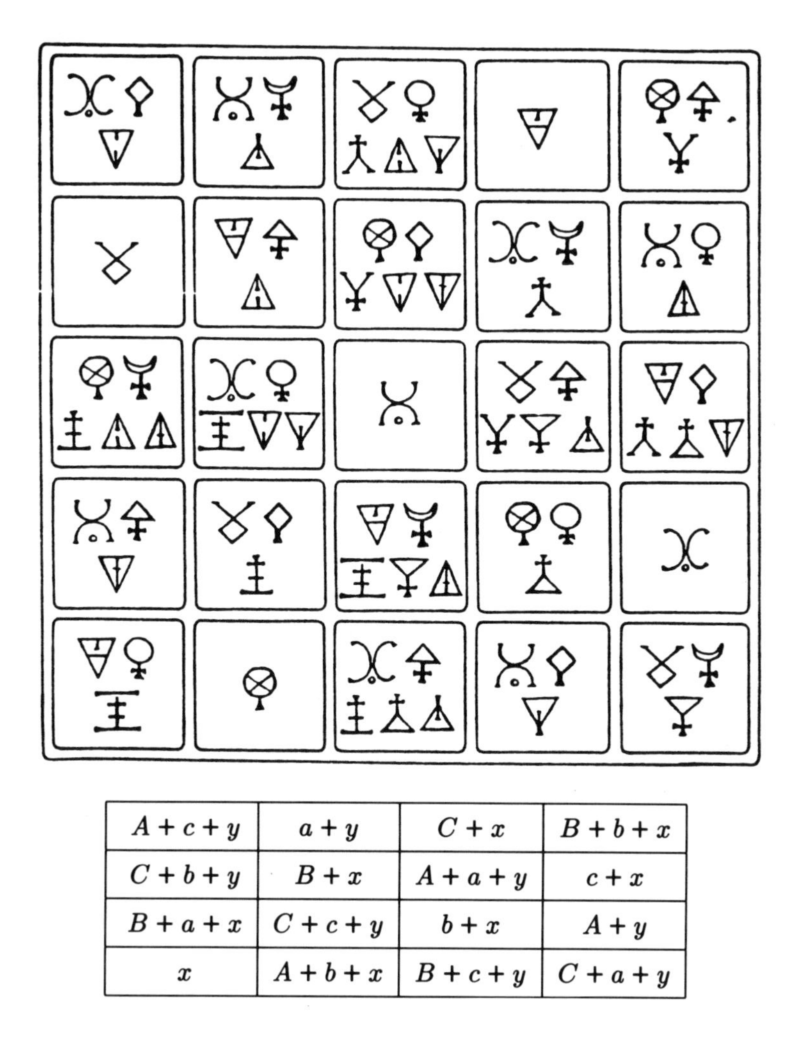

$A+c+y$	$a+y$	$C+x$	$B+b+x$
$C+b+y$	$B+x$	$A+a+y$	$c+x$
$B+a+x$	$C+c+y$	$b+x$	$A+y$
x	$A+b+x$	$B+c+y$	$C+a+y$

Figure 9: Near-Minimal Magic Square Formulae, Old and New

Above: A diagram found in Agrippa von Nettesheim's *De Occulta Philosophia* (Noviomagus edition), published in Lyons, 1533. Cabalistic symbols replace $a, b, c, \ldots$, in representing independent variables. Inverted versions of the same signs correspond to $-a, -b, -c, \ldots$ (compare the lower symbols among the cells). Astrological significance was attached to such magic diagrams. Rewriting the diagram in familiar notation produces what we recognize as an algebraic formula. Further elementary algebra is needed to show that the formula is a universal generalization, embracing all possible 5×5 magic squares. Even so, the number of variable occurrences is 81—12 more than the 69 required in a minimal formula for order 5.

Below: A modern near-miss at a 4×4 minimal formula due to John Horton Conway, a mathematical deity to whom we owe the gift of *Life* (taken from a letter to H. S. M. Coxeter, dated 7 March 1957). The number of variable occurrences in this square is 40—8 more than the 32 used in Figure 8, the minimal formula. To see that at least 32 must appear, note that Chernick's method (Figure 6) proves that 8 *independent* variables are involved. A little thought shows that, however arranged, each of these will have to appear 4 times in the square if it is to be magic.

8	1	3	5
3	5	8	1
5	3	1	8
1	8	5	3

Figure 10: Primary Latin Square

18	1	3	5
3	5	8	11
5	13	1	8
1	8	15	3

Figure 11: First Transform ($a = 10$)

18	12	3	5
3	5	19	11
16	13	1	8
1	8	15	14

Figure 12: Second Transform ($b = 11$)

18	12	23	5
3	25	19	11
16	13	1	28
21	8	15	14

Figure 13: Third Transform ($a = 10$)

12	23	8	15
18	5	21	14
25	19	13	1
3	11	16	28

Figure 14: Alphamagic Reshuffle of Figure 13

Noting that *eleven, thirteen,* ... can be substituted for *ten + one, ten + three,* ... without change to the logorithms, setting $a = 10$ in the formula produces a slightly less trivial alphamagic square, Figure 11. We can do better than this, however, as the linguistic coincidences associated with 1, 3, 5, 8 (arrived at through previous random experimentation) have not yet been exhausted:

$$\begin{aligned}\log 1 + \log 11 &= \log 12 + 3\\ \log 3 + \log 11 &= \log 14 + 3\\ \log 5 + \log 11 &= \log 16 + 3\\ \log 8 + \log 11 &= \log 19 + 3\end{aligned}$$

Here again, setting $b = 11$ in the formula, the constant difference between new entries and old at both numerical and logorithmic levels results in no adverse effects on alphamagic properties, as shown in Figure 12.

Only one more such trick and we shall have a matrix using sixteen distinct numbers. Falling back on an obvious standby, setting $c = 20$, our first nontrivial alphamagic square of order 4 will then be complete (Figure 13).

Since x in the formula is still equal to zero, this final matrix (Figure 7) is in fact patterned on a pure greco-latin square. As a consequence, it enjoys certain extra (alpha)magic properties special to that structure. In particular, the four cells in each quadrant and the four corner cells of each 3×3 subsquare also total to the magic constants 58;29. By happy coincidence, 58 is exactly twice 29. Moreover, cyclic permutations and other transpositions of its elements mean that the 16 cardinals here employed can be rearranged into no less than 144 distinct alphamagic squares (not counting rotations and reflections), every one of them displaying the 24 alphamagic constellations displayed by Figure 7. This ability to reshuffle can sometimes be used to manipulate elements into strategic positions. In Figure 14, for example, 1, 14, 18, and 25 have been maneuvered into the four cells occupied by x's in the general formula.

Thirty-one (9)	*Twenty-three* (11)	*Eight* (5)	*Fifteen* (7)
Seventeen (9)	*Five* (4)	*Twenty-one* (9)	*Thirty-four* (10)
Twenty-six (9)	*Thirty-eight* (11)	*Thirteen* (8)	*Zero* (4)
Three (5)	*Eleven* (6)	*Thirty-five* (10)	*Twenty-eight* (11)

Figure 15: A Non-Greco-Latin-Based Alphamagic Square

Having already established (again, by trial and error) that

$$\begin{aligned}\log 1 &= (\log 0) - 1\\ \log 14 &= (\log 15) + 1\\ \log 18 &= (\log 17) - 1\\ \log 25 &= (\log 26) + 1\end{aligned}$$

setting $x = 1$ leads to a square in which 0, 15, 17, 26 replace 1, 14, 18, 25—a transformation marred, however, by the double occurrence of 15 in the resulting matrix. Patience, though, discovers a way over the difficulty by adding 19 to the number represented by a in the formula, to produce a nontrivial alphamagic square no longer founded on a simple greco-latin square. The appearance of "zero" strikes me as an especially felicitous touch (see Figure 15).

Enough said about this somewhat makeshift method of construction, whose introduction has admittedly been very much a stopgap measure, primarily designed to smooth over an embarrassing semicompletion. Admittedly, systematic computer searches might go a long way toward displacing serendipity, although the factor not to be underrated here is the problem of stipulating exactly what properties are being sought. In any case, the need for a better algorithm becomes highlighted when we acknowledge the impossibility of assigning index numbers to squares formed in this fashion[1] Here we have a problem pleading for solution by computer, and I am curious to learn how better-equipped programmers will rise to the challenge, as doubtless some will. (Can someone produce English 4 × 4 Alphamagic Square No. 1? Remember, it need not be *attractive,* logologically speaking. And can anyone identify the index numbers of Figure 7 and 15?)

In the meantime, the manually-aided transmutation of greco-latin alloy into logological gold is a viable alternative, and I hope some readers will be encouraged to absorb the details above and go on to construct new squares of their own. For any who enjoy a puzzle, as well as the chance of making novel finds, it is an absorbing pastime, in some ways more akin to a skill than a science.

---

[1]Part I described a simple ranking system whereby every alphamagic is associated with a unique index number.

a	$-a-b$	b
$-a-c$	$a+b+c+d$	$-b-d$
c	$-c-d$	d

$-a+d$	$a+b-c-d$	$-b+c$
$a-b+c-d$	0	$-a+b-c+d$
$b-c$	$-a-b+c+d$	$a-d$

$-d$	$c+d$	$-c$
$b+d$	$-a-b-c-d$	$a+c$
$-b$	$a+b$	$-a$

Figure 16: General Formula for Order-3 Magic Cube

And the field is not limited to order 4, or even to alphamagic *squares.* Figure 16 gives a new general formula for a $3 \times 3 \times 3$ magic *cube,* of which even one alphamagic example has yet to be discovered.

Fifty-nine (9)	*Eighty-nine* (10)	*Seventeen* (9)	*Forty-four* (9)	*Sixty-one* (8)
Sixty-seven (10)	*Four* (4)	*One hundred one* (13)	*Fifty-seven* (10)	*Forty-one* (8)
Fifteen (7)	*One hundred seven* (15)	*Fifty-four* (9)	*One* (3)	*Ninety-three* (11)
Eighty-two (9)	*Fifty-one* (8)	*Seven* (5)	*One hundred four* (14)	*Twenty-six* (9)
Forty-seven (10)	*Nineteen* (8)	*Ninety-one* (9)	*Sixty-four* (9)	*Forty-nine* (9)

Figure 17: Concentric Alphamagic Square of Order 5

Lastly, in Figure 17, I beg to present a final specimen of the alphamagic art, the fruit of pensive nights and laborious days: a *concentric* alphamagic square of order 5, in which the outer layer of cells can be peeled off to leave a central alphamagic square of order 3 (No. 6). The formula for such a square is shown in Figure 18; let any who would improve upon this by all means try a hand.

$a+e$	$a+i$	$a-e-f-h-i$	$a+h$	$a+f$
$a+g$	$a+b$	$a-b-c$	$a+c$	$a-g$
$a-e+f-g-i$	$a-b+c$	a	$a+b-c$	$a+e-f+g+i$
$a+j$	$a-c$	$a+b+c$	$a-b$	$a-j$
$a-f$	$a-i$	$a+e+f+h+i$	$a-h$	$a-e$

Figure 18: General Formula for Concentric Alphamagic Square of Order 5

Conclusion

This has been a relatively brief reconnaissance in an unfrequented border country between the Mountains of Mathematics and the Lowlands of Logology, a hitherto unsuspected realm brought to light through *The Origin of Tree Worship.* (For an account of travels in some adjoining regions however, see the last three items on the reference list.) One unanticipated consequence of our alphamagic journey has been to discover how comparatively little is actually demanded of an arrangement of numbers in qualifying as an ordinary or, if you will, *beta*magic square.

Some may feel that here is something that Schroeppel's finding of 275-million-odd normal 5×5 squares should have made plain long since. Possibly so. Notwithstanding, innumerable publications in the field attest to a widespread, irrational susceptibility to traditional magic squares—a seemingly unflagging appetite for the cataloging of new specimens, no matter how inexhaustible in supply, how underwhelming and unworthy of attention they turn out to be on sober assessment. Many contributors avoid the worst excesses of this tendency, it is true, yet a surfeit of exclamation marks is almost a hallmark of the magic-square literature. It would be nice to think that the reference here was to factorials, the notation "$N!$" standing for $1 \times 2 \times 3 \times \cdots \times N$ (a common enough occurrence, as it happens, in formulae relating to enumerations). The truth of the matter, however, is less prodigious, the apparent superfluity resulting only from the too-often-encountered gasp of "... magic!!!!" expressed by authors moved to raptures over yet another find.

The advent of alphamagic squares promises a breath of fresh air in this respect, as their simultaneous compliance with magic requirements at two separate levels, outclasses the familiar prototypes and gives pause for reassessment in the field. And their unimposing, even whimsical exterior casts incidental light on the supposedly *mathematical* nature of ordinary magic squares—a view, I would suggest, as mistaken as it is pervasive.

The widely-held apprehension of magic squares as intrinsically mathematical objects is really a false impression encouraged by the sight of numbers in matrices. The genuine mathematical *problems* involved in their construction and enumeration reinforce this image. But rich as they are in mathematical connotations, the structures themselves—the completed squares—are not merely trivial but actually vacuous in any true mathematical sense: they impart no mathematical information, identify no mathematical relationships, possess no mathematical significance.

More essentially, the fascination they command, the interest they provoke, lies in the intriguing, counterintuitive quality of *co-incidence* they embody. Herein only resides the "magic"—a quality evinced to a greater degree by their alphamagic successors. it is their concomitant satisfaction of independent constraints that calls forth wonder, the role of the numbers as such being less central than first sight supposes: at root, these are simply a vehicle for the expression of the magical effect. Gardner's anagram, "*Eleven + two = twelve + one*," for instance, also exploits numbers to produce a magic constant, but we hardly see that as "mathematical." Likewise, the matrix or square arrangement has nothing to do with real matrices, being only a catchy device for marking off certain subsets.

For all that, one cannot ignore an important aspect of these structures that truly partakes of a mathematical nature: their abstract or Platonic status, independent of empirical reality, reflecting an absolute truth beyond all qualification of time or space. Should it emerge that intelligent creatures on some far-flung planet possess magic squares, we may be sure that theirs will be the same as ours.

Can the same be said of alphamagic squares? To the extent that these include the former, yes. But what of the mutability of number- representations? "Archimedes will be remembered when Aeschylus is forgotten, because languages die and mathematical ideas do not." Such is the opinion of G. H. Hardy, the great English number theorist, in his tender testimonial *A Mathematician's Apology*. Yet without notations, names, languages to remember them, how could there by any mathematical ideas? The duality of sign and signified is ineluctable. Doubtless the alphamagic squares of the Alpha Centaurians, say, will differ a trifle from terrestrial types, even as they remind us that the alphamagic *principle* remains transcendent.

We come thus to the close of this preliminary investigation touched off by the rediscovery of the *Li shu,* itself no mere example of its kind, but the great archetype of alphamagic squares in all the tongues of the globe, as the results of this research have shown. Perhaps we should not be surprised at the English origin of the legend, the alphamagic principle evidently finding its most perfect manifestation in the runes of Anglo-Saxon Northumbria. But whence came the magical formula? For, regretfully, the modern mind must reject the spirit of the yew tree as a primitive if colorful superstition. And yet, if we discount the supernatural agency of the tree spirit, who then was the human author? Alas, no name comes down to us from the veiled centuries of prehistory.

A Druid he was, no doubt, one high in the standing of King Mī, perhaps; a master of abacus and alphabet who commanded leisure for the pursuit of learning in the service of religious ritual and magic. Had this runemaster so chosen, can we doubt that he would have found means to pass on his name to posterity, so cunning a mind, a mixture of Merlin and Mycroft? Mayhap it was natural modesty. Or was he yet one who understood that where there is no mystery there can be no real magic?

References

S. Camman, The magic square of three in old Chinese philosophy and religion, *History of Religions* 1 (Summer 1961) 37–80.

J. Allardyce & M. Sandeford, New evidence for the Survival of Codex 221(b) [MSS. Cotton Catullus B XIV], *Journal of English and Germanic Philology* 87 (in press).

British Library Department of Occidental Manuscripts: Internal Report No. 2704/1729 (second series), 1985.

J. Denes & A. D. Keedwell, *Latin squares and Their Applications,* Academic Press, New York, 1974.

L. Sallows, In quest of a pangram, *Abacus* 2 (3) (Spring 1985) 22–40.

L. Sallows & V. Eijkhout, Co-descriptive strings, *The Mathematical Gazette* 86 (451)(March 1986) 1–10.

Z. Einschwein, *Tractatus Logologico Philosophicus,* Panjandrum Publications, Amsterdam.

Buurmansweg 30
6525 RW Nijmegen
The Netherlands

The Utility of Recreational Mathematics

David Singmaster

1 Introduction

This is written some $2\frac{1}{2}$ years after I gave the talk. Consequently, I will only sketch the points that I covered and I will generally write what I would be saying if I were giving the talk now. Although the title concerns the utility of recreational mathematics, I will devote much of my space to reporting on my Sources in Recreational Mathematics and on my visit to the Strens Collection in January 1985.

2 What is Recreational Mathematics?

To begin, I will try to define recreational mathematics. However, as with mathematics itself, it is not easy to describe in any exact way, e.g., a common way to define mathematics is that it is what mathematicians do. If anything, recreational mathematics must be mathematics that is fun. Unfortunately for the process of definition, but fortunately for mathematics, most mathematicians consider much of their work to be fun. So we must qualify the idea of fun in some way and I think that the correct qualification is that recreational mathematics must be popular. The average person ought to be able to understand the phrasing of a recreational problem and an average educated person should have the necessary knowledge to solve the problem, or at least to understand the solution. This attempt to define recreational mathematics provoked several other speakers to give their own definitions. One speaker said it was what mathematicians do even if they aren't being paid for it. In a report in the *Los Angeles Times*, Lee Dembart noted that most people think of recreational mathematics as a contradiction in terms!

In general, mathematicians are unaware of how popular mathematics is. But one recent phenomenon did point it out—something like 200 million Rubik's Cubes were sold in about three years. The previous classic example of a popular game was Monopoly, which sold about 80 million sets in fifty years. At one time, three of the top ten best selling books in the U.S.A. were books on how to do Rubik's Cube! In Hungary, there were more cubes sold than there are people!

Besides this extraordinary craze, there is a steady market for books on recreational mathematics. Several dozen such books appear every year. The occasion of this meeting is Calgary's acquisition of the Eugène Strens Collection of about 2200 books in the field, which gives some indication of the size of the field. I have also been collecting similar books and have about 3000, with perhaps a 20% overlap with the Strens Collection. [My catalogue is available.]

Many of the early Chinese, Indian and Arabic texts have many recreational problems, and there are texts entirely devoted to recreational problems by Alcuin (9th C) and Abu Kamil (c. 900). I should mention such perennially popular books as: Bachet's *Problèmes* of 1612 and 1624, which has been recently reprinted; Rouse Ball's (& Coxeter's) *Mathematical Recreations & Essays*, which has gone through 13 editions since 1892; Lucas's *Récréations mathématiques*, 4 volumes, 1882-1894, still being reprinted occasionally, and Schubert's *Mathematische Mussestunde*, now in its 13th edition. The works of Loyd, Dudeney and Phillips have been reprinted and edited several times. More mathematically, the work of Courant & Robbins, Hogben, Kasner & Newman, and Constance Reid have sold well.

Many newspapers and magazines carry mathematical puzzle columns, though this may be more popular in Europe than in the United States. Martin Gardner's columns in *Scientific American* deserved special mention as having introduced more people to more mathematics than any other means over the past thirty years. The books based on these columns are extremely popular. While looking up one of Dudeney's columns, I found his 500th column for the *Weekly Dispatch* and he later contributed a monthly column for about 20 years to the *Strand Magazine.* At present in London, the *Weekend Telegraph,* the *Sunday Times* and *New Scientist* carry regular puzzles and they appear in many of the popular computing magazines. I don't know the American scene so well, though J.A.H. Hunter has been running a column in Toronto for many years. As a result of this comment, the *Los Angeles Times* has begun a column conducted by Solomon W. Golomb and myself. Angela Newing and I are now providing half of the puzzles for the *Weekend Telegraph* in London.

3 Sources in Recreational Mathematics

In 1982, I began work on a project to produce a book giving the sources of classical problems in recreational mathematics. As I began to search, I found that there was a vast amount of material and that the sources of many problems are quite uncertain. (Even problems from the 1970's can be difficult to track down!) Consequently, I have shifted my aims toward producing an annotated historical bibliography, somewhat in the nature of Dickson's *History of the Theory of Numbers*. I have put this into a computer file and a first preliminary edition was printed for the Strens Conference. This had about 224 topics on 110 pages, but I tidied up the layout and added page numbers which spread it out to 129 pages. In 1987, I produced a second preliminary edition of 250 topics on 150 pages and in 1988, the third preliminary edition was 290 topics on 192 pages. The fourth preliminary edition of 1989 was 307 topics on 223 pages. The fifth preliminary edition should be completed in February, 1991, with 355 topics on about 310 pages. Copies can be obtained from me. The title has varied slightly — it is now *Sources in Recreational Mathematics — An Annotated Bibliography*. In addition, I have a file of *Queries*, currently 23 pages long. I have given several other presentations on this project which give more details [**4, 5, 6**].

I would be delighted to hear from anyone interested in this project, especially anyone with relevant information.

4 The Strens Collection

In January 1986, I was able to spend a week looking through the Collection, thanks to a grant from the Strens Bequest via Richard Guy. There are about 2200 items and some items are massively multiple — e.g., several boxes of clippings. The material is about equally distributed among Dutch, English, German and French, with a few percent of other languages. All of the standard books are present and a good number of the harder to get books. Below I list a few of the more unusual or rare items.

- Les amusemens mathematiques, 1749.
- W.W. Rouse Ball, translated and extended by J. Fitz-Patrick. Récréations mathématiques et problèmes, ..., three parts, Paris, Librairie Scientifique A. Hermann, 1926.
- Dictionnaire encyclopédique des amusemens des sciences, Paris, Panckoucke, 1792
- Divertissemens innocens, contenant les règles du jeu des échets, du billard, de la paume, ..., La Haye, Adrian Moetjens, 1696.
- Dudeney, 2 books of miscellaneous articles by him, from eight different magazines, apparently bound up by him. Included are two letters to Dudeney from Sir James Murray about the word 'tangram'. These are excerpted in *Amusements in Mathematics*, p. 44.
- H. van Etten (often attributed to J. Leurechon), revised by C. Mydorge, Les récréations mathématiques avec l'examen, ..., 6th ed, 1669.
- E. Hatton, A mathematical manual, London, 1728.
- Prof. Hoffmann, Puzzles old and new, London, Frederick Warne, 1893. And a dozen other books by Hoffmann.
- Edmond Hoyle, Mr. Hoyle's games of whist, quadrille, piquet, chess and backgammon, 12th ed., London, Thomas Osborne, n.d. [1751?]
- Indoor & outdoor games for boys and girls ..., c. 1859.
- John Jackson, Rational amusement for winter evenings, London, 1821. (Dudeney cites this as 'a rare little book I possess' – perhaps this was his copy?)
- C.F. de Jaenisch, Traité des applications de l'analyse mathématique au jeu des échecs, 3 vols., St. Petersbourg, 1862.
- Le jeu de Trictrac, Paris, Theodore Legras, 1757.
- M. Lebrun, Nouveau manuel complet des jeux de calcul et de hasard ou nouvelle académie des jeux, Paris, Encyclopédique de Roret, 1840
- J. Ozanam, Recréations mathématiques et physiques, 4 vols., Paris, Claude Jombert, 1735.
- Petite académie des jeux, ..., 5th ed., Metz, Verronnais, 1835.
- Hermann Schubert, Zwölf Geduldspiele, Berlin, Ferd. Dummlers, 1895. And a dozen other books.
- Ferdinand Winter, Das Spiel der 30 bunten Würfel, Leipzig, B.G. Teubner, 1934.

There are several relevant journals, e.g., there is a set of *Sphinx*, 1931-1939, together with the proceedings of the two Congresses on Recreational Mathematics that Kraitchik organized. The holdings of *Sphinx-Oedipe*, 1906-1926, seem to be more extensive than anything in France. There is a set of *Recreational Mathematics Magazine*, (part xeroxed). Sadly, the material on chess (5000(??) items) was disposed of separately, so the *Fairy Chess Review* is not present. [This has since been acquired – Ed.]

Strens's most obvious special interest was card games, especially Solitaire (U.S.A.) or Patience (U.K.). I have since discovered that this was a very popular pastime at the turn of the century. There is quite a bit on magic squares and on dominoes. There is some magic in general. A startling topic is the inclusion of a number of books with titles *Fun in Bed*. There are a fair number of catalogues of publishers and of magic and toy shops. There is some material on calendars and calendar reform and some material on computation, including tables and calculating machines.

5 The Utility of Recreational Mathematics

Though we may not be able to define recreational mathematics, I claim that it is remarkably useful. I have classified the utility in four categories.

1. Many fields of mathematics have arisen from recreational problems. Probability, graph theory, number theory and topology are the most obvious examples. More widely, many of the early problems for students had a recreational aspect. For example, Old Babylonian problems give a value for the area plus the difference between the length and the width of a rectangle, together with the value of the length plus the width. Obviously this is not a problem that arises in reality! (My thanks to Mogens Esrom Larsen for pointing out these problems.) One can also say that Greek geometry was not motivated by practical consideration, nor was the development of algebra by the Arabs and Italians, though neither subject would be called very recreational. Even the development of calculus had its recreational aspects, notably the study of new curves, such as the cycloid.

There are also many recreational topics which have led to rather specific applications. The Möbius strip and its use in magic and as a conveyor belt is one example. Knots are being applied in molecular biology. Prime factorization is the basis of public-key cryptography. Hyperbolic tessellations and the Penroses' 'Impossible Triangle' have been applied in the art of M.C. Escher. Art is not often recognized as a field for applying mathematics, but there is a great deal of interaction. The Penrose Pieces have created a new field of 'quasi-crystallography' and some 20 alloys are now known to exhibit such structure. It is even conjectured that some of these alloys may be stronger than other forms and so one day this flight of fancy may turn into real flights on real aircraft!

2. Recreational problems have immense pedagogic utility. Some of these problems have fascinated students for nearly 4000 years and should continue to do so indefinitely. I feel that a good problem is worth a thousand routine exercises. Recreational problems are interspersed in texts, from the earliest known examples, for the same reasons as we use them today – to give relaxation, variety and non-routine challenge.

I should add that many, if not all, of the recreational problems are open-ended. Often only the simplest situation has been completely solved and the study of the variations is very suitable for an undergraduate (or even graduate) student project. For example, the minimal number of moves for the Tower of Hanoi with 4 pegs remains unsolved. But I have recently learned about several variants of the 3-peg version and seen extended analyses of the 3-peg version which have revealed more of the structure and properties of this well known puzzle [**2**, **5**, **6**, & discussion with D.E. Knuth]. The mapping between binary and Gray coding is a related question and I have just seen some new questions and results on this [**1**]. A student and I have found a better solution for the river crossing problem with four jealous couples, a two-person boat and an island in the river [**3**]. The three jug problem (e.g. divide 8 pints in half using a 3-pint and a 5-pint measure) is easily solvable when the total amount is equal to the sum of the two smaller measures (provided their sizes are relatively prime), but one cannot divide 4 pints in half using a 4-pint and a 3-pint measure. I do not know which of these problems are solvable.

3. Recreational problems act as historical markers, showing the transmission of mathematics (and culture in general) in time and space. In particular, they illustrate the fact that most of the more algebraic and arithmetic parts of mathematics have their origins in the Orient, beginning with Babylonia and China and being transmitted through India and the Arabs. Hardly any Babylonian recreations are known, but from China we get magic squares, the Chinese rings, the coconuts problem and, possibly, overtaking problems and cistern problems. I would like to point out that this information has only emerged in recent years following the epochal work of Joseph Needham which has inspired a rising generation of historians of Chinese mathematics.
In [**4**], I have discussed this point in more detail.

With the current interest in multicultural aspects of mathematics, recreational problems are natural topics to consider. One must be a little careful with some of these problems, as past cultures were often blatantly sexist or racist, but such problems also show what the culture was like. For example, Bhaskara, in India in 1150, could say 'If a female slave sixteen years of age brings thirty two nishcas, what will one aged twenty cost?' The early forms of the Josephus problem usually involve Turks and Christians, but Turks and Jews or Christians and Jews also occur. Which group gets thrown overboard depends on who's telling the story. Other versions involve clever students and stupid students and a version from Florence about 1500 has two kinds of monks: Franciscans and Camoldensiens! The river crossing problem of the jealous husbands is quite sexist and transforms into masters and servants, which is classist, then into missionaries and cannibals, which is racist. With such problems, you can offend everybody!

4. Recreational mathematics is a channel of communication between the mathematical community and the general public. I have already mentioned this idea in maintaining that recreational mathematics is popular. But this channel is not used as much as it might be. With the increasing pressure on science to be accountable and the rise of pseudo-sciences, it behooves mathematicians (and scientists in general) to communicate their subject to others. The Strens Conference was remarkably successful in bringing together mathematicians from many fields, puzzle collectors, puzzle makers, reporters, and letting them communicate. Indeed, it was hard to get them to go to sleep!

At the Conference Dinner, I suggested that it would be nice to continue these conferences, perhaps every four years. The International Commission on Mathematical Instruction held a conference on the popularization of mathematics at the University of Leeds in September 1989.

Although different in aims, this might be considered a successor to the Strens Conference and I hope these will lead to a regular series of conferences on recreational and popular mathematics.

References

1. Francis Clarke, A theorem about the Gray code, Preprint, 1989.

2. Andreas M. Hinz, Pascal's triangle and the Tower of Hanoi, *Amer. Math. Monthly* **99**(1992) 538–544. (See also [**5** and **6**].)

3. Ian Pressman and David Singmaster, The jealous husbands and the missionaries and cannibals, *Math. Gaz.*, **73**(1989) 73–81. (A more detailed version, together with an extensive bibliography, is available from David Singmaster.)

4. David Singmaster, Some early sources in recreational mathematics, Presented at the Conference on Renaissance Mathematics, Oxford, 1984. In C. Hay, *et al.*, eds., *Mathematics from Manuscript to Print*, Oxford Univ. Press, 1988, 195-208. (The preprint is also available from David Singmaster.)

5. David Singmaster, Report of Topic Area 11: Mathematical Games and Recreations, *Proc. Sixth Internat. Congr. Math. Educ.*, Budapest, 1988, ICMI Secretariat and János Bolyai Mathematical Society, Budapest, 1988, 361-364. (An extended version is available from David Singmaster.)

6. David Singmaster, Old and new recreations in mathematics, *Proc. Conf. Math. Teaching*, 1988, Dept. of Mathematics, Univ. of Edinburgh, 1988, pp. 4–5.

Polytechnic of the South Bank,
London, SE1 0AA, U.K.

The Development of Recreational Mathematics in Bulgaria

Jordan Stoyanov

1 Introduction

In our opinion the following approximate definition could be accepted: recreational mathematics is mathematics dressed in recreational form. However, it is more important to emphasize that role which recreational mathematics plays in human society, in particular in the educational process, and in forming the thinking of people in general. The attractive form of the problems and great enjoyment when solving them successfully are the reasons ensuring the popularity of recreational mathematics all over the world. The wide circulation of books on recreational mathematics and the existence of special columns on this subject in many magazines and newspapers are a confirmation of that.

One of the peculiarities of recreational mathematics is its international character. It seems however, that each nation has some specific contribution to make in creating knowledge of this kind. Complete analysis of this complex process and of the collected material needs much time and effort. For this paper we decided to choose a few problems aiming to give a small but representative picture of recreational mathematics in Bulgaria. In general, there are too many problems. Some of them are relatively elementary, but others are serious enough. In all cases however, if we want to solve the problems, we sometimes have to demonstrate ingenuity and a good knowledge of special facts from mathematics.

In Section 2 we have formulated seven problems based on different ideas. Their solutions are given separately in Section 3. Thus the reader has an opportunity to enjoy himself finding out his own solution. Generalizations and versions of some of the problems from Section 2 are given in Section 4. In particular, this shows that there are recreational problems containing deep ideas which can be further developed. Some specific and general remarks are given in Section 5. Finally, the Reference list includes that minimum number of books on recreational mathematics with the maximum influence on the readers in Bulgaria.

2 Seven specific problems

Problem A. A bottle of "Bulgarian Cabernet" wine costs 1 Lev plus half a bottle of wine (the Lev (pl. Leva) is the official monetary unit in Bulgaria). How much do 3 bottles of that wine cost?

Problem B. A cat and a half eat a mouse and a half in an hour and a half. How many mice will 5 cats eat in 12 hours?

Problem C. To the question, how old his son was, a father answered: "If you subtract my son's trebled years six years ago from his doubled years now, you will obtain his years now." How old is the boy?

Problem D. The manager A and three brothers B, C and D live in a building, each occupying an apartment. It is known that:

(i) the apartment of A has 3 windows and 2 doors;

(ii) B has as many windows as the doors of C;

(iii) C has as many windows as the doors of B.

Besides, D has counted that in total his brothers have the same number of windows and doors. What is the color of the eyes of the manager's wife?

Problem E. Three men went to a barber. When the first man had been shaved, the barber told him: "Look how much money there is in the drawer, add the same amount, and take 2 Leva change." The barber told the same to the second man, and to the third man. When the three were gone, it turned out that there was no money left in the drawer. How much money was there in the drawer before the first man paid?

Problem F. (Old Russian problem) Two brothers have a flock of sheep. They decide to sell them, getting for each as many Roubles as the number of the sheep and then dividing the money equally. The older brother takes 10 Roubles for himself and gives 10 Roubles to the younger. He repeats such a distribution a few times more. In the end, the first brother takes 10 Roubles but it turns out that less than 10 Roubles remain for the younger brother. After a short reflection, the first brother says: "The division must be honest, so take what is left and I am giving you my pocket-knife to even it out." How much does the pocket-knife cost?

Problem G. A gold treasure is buried somewhere on an island. A man finds archives containing the following description: "Go from the stone to the paper-birch counting the steps. At the birch, turn to the right and walk the same distance. Thus you arrive at point M. Now start again from the stone, go to the oak-tree, measuring the steps. Being at the oak, turn to the left and walk the same distance. Thus you are at point N. The golden treasure is buried exactly in the middle between M and N." Reading this story you decide to try your fortune. You reach the island and find easily the birch and the oak. Unfortunately you discover that the stone has disappeared. How are you going to look for the treasure?

3 Solution of the seven problems

Solution of Problem A. It is easy to see that half a bottle of "Bulgarian Cabernet" cost 1 Lev. Thus 1 bottle of that wine costs 2 Leva and hence 3 bottles cost 6 Leva.

Solution of Problem B. Our reasoning comprises a few steps and we follow a rule called the proportionality principle. First, it easily follows that 3 cats eat up 3 mice in an hour and a half. Now doubling the time we conclude that 3 cats eat up 6 mice in 3 hours. This implies that 1 cat eats up 2 mice in 3 hours. Thus we find that in 12 hours, 1 cat eats up 8 mice. Therefore, if there are 5 times as many cats, they will eat up 5 times as many mice, i.e., 40 mice. Hence the answer is 40.

Solution of Problem C. If we denote the son's age by x, we have the relation

$$2x - 3(x - 6) = x$$

Thus $x = 9$, so the son is 9 years old.

Solution of Problem D. The question formulated in this problem is not mathematical. This would give a hint to us that something is wrong here. Indeed, let W_x and D_x denote respectively the number of windows and the number of doors of the person x. Then we have the equalities

$$W_B = D_C, \qquad W_C = D_B \tag{1}$$

However, according to the person D the following relation holds:

$$W_A + W_B + W_C = D_A + D_B + D_C \tag{2}$$

Now, introducing the notations $W_B = x$, $W_C = y$ and taking into account (1), (2) and the fact that $W_A = 3$, $D_A = 2$, we find

$$3 + x + y = 2 + y + x$$

which obviously is impossible. The only possible and reasonable conclusion from this contradiction is that the manager A is not a man, i.e., that A is a woman!

Solution of Problem E. *First solution*. Let x denote the amount of money in the drawer before the arrival of the men. Then, after the first man has paid, there is $x + x - 2$ in the drawer, i.e., $2x - 2$ Leva. Analogously, after the second man, the remainder is $4x - 6$ Leva and after the third man, the remainder is $8x - 14$ Leva. At this moment however, there is no money in the drawer. Hence $8x - 14 = 0$ and $x = 1.75$ Leva. This is the answer.

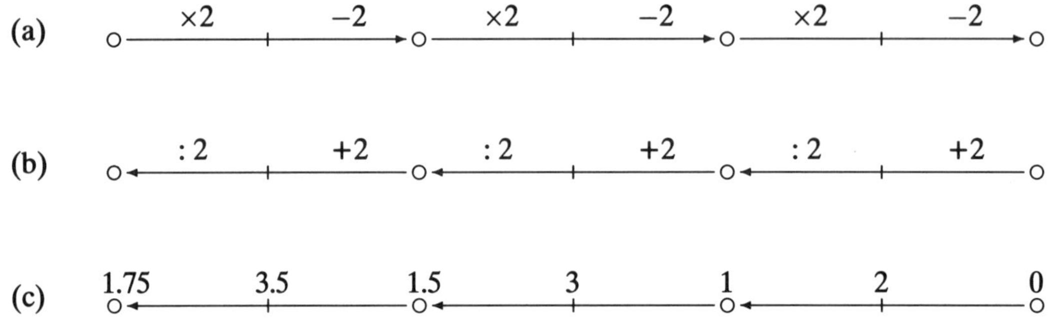

(a) ×2 −2 ×2 −2 ×2 −2

(b) : 2 +2 : 2 +2 : 2 +2

(c) 1.75 3.5 1.5 3 1 2 0

Figure 1:

Second solution. A somewhat different reasoning enables us to find the same answer but without using equations. The idea is clear if we look at Figure 1 where the operations "doubling" and taking 2 Leva are denoted by ×2 and −2, respectively. Case (a) represents the doubling of the amount of money and then taking 2 Leva consecutively by the three men. Case (b) shows the same operations but in the reverse order, i.e., from the natural end to the beginning. The operations "multiplication" and "subtraction" in case (a) change to their dual operations "division" and "addition", respectively. Finally, case (c) contains the answer to the question we are interested in. We start (from the right end) with the value 0 and following step by step the rules described in case (b), we easily find that 1.75 Leva is the content of the drawer at the beginning.

Solution of Problem F. The number of sheep can be written in the form $10a + b$ where a and b are positive integers and $b < 10$. Selling the sheep, the brothers get an amount of money denoted by M. Obviously, $M = (10a + b)^2$ Roubles. Hence $M = 2(5a^2 + ab) \cdot 10 + b^2$ and the first term $M = 2(5a^2 + ab) \cdot 10$ corresponds to an even number of payments, $(5a^2 + ab)$ times to each brother, it is obvious that everything further depends on the value of b^2. We know that in the end, the older brother takes 10 Roubles for himself and there is some remainder. This implies that b^2 contains an odd number of "tens", and that b^2 is an even number. Recalling that $b < 10$ we see that $b^2 = 16$ and $b^2 = 36$ are the only possible values of b^2. In both cases, after getting an equal number of "tens", the first brother takes another 10 Roubles and there is a remainder of 6 Roubles. This remainder is for the second brother, who also gets the pocket-knife from the first. Since this operation leads to an honest division, it follows that the pocket-knife costs exactly half of the difference (10 – 6) Roubles. Hence the pocket-knife costs 2 Roubles.

Solution of Problem G. First, it is not mentioned explicitly, but we always imply that a turn to the right or to the left means that we change the direction of motion by an angle of 90°.

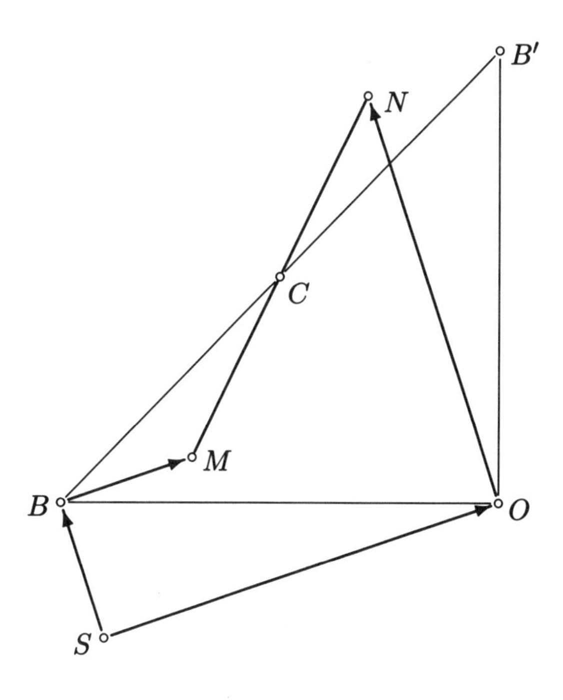

Figure 2:

Our reasoning uses the notation in Figure 2 where the letters S, B and O denote, respectively, the stone, the birch and the oak. We have $SB = BM$ and $SB \perp BM$, and also $SO = ON$ and $SO \perp ON$. It is easy to see that N is the image of M when applying two rotations with centers B and O and such that both are in a clockwise direction and both are under an angle of 90°. Thus, first M goes into S and then S goes into N. However, these two rotations result in one rotation of an angle 180° with respect to some centre, denoted by C. Now we easily see that C is just the middle of MN. Moreover, C does not depend on the position of the stone S, which implies that C can be determined uniquely by the two points B and O. Further, a rotation with respect to O under an angle 90° in a clockwise direction, transforms B into B' and the middle of BB' coincides with C. This is the key to your search. The birch and the oak are the only things you find on the island. However, following the above reasoning, you first find point B' and then the middle of

BB'. Thus you arrive at point C, i.e., the place where the golden treasure is buried. Therefore, fortune sometimes comes, but only if you know a little geometry!

4 Some Generalizations

4.1

Problem E and its two solutions enable us to formulate and solve other similar but more general problems. Here is one possibility.

Problem E1. Suppose that n clients, say $C_1, \ldots, C_n$, went to a barber. The client C_k first increases the money in the drawer p_k times and then takes q_k Leva, $k = 1, \ldots, n$. (In Problem E we have $n = 3$, $p_1 = p_2 = p_3 = 2$ and $q_1 = q_2 = q_3 = 2$.) After the last client C_n paid, the barber finds b Leva in the drawer. How much money was there in the drawer before the client C_1 paid?

The solution of this problem uses the same idea as that in Problem E. Indeed, if a denotes the amount of money before C_1 paid, then we easily derive the relation

$$(\ldots((b - q_n) : p_n - q_{n-1}) : p_{n-1} - \ldots - q_1) : p_1 = a.$$

One interesting particular case is when $p_1 = \ldots = p_n = \alpha$ and $q_1 = \ldots = q_n = \beta$. In this case the parameters a, b, α, β, n are connected by the relation

$$b = \alpha^n a - \beta \frac{\alpha^n - 1}{\alpha - 1}$$

Obviously, several useful conclusions can be deduced from here.

Another version of Problem E1 can be formulated if we change the mechanism of payment: suppose for example, that C_k first takes q_k Leva, and then increases the remaining amount p_k times, $k = 1, \ldots, n$.

Actually, Problems E, E1 and their versions lead to the consideration of numerical sequences defined in a special way. We can study different properties of the members of these sequences as well as to put interesting questions about their limit behavior when the number of clients, n, tends to infinity. A nice question is: what is the chance of the barber becoming a millionaire, if the clients follow a previously prescribed rule for payment?

4.2

Let us further discuss Problem F and its solution. We found that the pocket-knife costs 2 Roubles. However, the following situation also seems natural. Each of the brothers gets an equal number of "tens" and there is a remainder which is less than 10 Roubles. The first brother takes the whole remainder and gives his pocket-knife to the second. In this situation, what is the price of the knife?

The remainder is equal to b^2 and if the pocket-knife costs x Leva, then we have $b^2 - x = x$, implying that $b^2 = 4$ and $x = 2$. Surprisingly, we get the same answer: the pocket-knife costs 2 Roubles.

An interesting generalization of Problem F can be formulated if we change the number 10 to any positive integer m.

Problem F1. Suppose the brothers have n sheep and sell them, getting n Roubles for each. Now, instead of 10, the brothers take m Roubles each and this is repeated until the remainder becomes less than m Roubles. One of the brothers takes the whole remainder and gives his pocket-knife to the other. How much does the pocket-knife cost?

Obviously, there are two cases:

(a) The brothers get the same number of payments each of amount m.

(b) The first brother gets one payment more. In both cases we have to write the number n in the form $n = ma + b$, where a and b are positive integers and $b < m$, and then follow the same ideas as that in Problem F.

One interesting observation arises here and we should like to formulate it as an open problem.

Conjecture. The case $m = 10$ is the only one where the above problem has a unique solution.

5 Concluding remarks

Let us mention that the choice of the above problems was not such an easy task, e.g., two of them were chosen almost randomly among a large set of problems. It will not be surprising to learn that these problems or their versions are also widespread and popular in other countries. A nice illustration is the following. I knew Problem G several years ago (though not being able to identify the source), but in 1985 I learned that this is an "old Ukrainian problem." Surprisingly, almost the same problem was suggested to me by B. Brizeli during my visit in Montreal (October, 1988).

So, there are numerous and very interesting problems. We find them in books, journals, newspapers. In Bulgaria the books on recreational mathematics are always among the bestsellers. The "Matematika" magazine (published in Sofia since 1962) regularly includes an attractive rubric called "Recreational Mathematics". The people in Bulgaria like to ask mathematical riddles. In particular, if you are witty and ingenious and solve an interesting problem, you could drink a cup of nice Greek coffee or be shaved free of charge! Thus, it will be a correct statement to say that the Bulgarians like not only wine, songs and riddles, but they also like recreational mathematics!

In a wider perspective, looking at recreational mathematics round the world, we find much in common and naturally come to the biblical question: "What's new under the sun?". Well, the biblical answer is: "Nothing new under the sun!". But, ... new and nice recreational mathematical problems arise all the time!

References

1. I. Ganchev, K. Chimev & J. Stoyanov, *Mathematical Folklore*, 2nd ed., Narodna Prosveta, Sofia, 1987. (In Bulgarian; Russian Translation, Znanie, Moscow, 1987; Japanese Translation, Tokyo Tosho Publishing Co., Tokyo, 1990).

2. M. Gardner, *Mathematical Puzzles and Diversions*, volumes 1–3, Nauka i Izkustvo, Sofia, 1975–1980. (In Bulgarian; Translated from English.)

3. B. Kordemski, *Mathematical Quickness*, Narodna Prosveta, Sofia, 1964. (In Bulgarian; Translated from Russian.)

4. Ya. Perelman, *Living Mathematics*, Narodna Prosveta, Sofia, 1963. (In Bulgarian; Translated from Russian.)

5. H. Steinhaus, *Mathematical Kaleidoscope*, Technika, Sofia, 1974. (In Bulgarian; Translated from Polish.)

Institute of Mathematics,
Bulgarian Academy of Sciences,
BG-1090 Sofia, P.O. Box 373, Bulgaria

$V - E + F = 2$

Herbert Taylor

The sole purpose of this note is to describe a proof that $V - E + F = 2$ whenever conditions (1), (2), (3), (4), and (5) are satisfied. Considering a drawing of a graph on a surface we want to distinguish vertices, edges, and faces. Each vertex should be represented by a sizable dot. Let V denote the number of vertices. Edges should be simple paths which do not cross or touch each other. Let E denote the number of edges. Two points on the surface are in the same face if it is possible to travel on the surface from one point to the other without hitting any edge or vertex. Let F denote the number of faces.

Figure 1 shows an example of the kind of drawing under consideration.

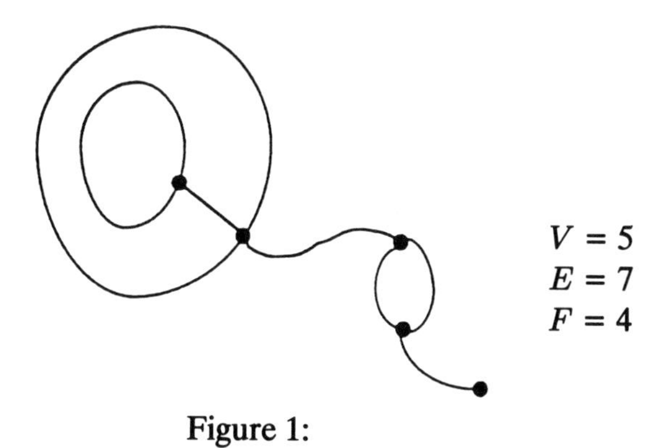

Figure 1:

It is worth special notice that conditions (1), (2), and (3) apply to the drawing, whereas conditions (4) and (5) refer to the surface on which the graph is drawn.

The Five Conditions

1. $V > 0$.
2. Every edge has two ends, and there is a vertex at each end.
3. Any two vertices are connected by a path consisting of edges and vertices.
4. If no edges are drawn the surface constitutes a single face.
5. Any edge that could be drawn with both ends on the same vertex would separate two distinct faces.

Theorem 1 *If conditions (1), (2), (3), (4), and (5) are satisfied, then $V - E + F = 2$.*

Proof. Starting on any surface for which conditions (4) and (5) hold, we may proceed by induction on E as follows.

First supposing $E = 0$, we use conditions (1), (3), and (4) to see that $V = 1$ and $F = 1$, and therefore $1 - 0 + 1 = 2 = V - E + F$.

Next, supposing $E > 0$ for a particular graph with parameters V, E, and F satisfying (1), (2), and (3), we make the inductive assumption that the theorem is true for any graph with $E-1$ edges. Choose any edge. Recalling (2) we may need to consider two cases.

Case I: The chosen edge has two different vertices at its two ends. In this case we shrink the edge to nothing and merge the two vertices into one. The reduced graph resulting from this operation still has F faces, but $V - 1$ vertices and $E - 1$ edges, while conditions (1), (2), and (3) still hold.

The truth of the theorem for this smaller graph tells us that $(V - 1) - (E - 1) + F = 2$ and therefore $V - E + F = 2$.

Figure 2: Illustration for Case I

Case II: The chosen edge has the same vertex at both ends. In this case we erase the edge. Because of condition (5) two faces become one as a result of this operation. Thus the reduced graph still has V vertices, but $F - 1$ faces and $E - 1$ edges, while conditions (1), (2), and (3) still hold. The truth of the theorem for this smaller graph tells us that $V - (E - 1) + (F - 1) = 2$ and therefore $V - E + F = 2$.

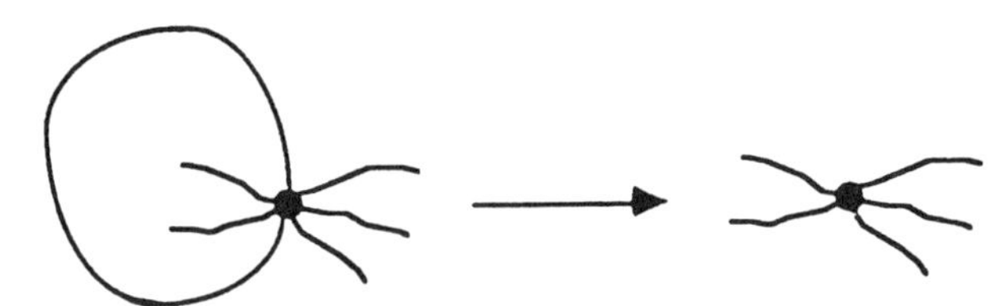

Figure 3: Illustration for Case II

The induction is completed by reasoning that since the theorem is true for $E = 0$, it must be true for $E = 1$, and therefore true for $E = 2$, and so on, that therefore it must be true when E is any positive integer. Q.E.D.

In conclusion it may be worth observing that this theorem incudes Euler's polyhedral formula for the sphere, or the plane, but covers a larger class of graphs. And other surfaces, for example a band of paper with a full twist, which could not be flattened on the plane, would nevertheless satisfy (4) and (5).

Center for Communications Research
4350 Executive Drive, Suite 135
San Diego, CA 92121-2116

Tracking Titanics

The late Samuel Yates

1 Introduction

There are 879 known primes, at the time that this is written, that have one thousand or more digits in their ordinary decimal form. There will undoubtedly be more such numbers when this is read, because hardly a month goes by when one or more of these **titanic primes**, as they are called, is not found [**13** and **8**, p. 120]. When the first table of titanics appeared in 1982, only 56 had been reported [**11**, p. 87]. As their number grows, we continue to be amazed at and filled with admiration for the achievements of those people who use their mental and physical resources so marvelously [**12**]. [The author told us that on 90-11-01 there were more than 1670 known titanic primes —Eds.]

In order to list titanics compactly, we use a set of special designations. Extracts from the first and last of the 17 pages of the current (Feb. 1989) table of titanics are displayed below and indicate how the tabulation is organized.

In general, the task of proving that a large number is composite — if it is composite — is not so difficult as proving primality. The "converse" of Fermat's "little" theorem is used.

Fermat's theorem states that if N is prime and does not divide b then N divides $A = b^{N-1} - 1$. If N does not divide A for some base b less than N, then N is composite. It is possible for a composite N to divide A for some bases and not for others. If the only bases smaller than N for which N fails to divide A are the divisors of N, then N is called a **Carmichael number**. When N divides A, further tests are needed to confirm or deny primality [**1**].

2 Single-Variable Functions

$R(N)$ and $U(P)$ are the only single-variable functions that we use. $R(N)$ is a **repunit** designation, standing for a string of N ones, $(10^N - 1)/9$. The only known prime repunits are $R(2)$, $R(19)$, $R(23)$, $R(317)$, and $R(1031)$. All other repunits less than $R(14500)$ [now $R(16000)$] have been shown by Dubner to be composite.

$U(P)$ is the product of all odd primes from 3 up to P. The largest known prime that is 1 more than a product of consecutive primes is the 5862-digit $2 \times U(13649) + 1$. Dubner has listed all of the 15 known primes of this form, and has found no others less than $2 \times U(17159) + 1$. [In 1989 he found the 8002-digit prime $2 \times U(18523) + 1$.] There are no known titanics of the form $2 \times U(P) - 1$, although testing up to $2 \times U(16699) - 1$, he reported eleven smaller primes and seven probable primes. The titanic table also lists **factorial primes**. There are 17 factorial primes of the form $N! + 1$ less than $2662! + 1$. The largest is the 4042-digit $1477! + 1$. There are 15 known factorial primes of the form $N! - 1$ less than $2063! - 1$. The largest is the 1051-digit $469! - 1$, and there are three larger probable primes of this form [**3**].

3 $A(K, N) = K \times 2^N + 1$

Most of the listed titanics are functions of two variables. The smallest recorded titanic of this form is the 1000-digit $A(25, 3314)$ of Cormack and Williams. The largest known non-Mersenne prime is 18029-digit $A(8423, 59877)$ of Buell and Young. [In 1989 the "Amdahl Six" (John Brown, Landon Curt Noll, Bodo Parady, Gene Smith, Joel Smith & Sergio Zarantonello) found five non-Mersenne primes larger than this. The largest is the 65087-digit prime $B(391581, 216193)$ (see next section for notation). It is presently (Nov. 1990) the largest known prime, having 37 more digits than the largest known Mersenne prime.]

In this form, an odd value of K indicates that N is the largest power of 2 in the expression. Where it is possible to get K to be the same as N by subtracting a from N and multiplying K by 2^a, the resulting $A(N, N)$ is called a **Cullen prime**. The largest such Cullen prime is the 5573-digit $A(18496, 18496)$ which, when written with an odd K is $A(289, 18502)$. Keller, who produced this and other very large Cullen primes, determined that there are no larger ones with N less than 20000 [**8**, p. 283].

$A(1, N)$ is composite whenever N is not a power of 2. It is called a **Fermat number**, Fm, when $N = 2^m$. The only known Fermat primes are $F0 = 3$, $F1 = 5$, $F2 = 17$, $F3 = 257$, and $F4 = 65537$. Every prime factor of a composite Fm is known to be of the form $A(K, m + 2)$. For instance, $F5 = 641 \times 6700417$, where $641 = A(5, 7)$ and $6700417 = A(52347, 7)$. The largest known prime Fermat factor is Keller's 7067-digit $A(5, 23473)$, which divides the Fermat number $F23471$. The size of this Fermat number is not readily conceivable: as Ribenboim points out [**8**, p. 74], the tiny $F37$ has more than thirty trillion digits!

4 $B(K, N) = K \times 2^N - 1$

When $A(K, N)$ and $B(K, N)$ have the same arguments and are both prime, they give examples of **twin primes** because their difference is 2. The largest known twins of these forms are 2003-digit numbers

$$A(520995090, 6624) \quad \text{and} \quad B(520995090, 6624)$$

found by Atkin and Rickert. [In 1989, the "Amdahl Six" found larger twin primes of the same forms, the 3389-digit primes $A(1706595, 11235)$ and $B(1706595, 11235)$.]

$B(1, N)$ can be prime only if N is prime. It is then known as a **Mersenne prime**. According to Colquitt and Walsh, who found $B(1, 110503)$, the 29th Mersenne prime in order of size, all Mersenne primes less in size than the 30th Mersenne prime $B(1, 132049)$ have been found. Slowinski discovered the 30th as well as the 65050-digit Mersenne $B(1, 216091)$ which is the largest that is presently known, but it has not yet been determined if there are others between these two.

Keller has extended the designation of Cullen primes to include $B(N, N)$, as well as $A(N, N)$. The largest presently known is his 5690-digit $B(18885, 18885)$.

5 Sophie Germain Primes

Sophie Germain, a contemporary of Gauss, showed [**8**, p. 261] that if p and $q = 2p + 1$ are both prime, there are no integers x, y, z, other than zero and multiples of p, such that $x^p + y^p = z^p$. Any reader who is familiar with the well-publicized efforts to prove the famous conjecture known as "Fermat's Last Theorem" can appreciate the contribution that she made in eliminating a large set of candidates for its possible contradiction.

The prime p is usually called a **Sophie Germain prime** when $q = 2p + 1$ is also prime but for convenience we refer to both p and q as Sophie Germains. It is clear that if p is $B(N, K)$, then q is $B(N, K + 1)$. The largest known Sophie Germain pair is Keller's 1812-digit $B(39051, 6001)$ and $B(39051, 6002)$. [In 1990 Dubner found the larger pair,

$$G(713851138, 1854) \quad \text{and} \quad G(1427702276, 1854) + 2$$

with 1863 and 1864 digits—see section 6 for notation.]

It is known that if p and q are Sophie Germains, and p is congruent to 3 (mod 4), then q is a divisor of the Mersenne number $B(1, p)$. The converse is also true. This is a convenient way to show that a particular Mersenne number is not prime. For example, 251 and 503 are a pair of Sophie Germain primes, with 251 congruent to 3 (mod 4). Then the Mersenne number $B(1, 251)$ is divisible by 503. Similarly, the large Sophie Germain prime just mentioned, $B(39051, 6002)$, divides the Mersenne number $B(1, B(39051, 6001))$.

Sophie Germain primes p and q have merited attention with respect to repunits and repetends [**11**, pp. 115–117]. If $q \equiv 3$, 27 or 39 mod40, then q is a **primitive divisor** of repunit $R(p)$, and the repetend length of q is p. For example, 7841 and 15683 are a Sophie Germain pair, and $15683 \equiv 3 \bmod 40$. Then 15683 divides $R(7841)$, but it divides no smaller repunit; and the number of digits in the repetend of any proper fraction $a/15683$ is 7841 [**10**].

If $q \equiv 7$, 19 or 23 mod 40, then q is a primitive divisor of $R(q - 1)$, so that q is a full period prime with period length $q - 1$. For instance, 99023 and 198047 are a Sophie Germain pair, and $198047 \equiv 7 \bmod 40$. The full period prime 198047 divides $R(198046)$, but it divides no smaller repunit, and the number of digits in the repetend of any proper fraction $a/198047$ is 198046. Every repetend is a cyclic permutation of each of the other 198045 repetends.

If we return to the titanic Sophie Germain primes mentioned above, we see that $B(39051, 6002)$ is congruent to 23 (mod 40). It therefore is also a full period prime, producing $39051 \times 2^{6002} - 2$ distinct repetends that are all members of one cyclic permutation set. Full period primes have only one cyclic permutation set. An example is 7, where the decimal equivalents of the six proper fractions with denominator 7 are cyclic permutations of $.\overline{142857}$ [**11**, pp. 34–55].

Some congruence subclasses are heavily laden with Sophie Germain primes. A subclass of primes congruent to 23 (mod 40) is that which contains the primes that are congruent to 23 (mod 120). Among the 556 primes less than 200000 which are congruent to 23 (mod 120), there are 533 full period primes, mostly Sophie Germains each of which is 1 more than twice a prime.

There exist sequences or chains of Sophie Germain primes in which each prime is 1 greater than twice the one before it. Examples of such n-tuples or ordered sets are

$$(2, 5, 11, 23, 47) \quad \text{and} \quad (89, 179, 359, 719, 1439, 2879).$$

We have shown that no such sequence can be infinite, and that the maximum possible length of a sequence whose initial prime is an odd a is $a-1$ terms. There is only one such sequence with as many as eight terms when the initial prime is less than 20000000 [**11**, p. 116]. Löh has found larger chains, including several with 12 terms and with initial values being 12 to 15 digits long [**7**].

6 $G(K,N) = K \times 10^N + 1$

The largest known prime of this form is Dubner's 8006-digit $G(150093, 8000)$ which ends with 7999 zeros and a one.

$G(K,N)-2$ is written as $K-1$ followed by N nines. Dubner's 4333-digit $G(6,4332)-2$ consists of a five followed by 4332 nines. A titanic with no even digits and with all odd digits appearing at least once is his 3825-digit $G(1358,3821)-2$. The largest known prime written as a 1 followed only by nines is Williams's 3021-digit $G(2,3020)-2$. Dubner's $G(2 \times R(1439), 1440)$ is written as 1439 twos and 1439 zeros followed by a one. The largest known twin primes are Dubner's 2259-digit

$$G(107570463, 2250) \quad \text{and} \quad G(107570463, 2250) - 2.$$

A titanic written with only ones and zeros is his 2125-digit $G(R(820), 1305)$. [In 1989 Dubner also found the 2309-digit pair

$$G(75188117004, 2298) \quad \text{and} \quad G(75188117004, 2298) - 2,$$

and see the note added in section 4.]

It probably goes without saying that the prime instigator of many of Dubner's rare forms is that well-known prime mover, prime collector, and prime minister, Rudolf Ondrejka. Typical of his involvement is his urging Dubner to seek out **pandigits**, numbers in which a sequence of decimally ordered digits stands out in interesting ways. As an example, the fraction $R(mn)/R(m)$ produces a set of n ones, each consecutive pair of which is separated by $m-1$ zeros. If $R(mn)/R(m)$ is multiplied by an m-digit string, the result is n consecutive m-digit strings. For instance, $R(12)/R(3)$ is 1001001001, which when multiplied by 157 is 157157157157. This device is used several times in the titanic table. For instance, the 3284-digit pandigit

$$G(123456789 \times R(999)/R(9), 2285)$$

is written as 111 consecutive strings of 123456789 followed by 2284 zeros and a one.

Another kind of pandigital prime is of the form $Q(K,N)$ whose digits are a sequence of K ones, K twos, K threes, K fours, K fives, K sixes, K sevens, K eights and K nines followed by $N-1$ zeros and a one. The largest known prime of this type is Dubner's 2917-digit $Q(111, 1918)$.

7 $T(K,N) = K^N + 1$

It was noted above that a Fermat number $A(1,N)$ is 1 more than the Nth power of 2, where N itself is a power of 2. $A(1,N)$ cannot be prime unless N is a power of 2. Every prime factor of a Fermat number $A(1,N)$, where $N = 2^m$, must be of a form $A(K, m+2)$. Dubner has investigated "generalized Fermat primes" [4], denoted in our list by $T(K,N)$, each of which is 1 more than the N-th power of K, where N is a power of 2. As with regular Fermat numbers, such a number cannot be prime unless $N = 2^m$. Also, any factor of $T(K,N)$ must be of the form $A(J, m+1)$. For instance, the generalized Fermat number $T(6,8) = 1679617$ has two factors, $17 = A(1,4)$ and $98801 = A(6175,4)$.

The largest known generalized Fermat prime is Dubner's 8556-digit $T(15048, 2048)$.

8 Palindromes and Reversibles

Let K be a number that has k digits. If K is preceded and followed by $M-1$ zeros and a one, then the resulting number, designated $S(N,K,M)$, where $N = M+k-1$, is a palindrome if K is a palindrome. Dubner's 5031-digit $S(2520, 11101310111, 2510)$ is a titanic prime palindrome. To intrigue numerologists and cabalists, he found other prime palindromes with 666, the number of the Beast [5], in the centre. The largest that he discovered is the 5251-digit $S(2626, 666, 2624)$. His largest pandigital palindrome is the 4955-digit

$$S(2484, 976543282345679, 2470).$$

Another different kind of palindrome is his 3159-digit $9 \times R(3159) - 8 \times 10^{1579}$, which is two strings of 1579 nines each, with a one between them.

If K is not a palindrome, $S(N,K,M)$ is used for reversible primes, pairs of numbers that are not palindromes but that are primes when read from left to right as well as when read from right to left. One such titanic pair is Dubner's 1015-digit

$$S(510, 2888601, 504) \quad \text{and} \quad S(510, 1068882, 504).$$

9 Largest Smith Number

A Smith number is a composite number the sum of whose digits is equal to the sum of the digits of its prime factors. For instance, Dr. Smith's phone number 4937775 is a Smith number because the sum of its digits, 42, is the same as the sum of the digits of its prime factors which are 3, 5, 5, and 65837. Without resort to a large computer, we are able to construct and to confirm the largest known Smith number, one which has 10694985 digits, by using two titanic primes, a repunit and a palindrome, as the kernels [9]. The large Smith number may be written as

$$9 \times R(1031) \times (G((S(2297, 3, 2297))^{1476}, 3913210) - 1).$$

[Dubner's 1990 discovery of the large palindromic prime $S(3286, 3, 3286)$ enabled the author to modify this to give a 13614513-digit Smith number.]

10 Carmichael Numbers

It was mentioned earlier that a Carmichael number is a composite number that satisfies the "converse" of Fermat's theorem for all smaller numbers except its divisors. A Carmichael number must have at least three prime divisors. The smallest Carmichael number is 561. Several approaches have been taken to generate large Carmichael numbers. Dubner has devised a method [2] which culminates in obtaining a Carmichael as the product of three inter-related prime numbers called Carmichael parameters, denoted by $V(P,K,N)$, $W(P,K,N)$ and $Y(P,K,N,M)$ on the titanic function list. His largest Carmichael number has 4407 digits. It is obtained using $P = 47$, $K = 10233053$, $N = 45$ and $M = 12441$. Löh has succeeded in finding much larger Carmichaels, the largest having more than 2300 smaller prime factors and over 15000 digits, but Dubner's are the largest known Carmichaels with only three prime factors [6].

11 Concluding Remark

Tracking titanic primes has allowed this grateful viewer to cast his eyes and mind on a picturesque landscape drawn by a variety of number theoretical artists with great talent, vivid imaginations, and a plethora of paints and brushes.

REFERENCES

1. J. Brillhart, D.H. Lehmer, J.L. Selfridge, B. Tuckerman & S.S. Wagstaff, *Factorizations of $b^n \pm 1$, $b = 2, 3, 5, 6, 7, 10, 11, 12$ up to high powers, Contemporary Math.* **22**, Amer. Math. Soc., Providence, 2nd edition, 1988.

2. H. Dubner, A new method for producing large Carmichael numbers, *Math. Comp.*, **53**(1989) 411–414; MR **89m**:11013.

3. H. Dubner, Factorial and primorial primes, *J. Recreational Math.*, **19**(1987), 197–203.

4. H. Dubner, Generalized Fermat primes, *J. Recreational Math.*, **18**(1985–86) 279–280.

5. M. Keith, The number 666, *J. Recreational Math.*, **15**(1982–83) 85–87.

6. W. Keller, Letter to S. Yates, 88-09-21.

7. G. Löh, Long chains of nearly doubled primes, *Math. Comp.*, **53**(1989) 751–759; MR **90e**:11015.

8. P. Ribenboim, *The Book of Prime Number Records,* Springer, New York, 1988.

9. S. Yates, Digital sum sets, in Number Theory (Proc. 1st Canad. Number Theory Assn. Conf., Banff, 1988), de Gruyter, New York, 1990, pp. 627–634.

10. S. Yates, *Prime Period Lengths,* New Jersey, 1975.

11. S. Yates, *Repunits and Repetends,* Florida, 1982.

12. S. Yates, Sinkers of the Titanics, *J. Recreational Math.*, **17**(1984–85) 268–274.

13. S. Yates, Titanic primes, *J. Recreational Math.*, **14**(1983–84) 250–262.

KNOWN PRIMES WITH 1000 OR MORE DIGITS

Compiled by Samuel Yates, February 1989

$A(K,N) = K \times 2^N + 1$, $B(K,N) = K \times 2^N - 1$, $G(K,N) = K \times 10^N + 1$
$R(N) = (10^N - 1)/9$, $S(N,K,M) = (10^N + K) \times 10^M + 1$, $T(K,N) = K^N + 1$
$U(P) = 3 \times 5 \times 7 \times 11 \times \ldots \times P$, $W(P,K,N) = 1.5 \times (U(P) \times K - 1)^N + 1$
$V(P,K,N) = 2 \times W(P,K,N) - 1$, $Y(P,K,N,M) = (V(P,K,N) \times W(P,K,N) - 1)/M + 1$

	Prime	Digits	Discoverer	Year	SpecialType
1	$B(1,216091)$	65050	S	1985	Mersenne
2	$B(1,132049)$	39751	S	1983	Mersenne
3	$B(1,110503)$	33265	WCW	1988	Mersenne
4	$B(1,86243)$	25962	S	1982	Mersenne
5	$A(8423,59877)$	18029	BY	1988	
6	$A(8423,55157)$	16608	BY	1988	
7	$A(7,54486)$	16403	Y	1987	
8	$B(1,44497)$	13395	SN	1979	Mersenne
9	$A(32161,43796)$	13189	BY	1988	
10	$A(77899,43194)$	13008	BY	1988	
11	$A(16817,42155)$	12695	BY	1988	
12	$A(14027,40639)$	12238	BY	1988	
13	$A(36983,38573)$	11617	BY	1987	
14	$A(53941,36944)$	11126	BY	1987	
15	$A(3061,33288)$	10025	BY	1987	
16	$A(75841,31220)$	9404	BY	1987	
17	$A(69107,29175)$	8788	BY	1987	
18	$A(71671,28884)$	8700	BY	1987	
19	$T(15048,2048)$	8556	D	1987	Generalized Fermat
20	$T(11272,2048)$	8299	D	1986	Generalized Fermat
21	$A(71417,26807)$	8075	BY	1987	
22	$G(150093,8000)$	8006	D	1986	
23	$A(39079,26506)$	7984	K	1985	
24	$A(64007,26015)$	7837	BY	1987	
25	$A(7651,25368)$	7641	BY	1987	
26	$A(67193,24297)$	7319	BY	1987	
27	$G(217833,7150)$	7156	D	1985	
28	$G(6006,7090)$	7094	D	1985	
29	$A(5,23473)$	7067	K	1984	Fermat F23471 factor
30	$B(1,23209)$	6987	N	1979	Mersenne
31	$A(5897,22619)$	6813	K	1985	
32	$A(78181,22024)$	6635	BY	1987	
33	$B(1,21701)$	6533	NN	1978	Mersenne
34	$A(18107,21279)$	6410	BY	1987	
35	$A(3,20909)$	6295	K	1985	
36	$T(1000174,1024)$	6145	D	1986	Generalized Fermat
37	$B(1,19937)$	6002	T	1971	Mersenne

...

...					
850	$G(3, 1020)$	1021	D	1984	
851	$G(6, 1019)$	1020	D	1984	
852	$A(69, 3379)$	1020	K	1983	
853	$A(141, 3375)$	1019	K	1983	
854	$A(77521, 3360)$	1017	K	1983	
855	$A(34565, 3361)$	1017	J	1982	
856	$S(509, 666, 507)$	1017	D	1987	Beastly Palindrome
857	$S(510, 2888601, 504)$	1015	D	1988	Reversible
858	$S(510, 1068882, 504)$	1015	D	1988	Reversible
859	$A(9, 3354)$	1011	CW	1979	
860	$S(510, 30555955503, 500)$	1011	D	1985	Palindrome
861	$S(510, 30555855503, 500)$	1011	D	1985	Palindrome
862	$S(506, 171, 504)$	1011	D	1987	Palindrome
863	$G(6, 1009)$	1010	D	1984	
864	$A(77521, 3336)$	1010	J	1982	
865	$B(9, 3349)$	1010	BB	1980	
866	$A(217, 3344)$	1009	SU	1983	
867	$A(209, 3343)$	1009	SU	1984	
868	$S(512, 23456789598765432, 496)$	1009	D	1987	Pandigital Palindrome
869	$S(510, 234567890629590, 496)$	1007	D	1988	Pandigital Reversible
870	$S(510, 095926098765432, 496)$	1007	D	1988	Pandigital Reversible
871	$G(470796, 1000)$	1006	D	1987	
872	$G(235398, 1000)$	1006	D	1987	
873	$A(105, 3331)$	1005	K	1983	
874	$A(61, 3328)$	1004	K	1983	
875	$(2^{1666} + 2^{834} - 1)^2 + 12769 \times 2^{1668}$	1004	SU	1989	
876	$A(27, 3322)$	1002	CW	1979	
877	$(2^{1662} + 2^{832} - 1)^2 + 9025 \times 2^{1666}$	1001	SU	1989	
878	$S(502, 81918, 498)$	1001	D	1988	Palindrome
879	$A(25, 3314)$	1000	CW	1979	

KEY TO DISCOVERERS

BB	Walter Borho & Jurgen Buhl	NN	Landon Curt Noll & Laura A. Nickel
BY	Duncan A. Buell & Jeffrey Young	S	David Slowinski
CW	G.V. Cormack & Hugh C. Williams	SN	David Slowinski & Harry L. Nelson
D	Harvey Dubner	SU	Hiromi Suyama
J	Gerhard Jaeschke	T	Bryant Tuckerman
K	Wilfrid Keller	WCW	Walter N. Colquitt & Luther Welsh
N	Landon Curt Noll	Y	Jeffrey Young

The Eugène Strens Memorial Conference
List of Participants

Vaughan Aandahl
1228 Jasmine Street
Denver, CO 80220

Karen Balcombe
406 5204 Dalton Drive NW
Calgary, Alberta T3A 3H1

Leon Bankoff
471 Rodeo Drive
Beverly Hills, CA 90212

Elwyn Berlekamp
University of California
Berkeley, CA 94720

Stanley Bezuszka
Department of Mathematics
Boston College
Chestnut Hill, MA 02167

Tibor Bisztriczky
Department of Mathematics and Statistics
The University of Calgary
Calgary, Alberta T2N 1N4

Lawrence S. Braden
1005 583 Kamoku Street
Honolulu, HI 95826

David Bruce
7760 SW Miner Way
Portland, OR 97225

Simcha Brudno
204 318 South Throop Street
Chicago, IL 60607

Sonia Cantu
526 West Surf
Chicago, IL 60607

W. W. Chernoff
Department of Mathematics and Statistics
University of New Brunswick
Fredericton, New Brunswick E3B 5A3

Stewart T. Coffin
79 Old Sudbury Road
Lincoln, MA 01773

Curtis N. Cooper
Department of Mathematics and Computer Science
Central Missouri State University
Warrensburg, MO 64093-5045

Prof. Emer. H. S. M. Coxeter*
Department of Mathematics
University of Toronto
Toronto, Ontario M5S 1A1

Donald W. Crowe
Mathematics Department
University of Wisconsin
Madison, WI 53706-1380

Lee Dembart
290 Lombard Street
San Francisco, CA 94133-2402

Dr. M. Anne Dow
MIU
Box 1127
Fairfield, IA 52557-1380

Andrejs Dunkels
Department of Mathematics
University of Lulea
S-951 87 Lulea SWEDEN

*People who could not attend the conference but are eager to correspond.

Roger B. Eggleton
Department of Mathematics
Illinois State University
Normal, IL 61761

Jack A. Eidswick
Department of Mathematical Sciences
University of Montana
Missoula, MT 59812-1032

Douglas A. Engel
2935 W. Chenango
Englewood, CO 80110

Ernest G. Enns
Department of Mathematics and Statistics
The University of Calgary
Calgary, Alberta T2N 1N4

Kenneth Falconer
Department of Mathematics
University of St. Andrews
St. Andrews, Fife KY16 9SS
SCOTLAND

Kenneth J. Ferguson
Department of Mathematics
University of Utah
Salt Lake City, UT 84112-1107

David J. Ferguson
Department of Mathematics
Boise State University
Boise, ID 83725-1555

J. Chris Fisher
Mathematics Department
University of Regina
Regina, Saskatchewan S4S 0A2

Aviezri S. Fraenkel
Department of Applied Mathematics
Weizmann Institute of Science
P. O. Box 26
Rehovot 76100
ISRAEL

David Gale
Department of Mathematics
University of Calfornia
Berkeley, CA 94720-0001

Tony Gardiner
Department of Pure Mathematics
University of Birmingham B15 2TT
ENGLAND

Martin Gardner*
Woods End Inc.
110 Glenbrook Drive
Hendersonville, NC 28739-4070

Solomon W. Golomb
Department of Mathematics
University of Southern California
Powell Hall 506, University Park
Los Angeles, CA 90089-0272

Stephen B. Grantham
Mathematics Department
Boise State University
Boise, ID 83725

Tom Griffiths
30 Conifer Court
London, Ontario N6K 2X4

Branko Grünbaum
C138 Padelford Hall, GN-50
University of Washington
Seattle, WA 98195-0001

Richard K. Guy
Department of Mathematics
The University of Calgary
Calgary, Alberta T2N 1N4

Denis Hanson
Mathematics Department
University of Regina
Regina, Saskatchewan S4S 0A2

Heiko Harborth
Bienroder Weg 47
D-3300 Braunschweig
GERMANY

E. O. Hare
Department of Computer Science
Clemson University
Clemson, SC 29634

Graham Hoare
Dr. Challoner's Grammar School
Amersham, Buckinghamshire HP6 5HA
ENGLAND

Stanley S. Isaacs
210 E. Meadow Drive
Palo Alto, CA 94306-4211

W. H. Jamison
Physics and Mathematics Department
Rocky Mountain College
Billings, MT 59102

K. B. Jones
Kadon Enterprises Inc.
1227 Lorene Drive, Suite 16
Pasadena, MD 21122-4645

Robert E. Kennedy
Department of Mathematics and
Computer Science
Central Missouri State University
Warrensburg, MO 64093-5045

Margaret J. Kenney
Department of Mathematics
Boston College
Chestnut Hill, MA 02167-3809

Ronald E. Keutzer
592 Cypress Avenue
Batavia, IL 60510-1131

Scott Kim
Box 9414
Stanford, CA 94305

Prof. Emer. Joseph D. E. Konhauser
Mathematics Department
Macalester College
ST. Paul, MN 55105

Carole Lacampagne
4530 Connecticut Avenue, NW #701
Washington, DC 20008-4316

Evert Lamfers
Buurmansweg 30
Nymegen 6525
THE NETHERLANDS

Mogens Esrom Larsen
Københavns Univ. Mate. Inst.
Universitetsparken 5
2100 København
DENMARK

Michel Las Vergnas
Univ Pierre et Marie Curie
U.E.R. 48
4 Place Jussieu
F-75230 Paris Cedex 05
FRANCE

Hendrik W. Lenstra
University of Amsterdam
Math Institute Spui 21
Amsterdam 1012 THE NETHERLANDS

Matti Linkola*
Pakkaajankatu 13
87150 Kajaani 15
FINLAND

Andrew C.-F. Liu
Department of Mathematics
University of Alberta
Edmonton, Alberta T6G 2G1

Jeremiah J. Lyons
Springer-Verlag
75 Fifth Avenue
New York, NY 10010

John McCallum
186 Mitchell Crescent NW
Medicine Hat, Alberta T1A 6V4

Douglas McKenna
93 Wolcott Road
Brookline, MA 02167-3108

P. McMullen
Department of Mathematics
University College
Gower Street
London WC1E 6BT
UNITED KINGDOM

Victor Meally
Deceased

L. Harding Migotti
T. A. Associates
550 999-8 Street SW
Calgary, Alberta T2S 1Y5

Eric C. Milner
Department of Mathematics and Statistics
The University of Calgary
Calgary, Alberta T2N 1N4

Barry Monson
Department of Mathematics
University of New Brunswick
P. O. Box 4400
Fredericton, New Brunswick E3B 5A3

William O. J. Moser
McGill University
Burnside Hall
805 Sherbrooke Street West
Montreal, Quebec H3A 2K6

Mrs. Angela Newing
The Rectory
Brimpsfield, Gloucester GL4 8LD
ENGLAND

Richard J. Nowakowski
Department of Mathematics
Dalhousie University
Halifax, Nova Scotia B3H 4H2

Peter O'Halloran
Department of Mathematics
Canberra College of Advanced Education
Belconnen, ACT 2616
AUSTRALIA

Victor Pambuccian*
Department of Mathematics
The University of Michigan
Ann Arbor MI 48109-1003

Robert Peard
Brisbane College of Advanced Education
Kelvin Grove, Queensland 4059
AUSTRALIA

Ivars Peterson
Science News
1719 N Street NW
Washington, DC 20036

James Propp
Department of Mathematics
MIT
Cambridge, MA 02139-4307

Tom Ransom
37 Silver Birch Avenue
Toronto, Ontario M4E 3L1

John F. Rigby
University of Wales, College of Cardiff
School ofMathematics
Senghennydd Road
Cardiff CF2 4AG
Wales, United Kingdom

Nicholas J. Rose
Department of Mathematics
North Carolina State University
Raleigh, NC 27695-8205

Lee C. F. Sallows
Buurmansweg 30
6525 RW Nijmegen
THE NETHERLANDS

Nola Samson
Box 102
Carbon, Alberta T0M 0L0

Bill Sands
Department of Mathematics and Statistics
The University of Calgary
Calgary, Alberta T2N 1N4

Norbert W. Sauer
Department of Mathematics and Statistics
The University of Calgary
Calgary, Alberta T2N 1N4

Jonathan Schaer
Department of Mathematics and Statistics
The University of Calgary
Calgary, Alberta T2N 1N4

Doris J. Schattschneider
Moravian College
Bethlehem, PA 18018

James L. Scott
4955 Viceroy Drive NW
Calgary, Alberta T3A 0V2

John Selfridge
Department of Mathematics
Northern Illinois University
DeKalb, IL 60115

Leslie Shader
Department of Mathematics
University of Wyoming
Laramie, WY 82071

Joan Shields
Box 355
Picture Butte, Alberta T0K 1V0

David Singmaster
Department of Comp. and Mathematics
Polytechnic of the South Bank
London SE1 0AA
ENGLAND

Gerald K. Slocum
257 South Palm Drive
Beverly Hills, CA 90212

Florentin Gh. Smarandache*
P. O. Box 42561
Phoenix, AZ 85080-2561

Bruce Smith
57 Glenside Way
San Rafael, CA 94903

Athelstan Spilhaus
Box 1063
Middleburg, VA 22117

M. G. Stone
Department of Mathematics and Statistics
The University of Calgary
Calgary, Alberta T2N 1N4

Jordan Stoyanov*
Bulgarian Academy of Sciences
Math Institute
P. O. Box 373
BG-1090 Sofia
BULGARIA

Michael Stueben
4651 Brentleigh Court
Annandale, VA 22003

Robert Sulanke
Department of Mathematics
Boise State University
Boise, ID 83725

Herbert Taylor
Center for Communications Research
4350 Executive Drive, Suite 135
San Diego, CA 92121-2116

Prof. Emer. Charles W. Trigg*
2404 Loring Street
San Diego, CA 92109

Daniel Ullman
The George Washington University
2130 H Street, NW
Washington, DC 20052-0001

Stan Wagon
Department of Mathematics
Macalester College
St. Paul, MN 55105-1899

Asia Ivic Weiss
Department of Mathematics
York University
4700 Keele Street
North York, Ontario M3T 1P5

John B. Wilker
Department of Mathematics
University of Toronto
Scarborough College West Hill, Ontario M1C 1A4

Harald Wimmer
Math Inst der Universitat
D-8700 Wurzburg
GERMANY

Robert E. Woodrow
Department of Mathematics and Statistics
The University of Calgary
Calgary, Alberta T2N 1N4

Ken Yanosko
Department of Mathematics
Humboldt State University
Arcata, CA 95521-5999

Samuel Yates
Deceased

Nobuyuki Yoshigahara
4-10-1-408 Iidabashi
Tokyo 102
JAPAN

Joseph Zaks
Department of Mathematics
University of Haifa
Mt. Carmel, Haifa 31999 ISRAEL